Fractal Surfaces

Fractal Surfaces

John C. Russ

North Carolina State University
Raleigh, North Carolina

Plenum Press • New York and London

Library of Congress Cataloging-in-Publication Data

Russ, John C.
Fractal surfaces / John C. Russ.
p. cm.
Includes bibliographical references and index.
ISBN 0-306-44702-9
1. Surfaces (Physics)--Measurement. 2. Fractals. I. Title.
QC173.4.S94R88 1994
514'.74--dc20 93-45023
CIP

ISBN 0-306-44702-9

A Division of Plenum Publishing Corporation
233 Spring Street, New York, N.Y. 10013

Printed in the United States of America

Preface

Fractals occupy a borderline between Euclidean geometry and complete randomness. In terms of the frequency distribution, one method that can be used to measure the fractal dimension, this is the space between a pure tone and white noise. This is similar, and perhaps related to, the fact that chaos theory (also called complexity theory) in mathematics lies in the critical boundary between pure Newtonian deterministic physics and complete random unpredictability. It appears that Nature tends toward this "self-organized criticality" as an attractor.

Like all new paradigms, the notion of fractals has been as much abused as used. It remains to be seen how broadly useful it will become in many different fields of application, or whether it really does offer any explanation of surfaces, their history, and properties. But as a phenomenological description of surface geometry, fractal dimensions "work" in a surprising number of instances. Since this is the way Nature has chosen to behave, it will be important for researchers to apply the new methods for characterization discussed in this book, and for models that attempt to describe the generation and behavior of surfaces to incorporate this geometry.

In writing this book, I have attempted to merge several different facets of the overall problem. Most of the early chapters are devoted to discussions of fractal geometry, and the methods available for the measurement of the dimension of surfaces. This can be done by a few direct methods and many indirect ones, either dealing with the surface in its entirety or by working in a lower dimension by intersection of the surface with a horizontal or vertical plane. Each of these choices presents some difficulties of technique or interpretation, and in any case the various "dimensions" that are measured are not all equivalent, a point that has been seriously under-appreciated by many workers in this young field.

A second section of the book discusses the modeling of fractal surfaces. This is not done with an eye to making pretty images for magazine covers or sets for movies, but to duplicate real surfaces produced by various processes, and to provide a basis for understanding the various measurement tools and the role of instrument noise and other limits to performance.

The final chapter attempts to compile some of the applications of fractal geometry to real surfaces, produced by a variety of operations such as machining, fracture, deposition, and so forth. Some of these examples are also used to illustrate the earlier sections on methods, and some duplication of information between the various chapters has been unavoidable. The breadth of applications of fractal geometry is astounding, but the depth of the literature is still rather shallow in most places. Much more work remains to be done before conclusions about the relationship(s) between surface fractal dimension and the history or properties of the surfaces can be reached.

The wide variety of applications has two consequences. Publication of preliminary results showing that some particular surfaces can be described by fractal geometry has taken place in

a great many different journals, reaching different audiences. This entails a certain duplication of effort and rediscovery or reinvention of principles. Secondly, the follow-on work to find the quantitative relationships and model the underlying physics, in an attempt to "explain" the observations, has been quite spotty. In many fields, this has not yet taken place.

One of the frustrations in writing the book has been keeping up with the ongoing publication of new results and methods in the field. I will apologize in advance for missing some papers. A second difficulty has been finding an organization that makes it possible to read about a particular application or technique without jumping all over the book. There is a certain amount of necessary duplication as a result. Some ideas appear and are at least briefly described in several places; the alternative was to force the reader to skip back and forth.

It is hoped that this book will stimulate more work, provide some access to the basic tools for measurement and interpretation of data, and encourage more researchers to apply fractal geometry to their individual problems. As more quantitative data are accumulated, and more correlations between fractal parameters and surface and material properties are discovered, it may lead to an understanding of the reasons that Nature has adopted this geometry to shape so many parts of our world.

It is a pleasure to acknowledge the contributions of coworkers to this book. Ron Scattergood, Tom Hare, and Mark Ray have all endured head-banging sessions at the blackboard helping to find ways through strange territory. Several collaborators have provided data and feedback to questions, and exchanged papers before publication, particularly Miguel Aguilar, Johannes Bueller, Brian Kaye, Michael Hamblin, and Paul Scott. Carl Zanoni (Zygo Corporation) and George Collins (Topometrix) have kindly provided data on specimens from state-of-the-art high-resolution surface characterization tools. A number of graduate students, particularly Yusef Fahmy, Sreeram Srinivasan, Michael Tidwell, and John Tyner, have obtained data as part of their thesis programs which are incorporated here. Mark Schaffer has been a willing and able "gofer" to track down elusive and often incomplete references. Chris Russ wrote the original Macintosh program shell and contributed portions of the code. Marty Esterman translated the finished program to run under Windows. Helen Adams has been a willing ear, a sounding board to see whether my explanations make sense, and has made possible a program to teach these concepts to 4th and 5th grade children (Adams and Russ 1992). She has also provided the moral support and enthusiasm necessary to write the book.

John C. Russ
October, 1993

Contents

1. Introduction 1

A Little History 1
Classes of Fractal Surfaces 2
Monster Curves 3
Random Fractals 6
Dusts 10
Zerosets 13
Korcak Islands 15
Deposited Surfaces 16
Pore Structures 17
L-Systems 19

2. Measuring the Fractal Dimension of Boundary Lines 27

Richardson Plots 27
Using a Computer 31
Digitized Images 32
Mosaic Amalgamation and the Kolmogorov Dimension 38
The Minkowski Sausage 41
Implementation with the Euclidean Distance Transform 42
Fitting Lines to Data 48
Other Methods 51
Comparison of Dimensions 55

3. The Relationship between Boundary Lines and Surfaces 59

Analogies 59
Direct Methods 61
Zerosets 64
Dimensional Analysis 66

The Minkowski Comforter 71
Data Formats 73
Effects of Noise 75
Simulated Noise 77

4. Hurst and Fourier Analysis 83

Time-Based Data 83
The Hurst Plot 83
Fractal Brownian Motion 87
Elevation Profiles 88
Hurst Analysis of Range Images 89
The Hurst Orientation Transform 95
Fourier Analysis 97
Fourier Analysis of Boundary Lines 101
White and 1/*f* Noise 102
Fourier Analysis in Two Dimensions 104
Anisotropy 107
Characterizing the Magnitude of the Roughness 110
The Topothesy 112

5. Light Reflection and Scattering 115

Visual Appearance 115
Fractal Brownian Profiles 116
Electrons, Radar, etc. 117
Local Texture Measurement 121
Brightness Patterns from Rough Surfaces 122
Relating the Surface and Brightness Dimensions 129
Light Scattering 138
Scattering of Diffuse Light from Rough Surfaces 139
Range Measurement Methods 141
Range Images and Surface Parameters 144

6. Modeling Fractal Profiles and Surfaces 149

Fractal Profiles 149
Particle Aggregation 153
Deposited Surfaces 157
Modeling a Fractal Surface 159
Fractal Brownian Surfaces 165
Mandelbrot–Weierstrass Functions 167

Comparing Models and Measurements 172
Takagi Functions 177
Weak and Strong Anisotropy 181
Modeling an Anisotropic Surface 183

7. Mixed Fractals 191

Limited Self-Similarity 191
Kaye's Definition 193
Simulating the Projection 195
Vicsek's Fat Fractal 197
Mixed Fractals 200
The Mandelbrot Conjectures 201
Measurement of Fractal Dimensions 203
Addition of Fractals 204
Variation with Scale 206
Splicing Fractals Together 208
Directionality 212
Tentative Conclusions 224

8. Examples of Fractal Surfaces 227

Brittle Fracture 229
Machining and Wear 238
Deposited Surfaces 244
Pore Networks 248
Other Surface Applications 250
Very Flat Surfaces I: Interferometry 251
Very Flat Surfaces II: Atomic Force Microscopy 259

References 267

Appendix 285

Index 301

1

Introduction

A Little History

It is impossible to begin a book on fractals without first acknowledging the debt owed to Benoit Mandelbrot. Fractal surfaces, and the many other natural manifestations of fractal behavior (only a few of which will be described here), existed long before Mandelbrot described them, but his 1982 book *The Fractal Geometry of Nature* collected together many ideas in one place, brought them with the aid of striking graphics to a general audience, and placed the made-up word "fractal" firmly in the modern vocabulary. His popularization of the concepts of fractal geometry has spurred many other workers to explore various aspects of this newly revealed subject.

Other workers had published observations and speculations before 1982, and indeed Mandelbrot cites hundreds of references to this prior literature. He has himself published other papers of a more traditional scientific nature before and since, dealing with particular aspects of fractals from stock market fluctuations to metal fractures. *The Fractal Geometry of Nature* is not an easily classified book. It is not a tutorial guide or textbook, nor a conventional technical or scientific exposition, nor a picture book, although it contains pieces of each. Mandelbrot himself describes the book as an essay.

The ideas follow each other so rapidly that none is fully developed but serves as a kind of guidepost to a rich lode that can be mined deeply by those interested. The descriptive concepts provided by fractals have proven to have a deep resonance with the observed behavior of many aspects of the natural world. In the often over-used parlance of current science, fractal geometry provides a new paradigm for understanding many physical phenomena. This is timely and fortunate, because studies of many of the phenomena which now appear to be well described by fractal geometry are becoming increasingly important in various aspects of technology, and are not satisfactorily explicable by conventional means.

The significance of Mandelbrot's book is threefold: First, the book introduced the vocabulary of fractals to a broad audience. This is important because it is not possible to describe or perhaps even to recognize properties or behavior for which we have no language. Words such as "self-similar" and "fractal" became available, and the concept of noninteger dimensions was given a clear and graphic meaning. Second, the book collected together in one place some of the mathematical ideas which had been around since the beginning of the twentieth century. Several mathematicians and geometers had pointed out that the Euclidean definition of a line was not the only possible one, and that "monster" curves could be easily defined which had properties unreconcilable with "conventional" geometry, but the similarities between these formulations were not particularly obvious. Third, a number of natural phenomena are cited which exhibit the kind of behavior which this mathematics describes. It is

interesting that much of the natural world seems to obey fractal geometry, rather than Euclidean geometry. It is only over a very small range of dimensions, from submillimeter to a few meters, where man has exerted influence to create "flat" planes and "straight" lines. Beyond that scale, nature reasserts its own natural geometry, and that is often fractal.

There are many "kinds" of fractals. Some are primarily of interest to mathematicians and physicists. They describe the behavior of attractors in high dimensionality spaces which control the chaotic behavior of variables in equations, which may have important connections to physical behavior but are not particularly obvious to anyone without the skills to follow the math. The fact that fractals do seem to describe some aspects of the natural world has also encouraged artists to use them as an aesthetic device (Peitgen and Saupe 1988; Jürgens, Peitgen et al. 1990; Pickover 1990; Peitgen, Jürgens et al. 1992). At a very different end of the spectrum are simple numerical descriptors which compactly characterize boundary lines, surfaces, clusters of particles, and other "real-world" objects. It is this latter use of fractals which is most important in this book.

Classes of Fractal Surfaces

Even within the topic of fractal surfaces, there are (at least) three types which can be distinguished (Pfeifer and Obert 1989; Gouyet, Rosso et al. 1991). These are illustrated schematically in Figure 1. The first is a dense object with a fractal surface, which we will see in later chapters is a self-affine fractal rather than self-similar. The second type is a fractal object such as a network or cluster, whose surface will also be fractal (called a mass fractal). The third is a dense object within which there exists a distribution of holes or pores with a fractal structure (called a pore fractal). In the first category are erosion surfaces (materials or mountains), most deposited thin films (some deposition processes produce the second type of fractal surface), fractures, machined surfaces, corrosion and wear surfaces, most chemically dissolved surfaces (some dissolution processes produce porous structures of type 3), and many more. Most of the interest in this book will center on the first type.

One of the particularly compelling features of fractal geometry is that it becomes possible to construct objects or surfaces of dazzling and in some senses unlimited complexity from extremely compact and simple mathematical relationships. It seems natural to expect simple objects to have simple arithmetic descriptions, but perhaps to expect complex objects to require

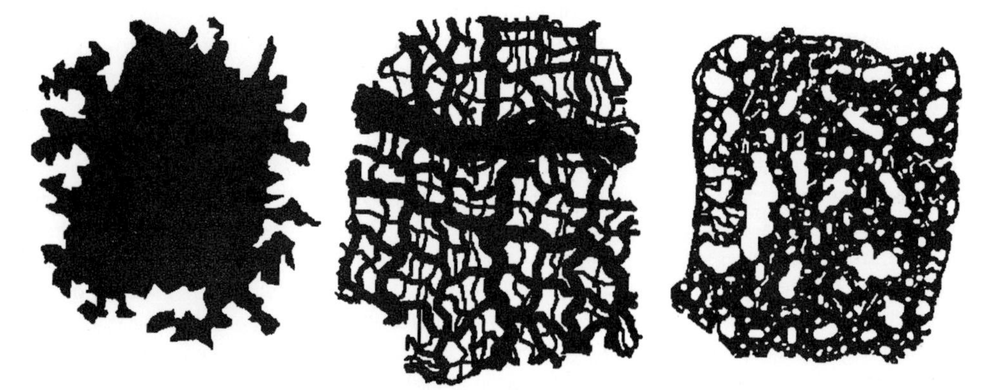

Figure 1. Schematic diagram of three classes of fractals. Type 1: a dense object with a fractal surface; Type 2: a network or cluster; and Type 3: A dense object containing pores.

more complex generating functions. Fractal geometry is a tool by which simple rules can be used to construct realistically complex objects. In a few cases, the simple rules can furthermore be related to physical processes that operate to produce the real objects and surfaces. These processes are typically iterative and always operate at many different dimensional scales, which is the hallmark of fractal geometry.

While some mathematical expressions will be used to describe the fractal concepts discussed in this book, they will be used with restraint, and comparison to tangible objects and experiments will be employed to help explain the relationships. Likewise, computers are a powerful tool for simulating and measuring fractal behavior, but the concepts are simple enough that stepping through the algorithms by hand can provide a more direct experience for the reader. The algorithms for the modeling, measurement, and graphical display of fractals are simple enough that they can be run on a desktop computer. An appendix presents a program that can be run on desktop computers to generate and to measure images of fractal surfaces.

No rigorous definition of a fractal is needed as yet. Several will evolve in the course of the various chapters. Most of these are operative definitions, rather than formal ones. The basic idea of a fractal is well summed up by Jonathan Swift:

> So, Naturalists observe, a Flea
> Hath smaller Fleas that on him prey,
> And these have smaller yet to bite 'em,
> And so proceed ad infinitum.

This idea, namely that there exists a hierarchy of ever-finer detail in the real world, is familiar to most observers. Conventional Euclidean geometry does not deal adequately with such a situation. The straight lines and circular arcs of the geometry taught to high school students do not actually describe the objects found in nature. Man-made objects have such shapes, at least over a narrow range of dimensions. But apparently smooth, geometrical forms become rough and irregular at small dimensions and become insignificant at larger scales. Clearly, Euclidean geometry is not able to characterize the natural world. Mandelbrot has often said, "Mountains are not cones, clouds are not spheres."

Monster Curves

Mathematicians have been aware of this limitation for some time, and several "non-Euclidean" geometries have been developed to deal with various shortcomings in the conventional tools. About 100 years ago, several mathematicians found functions which violated some of the common assumptions of Euclidean definitions of a line. These curves were known as "monsters" which violated accepted precepts of mathematics, and were generally put aside from the main stream of mathematical development. It was further assumed that they had nothing to do with anything real. Only recently has interest revived as the utility and relevance of these monsters became known.

One of the non-intuitive characteristics of these monster curves was that they had an indeterminate and perhaps infinite length. While continuous, they were nowhere differentiable, meaning that they had no easily definable local direction or variation of direction. An example of such a curve is the Koch curve (Koch 1904) shown in Figure 2. This curve is defined by iteration. The initial figure consists simply of a square, with four straight sides. The rule for elaboration of the boundary line is to replace each straight line segment with the shape shown, which has twice the length. This produces a new figure with 32 straight line segments for sides, each one-quarter as long as the original.

Figure 2. The first and third iterations of the Koch Quadratic curve ($D = \log(8)/\log(4) = 1.500$).

If each of these straight lines is likewise replaced with a reduced version of the same shape, another figure appears with 256 shorter straight sides. At each step, the total boundary length increases by a factor of 2. If this process is continued *ad infinitum*, the boundary length increases without limit. However, the area bounded by the curve remains constant at each step, since the irregular shape removes as much area as it adds. The final result is a boundary line which encloses a finite area but has an unbounded or infinite length. As noted above, it is everywhere continuous but nowhere differentiable. It was this nonintuitive behavior that led to the classification of such curves as monsters.

This particular figure has a very regular pattern to it. We will shortly encounter lines which have the same property of increasing length as finer details are considered but do not have such regularity in the process. These are generally called random or natural fractals, as compared to the regular or structured ones like the Koch curve. Before leaving the regular Koch curves, we can examine the effect of changes which alter the rate at which the length of the boundary line increases. The Koch curve illustrates the concept that the length of a fractal line does not possess a well-defined value. A Euclidean or "rectifiable" curve is one whose length approaches a finite limit as step length approaches zero, and perhaps fits better our everyday notion of a line, but these exist only rarely in nature. The Koch curve was originally defined as being everywhere continuous but nowhere differentiable.

The shapes shown in Figure 3 will also increase the length of the boundary at each step of refinement without altering the enclosed area. But the rate of increase of length is not the same. For the square steps shown in Figure 2, the rate of increase is described by the ratio of the logarithm of the lengths at each step, or $\log(8) / \log(4) = 1.5$. The shapes in Figure 3 have ratios of 1.262 and 1.465. This dimension value will become a measure of the irregularity of the resulting boundary line, and will be described as a fractional or "fractal" dimension for the line. Technically, this is the Hausdorf or similarity dimension, but for our purposes at the moment, all of the various measures of the degree of irregularity can be considered as interchangeable.

The fractal dimension is a value greater than the topological dimension of a line (exactly 1). Fractal surfaces have a dimension which is greater than 2, which is the topological dimension of a surface. For the fractal line, the dimension greater than 1 is in some senses the extent to which the line "spreads out" into the plane, or the difficulty of telling whether

a particular location is on one side or the other of the boundary. As the fractal dimension increases toward 1.999..., the line covers more and more of the plane. A number of "monster" curves have been defined, which have a dimension of 2.0 and in the limit pass through every point in the plane. Figure 4 shows the evolution of the Hilbert curve, one of this set of monsters.

The idea that dimension need not be an integer may seem disturbing at first, since it is conventional to deal with the topological values of 0 (point), 1 (line), 2 (surface), and 3 (volume). But many real-world lines and surfaces have a roughness that reaches out into a higher dimension, and expressing that as a fractional value may be understood as a qualitative measure of the roughness. From a mathematical point of view, the meaning of dimension is both more general and more specific than this notion. And just as the topological (integer) concept of dimension generalizes to fractal dimension, so do other math operations (Le Méhauté 1990; Le Méhauté 1991). For example, $n!$ generalizes to the gamma function $\Gamma(n)$, going from discrete to continuous. So why not have a fractional derivative (and integral) at least with respect to time? Consider motion along a path which may be either Euclidean or fractal, and for which the integer orders of a generalized D (time derivative) operator are: D_2 = acceleration, D_1 = velocity, D_0 = position, D_{-1} = distance. Thus the second derivative

Figure 3. Two more Koch curves. The Koch Triadic curve ($D = \log(4)/\log(3) = 1.262$) and the "spike" curve ($D = \log(5)/\log(3) = 1.465$). The second and fifth iterations of each are shown.

Figure 4. A space-filling (fractal dimension 2.0) "monster" curve; the first and fifth iterations are shown.

is acceleration and the integral is distance. What would be the meaning of fractional derivatives?

Random Fractals

Structured fractals have some properties disturbing for the mathematician, but do not seem to be closely connected with many real-world situations. It is not difficult, however, to generate a random fractal line which closely models or simulates real-world processes. We will later encounter several ways to do this, but a simple one can be constructed in a way that is quite similar in principle to the Koch curve. Start with a straight (Euclidean) line between two points. Displace the midpoint of the line by some random amount, either up or down with equal probability. This creates a line with two straight segments as shown in Figure 5. Repeat the operation for each of their midpoints. Continue the process, at least until the line segments become shorter than can be drawn or seen as part of the profile.

It is most common to use a Gaussian random number generator to vary the amount of the displacement. Instead of reducing the mean value of the deflection in proportion to the halving of the length of each segment in the curve with each iteration, a different relationship may be chosen. For reasons that will make sense in the context of a later chapter, we will choose here to have the mean magnitude of the deflection decrease as the segment length to a power, alpha. This coefficient will turn out to control the fractal dimension of the line.

In Figure 5, several successive iterations of the algorithm are shown. Since the length of each line segment is halved at each step, it takes only a few iterations to produce segments so short that they cannot be printed with enough resolution to show the details. In this particular example, the program shown in Listing 1 was used. It creates arrays of numbers 1024 points long for the coordinates of the line, which takes 10 steps ($2^{10} = 1024$) of successive halving to produce displacement of the individual points.

The final plot looks "natural" and shows the kind of random self-similarity which we associate with the word "fractal." In a formal sense, the length of the line would be indeterminate if we had enough resolution to continue the iteration. The length of the line increases at a constant rate with each iteration, which depends on the value of alpha.

Listing 1

```
PROCEDURE MidpointDisplacement;
{One of the most straightforward methods for generating a fractal profile.
Starting with the entire length of an array, the midpoint is displaced up
or down by a random amount. This procedure is repeated for each of the two
segments to produce four, and continues down to individual points. The
magnitude of each displacement is reduced as the length of the segment is
reduced, so that for a length scale r=1/2^n, the mean square variation is
r^2*alpha, and the fractal dimension is 2-alpha.}

CONST
      NumP=1024; {number of points in the array}
      WindWidth=500; {dimensions of the window for plotting}
      WindHeight=400;
VAR
      Data : Array[0..NumP] of real;
      i, j, k, steps, depth, h, v: integer;
      alpha, temp, value, increment, hscale, vscale: extended;

      PROCEDURE InitRand;
      BEGIN
            GaussFac := sqrt(4 / 12); ranscal := scalb(31, 1) - 1;
            seed := num2extended(TickCount) / 1000;
      END;

      FUNCTION Gauss: extended;
      {Gaussian random number generator, mean value=0, std. dev:=1}
      VAR
            gsum: extended;
            ii: integer;
      BEGIN
            gsum := 0;
            FOR ii := 1 TO 4 DO
                  gsum := gsum + RandomX(seed) / ranscal; {uniform rand 0…1}
            Gauss := ((gsum / 4) - 0.5) / GaussFac;
      END; {Gauss}

BEGIN{Midpoint}
      InitRand; {set up the random number generator}
      FOR i := 0 TO NumP DO {clear the initial array values}
            Data[i] := 0;
      value := 0.5; {each step halves the base width}
      steps := round(ln(num2extended(NumP)) / ln(2)); {for the fixed case
            of NumP=1024, the number of steps is 10 since 2^10=1024}
      alpha := GetAlpha('Alpha=', 0.5, 0, 1);
            {This function  allows the user to type in a real value between
             0 and 1, with the default initial value set to 0.5. It uses the
             Macintosh modal dialog routines.}
      ShowWatch; {change the cursor to the wristwatch while we work}
      j := NumP;
      FOR depth := 1 TO steps DO
            BEGIN
                  value := value * XpwrY(0.5, alpha);
                  j := j DIV 2;
                  i := j;
                  REPEAT
                        temp := Gauss;
                  increment := value * temp;
                  FOR k := 0 TO (j DIV 2) DO
                        BEGIN {interpolate a line between the new value
                              and the previous neighbor points}
                              temp := increment*2.0 * num2extended(k) /
                                          num2extended(j);
                              Data[i - j + k] := Data[i - j + k] + temp;
                              IF k < (j DIV 2) THEN
                                    Data[i - k] := Data[i - k] + temp;
                        END;
```

Continued

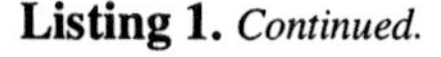

Listing 1. *Continued.*

```
                        i := i + j;
                 UNTIL (i > NumP);
                 IF (depth = steps) or ShiftKeyDown THEN
                     BEGIN {Plot the array on the screen at the end, or
                            after each iteration if the user requests it}
                           hscale := WindWidth / num2extended(NumP);
                           vscale := WindHeight / 2.0;
                           {sets arbitrary vertical plotting scale of ±1}
                           v := trunc(vscale * (Data[0] + 1.0));
                           MoveTo(0, v);
                           FOR k := 1 TO NumP DO
                               BEGIN
                                   v := trunc(vscale*(Data[k] + 1.0));
                                   h := trunc(hscale * k);
                                   LineTo(h, v);
                               END;
                     END;
         END; {next depth}
END; {MidpointDisplacement}
```

A few words are appropriate here with regard to the listing. Pascal has been used for several reasons. At one time, most scientific calculations were performed in Fortran. However, this language is not widely available on desktop computers, and does not naturally support some of the graphics displays which are now available on most computers and workstations. Also, Fortran programs can become hard to read if many variables are used. Basic is probably the most widely used small-computer language, and would perhaps make these programs directly usable by the largest number of readers. However, Basic is not well standardized on different computers, and the short variable names and limited structure to the code are less than ideal for tutorial purposes.

Pascal has become a standard language for instruction in computer science because it is well structured, and is easy to document and read. Pascal compilers are available for most small computers, and graphics are usually well supported. Other small computer languages such as C are quite efficient from a programming point of view, but are more difficult to read and can become quite cryptic in their brevity. The procedure listings shown here and in the appendix are short enough that you can actually try them out yourself.

The program in the appendix was written to analyze many of the images shown in this book, and to generate others. It was compiled and run on a Macintosh, using the MPW Pascal compiler. However, only the commands having to do with file access or graphics displays are in any way machine-specific and they are commented so that you may modify them as needed for your particular system. The program shown in Listing 1 uses the system clock to initialize

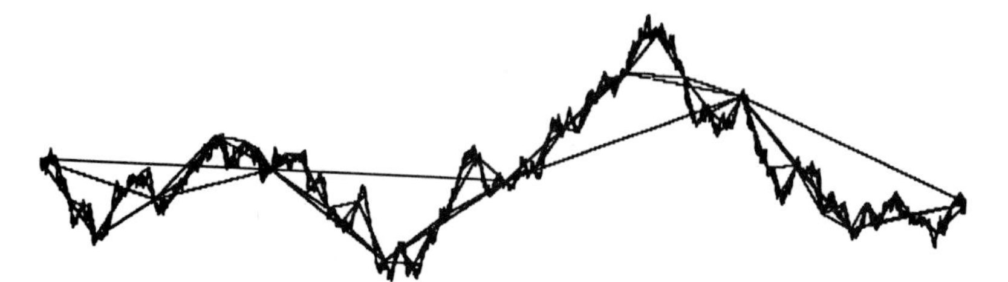

Figure 5. Generation of a fractal line by iterative random midpoint displacement.

the random number generator, and so this program will produce a different sequence of numbers and consequently a different fractal profile each time it is run. Only the average degree of self-similar irregularity (and the fractal dimension) will remain approximately the same. This is visible to a human as the perceived "roughness" of the line.

This listing creates the numbers which represent the z elevation of the line as a function of x position, and saves them in array so they may be written to a disk file, plotted on the screen, or otherwise analyzed. A simple plotting routine is included in Listing 1 which can display the results of each iteration. A more versatile plotting routine is shown in the more complete program in the Appendix. This also shows the particular commands needed to create a window on the Macintosh, allow the user to input values, perform the plotting, and read and write data files.

Varying the alpha coefficient in the algorithm used to produce Figure 5 alters the self-similar irregularity of the resulting boundary lines. Figure 6 shows several examples. When asked to rank the lines in order of apparent or visual roughness, people have no difficulty in doing so and the results agree with the fractal dimension. This parameter does indeed summarize much about our experience with real, irregular boundaries. On the other hand, changing the magnitude of the displacement or the probability distribution function for the displacement from a Gaussian one to some other function makes no change in the fractal dimension, although it does alter the visual appearance of the curve somewhat. This is illustrated in Figure 7.

There are other types of fractal structures which can be produced by iterating complexity in a self-similar way. The examples which follow all use a structured self-similarity, but just as for the Koch curve it is possible to substitute random events which produce results that appear more natural.

Readers interested in generating various kinds of fractals, particularly those which mimic natural terrain and so can be used to produce images of artificial (but realistic-looking) worlds, should read Peitgen and Saupe (1988), where several such program fragments are presented. You may have seen quite a few such examples without realizing it, since these methods are now used extensively in the motion picture industry to produce artificial landscapes, plants, etc., which appear natural but strange, by slightly altering the fractal dimension from those typical of our world. It is generally less expensive to generate such landscapes in the computer than to physically construct sets. This technology has become common, particularly in science fiction movies, such as *Alien*, *Star Trek II*, etc.

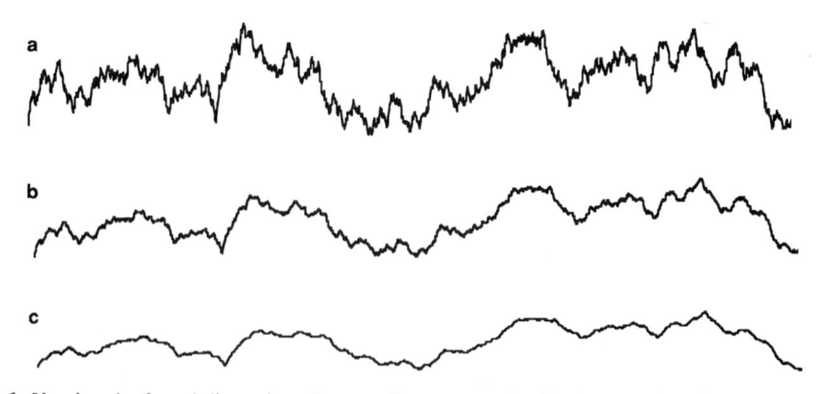

Figure 6. Varying the fractal dimension: three profiles generated with the same list of random numbers—midpoint displacement fractals generated with α = **(a)** 0.5, **(b)** 0.7, and **(c)** 0.9.

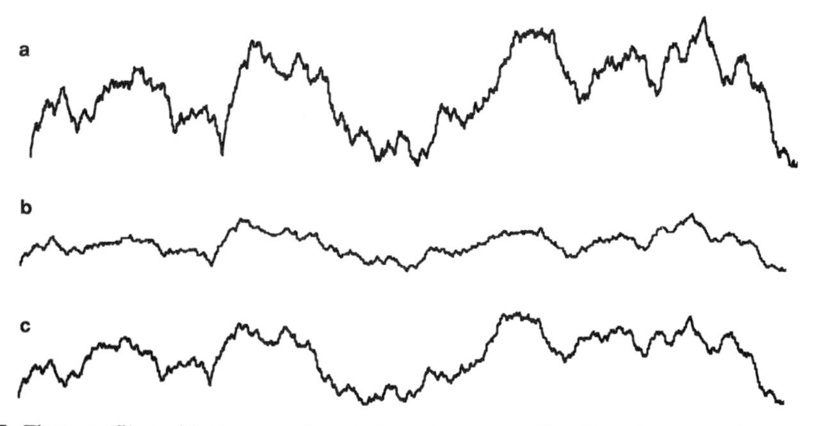

Figure 7. Three profiles with the same fractal dimension ($\alpha = 0.7$), although generated with different displacement magnitudes or with displacement values taken from distributions which are not Gaussian (compare to Figure 6b): **(a)** magnitude × 2, **(b)** squared distribution, and **(c)** square-root distribution.

Dusts

While the field of interest in this book is primarily that of fractal surfaces and the various ways that they are produced by deposition, fracture, machining, and so forth, there are other fractals which may be related in some instances. One of the simplest mathematical fractals is a Cantor dust. Start with a Euclidean line segment one unit long, containing an infinity of points. Remove the central third of the points (from 1/3 to 2/3). Of course, the number of points is still infinite. Now remove the central third of each remaining segment, as shown in Figure 8. Repeat this process endlessly. The result is an infinite number of isolated points in the interval from zero to one. The dimension of this dust is clearly less than 1.0, since the coverage is less than that of the original line. Defined by the rate at which the coverage decreases with each iteration, the dimension of this dust is given by log(2)/log(3) = 0.631.

The same process of removing pieces in either a structured way, as used for the Cantor dust example, or using a random number generator, can be extended to planes and solids. A Cantor dust in two-dimensional space describes, as an example, the distribution of mineral resources on the surface of the earth. In the case of the three-dimensional solid, removing portions will produce a Menger sponge—a construction which extends through three-dimensional space without filling it. In fact, the volume of the sponge vanishes in the limit while the extent is unchanged, the structure remains a single continuous object, and the surface area becomes infinite. It appears that the distribution of galaxies is an example of such a structure, as is the distribution of metal grains in an ore deposit.

The variety of fractal constructs which exist in Nature and/or can be generated by simple application of the geometric principle of iterated self-similarity is very great. Some of the relatively familiar examples include Menger sponges, Cantor dusts, Korcak islands, and L-systems (Lindenmayer systems), each named after someone who studied them in detail.

Figure 8. Producing a Cantor dust by iterative removal of the central 1/3 of line segments. Five iterations are shown from top to bottom.

These have at best an indirect relationship to our interest here in fractal surfaces, but it is worthwhile to have some familiarity with each.

The Sierpinski fractal is a structure that occupies a two-dimensional space and has an unbounded perimeter, but in the limit has no area. It is shown in Figure 9. The extension of this to a volume results in a sponge, as described above, that can be modeled as a solid block from which holes are removed. If a structured approach to the removal is used, the result can be as shown in Figure 10. There is only one large hole centered in the block, which in this example removes 7 of the 27 smaller cubes into which the original block can be subdivided. If each of the smaller cubes is then cut up in the same way, the volume remaining after two iterations would be $(20/27)^2$. As the process is repeated, the remaining volume decreases as shown in Figure 11. At the same time, the surface area increases. The fractal dimension for the volume is $D = \log(20)/\log(3) = 2.7268$. This is less than three because the object does not fill space, although it occupies it. Other structuring elements can be devised which will give different fractal dimensions. Not all of these will have another interesting property of the Menger sponge shown here, which is that both the solid and the pore are connected sets. That is, there is only a single object present with all points connected, and the same thing is true of the pore space.

The surface area of the pore space is also interesting. The initial cube has a surface area of 6. After the initial step, 24 new faces have been created, each with an area of 1/9. However, size squares of area 1/9 have been lost on the faces, so the net gain is 18/9 = 2 square units. On the next iteration, there are 20 small cubes in which the pores are created, and the area of each tiny face is now 1/81, so the area gain is $20 \cdot 18/81$. The formula for the area after n generations is

$$\text{Surface} = 6 + (18/20) \cdot (20/9)^n$$

Hence the fractal dimension is $D = \log(20)/\log(3) = 2.7268$, the same as the volume dimension.

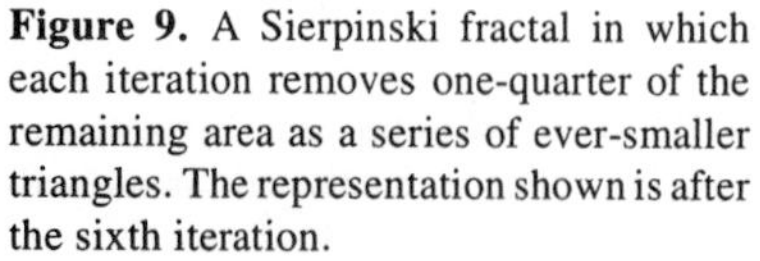

Figure 9. A Sierpinski fractal in which each iteration removes one-quarter of the remaining area as a series of ever-smaller triangles. The representation shown is after the sixth iteration.

The pores in a Menger sponge represent one realization in three dimensions of a Cantor set. The simplest example of a Cantor set occupies a space in one dimension—a line. As described above, the iterative removal of one third of the points leaves a "dust" of points whose dimension is less than one, the dimension of the space occupied by the dust, and is a measure of how well the dust fills the space. Using similar iterative rules to reduce planes or solids to a dust produces differing dimensions, but always less than 1.0 because the dust is a series of disconnected points.

Of course, using appropriate randomization rules for either the Menger sponge or the Cantor dust can produce more "natural" fractals. Mandelbrot has suggested that a cross section

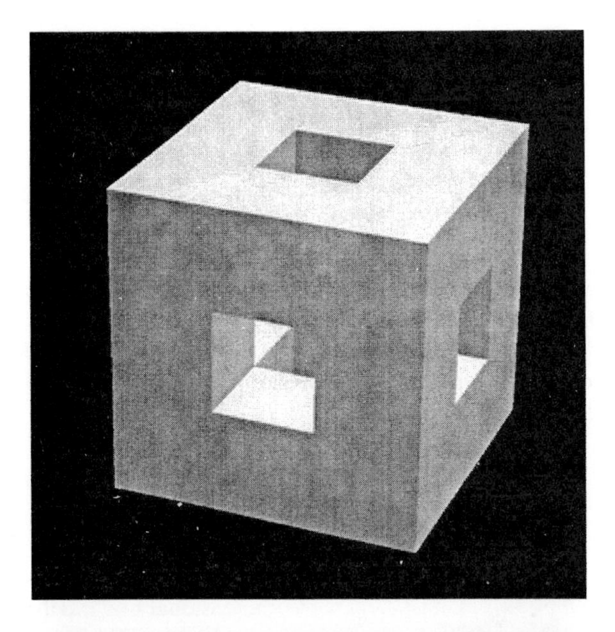

Figure 10. Iterative generation of a Menger sponge by removing holes from a block. Only the first two iterations are shown.

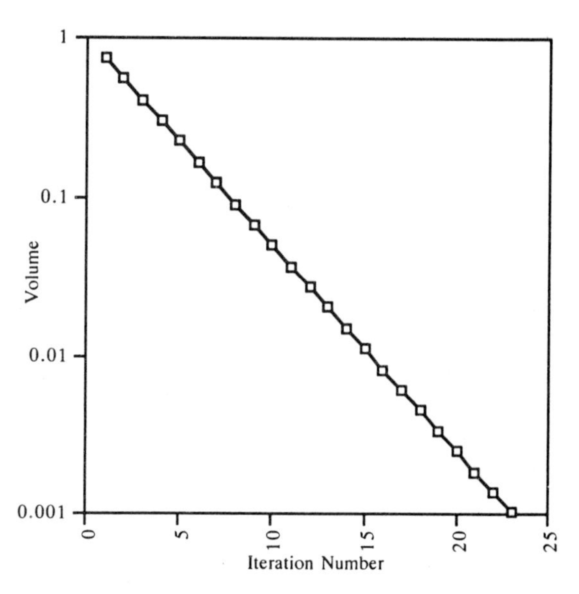

Figure 11. Plot of the change in volume of the Menger sponge with each iteration.

of Saturn's rings may be one example of a Cantor dust, in which each orbit is seen to divide into several separate ones as the magnification increases. Certainly, many of the chaotic attractors, such as the well known Henon maps or the Lorenz function, show this kind of behavior. A cross-section across such orbits (known as a Poincaré section) shows points which can be magnified to reveal finer detail.

Zerosets

This idea of a cross-section across orbits provides a way to evaluate fractal dimensions and is an important relationship between the Cantor dust along a one-dimensional line and the line itself. Consider a line which is position versus time for a particle undergoing Brownian motion. The particle itself wanders up and down, sometimes close to the initial starting position and sometimes far from it. The characteristic of such a "drunkard's walk" is that the moving point may wander off for a time, but it always returns to the starting point. ("The drunk always comes back to the bar.") A line which traces this motion as a function of time is fractal with dimension $D = 1.5$, and such line profiles will be discussed in detail in subsequent chapters. The times at which the particle returns to the initial position (or to any other fixed position chosen) correspond to points where the fractal line crosses the straight line which is the time axis, as shown in Figure 12. The axis is in effect a Poincaré section through the fractal. The intersection is also called a zeroset of the fractal. The points where the fractal path crosses the axis are a Cantor dust, and the dimension of the dust is $D = 0.5$, exactly 1 less than the dimension of the line of Brownian motion.

Exploiting this effect as a means of measurement provides one set of tools to measure fractal dimensions. The dimension of a Cantor dust can be determined by plotting the frequency distribution of the distances between the points. Going back to the fractal line shown in Figure 12, there are some periods of time when the moving particle is quite far from the axis, and so the time interval between crossing points is large. However, when the line does return to the axis it is likely that near each crossing point there are many others, given the tendency of the line to wander up and down with many small deviations. Making a plot of the number of time

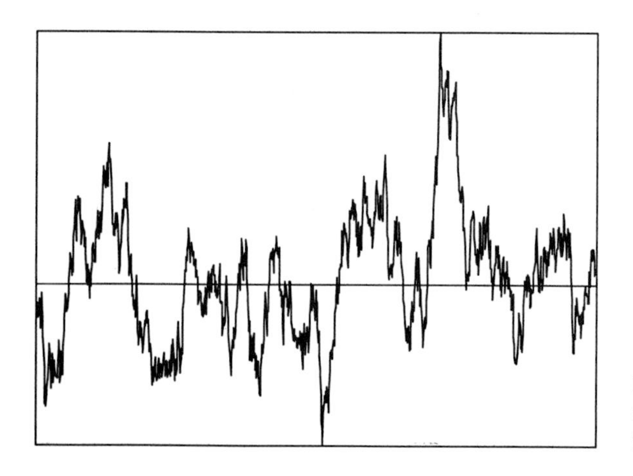

Figure 12. Intersection of a Brownian motion trace with a zeroset. The intersection points form a Cantor dust.

intervals which exceed a value *t* as a function of *t* produces a linear result on a log–log graph, and the slope of that plot is 0.5, or exactly 1 less than the fractal dimension of the line. This is known as the Korcak relationship, and can be written as

$$N(T \geq t) \propto t^{-(D-1)}$$

Actually, measuring the distribution of intervals between zero crossings is a very clumsy process that is not often used for line profiles, since other techniques are available that make more efficient use of the data in the profile. However, it has one advantage. The largest intervals are the easiest to measure, while the small ones are limited by the resolution of the measuring tool. The nature of the cumulative plot is such that the small ones are not needed. A measurement of enough relatively large intervals can define the line and hence the dimension with good precision.

However, for any finite extent of the profile, the largest intervals are systematically undersampled because they cannot fit within the extent. This effect is shown in Figure 13. This is a cumulative plot of the number of horizontal intercepts along many horizontal lines through the profile. The plot departs from linearity and bends down for longer intervals. This complicates fitting a straight line to the data to determine the slope and hence the fractal dimension.

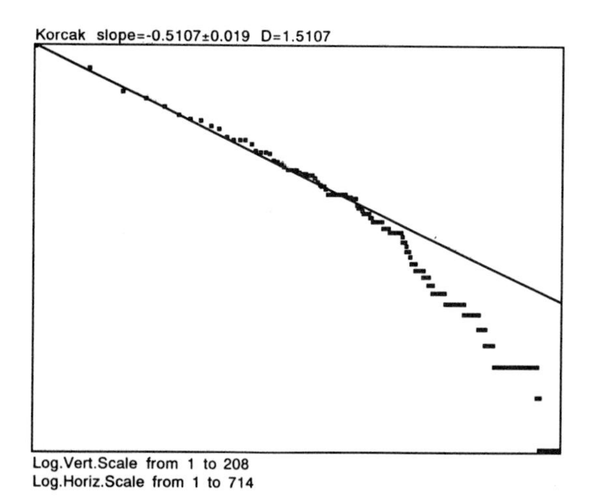

Figure 13. Korcak plot of cumulative frequency of intercept lengths for a 2048 point Brownian profile (D = 1.5), showing the drop at long length values.

Korcak Islands

This idea of the zeroset or Poincaré section can be extended to surfaces as well. Korcak's original application was to the surface of the earth. Sea level can be thought of as a zeroset for that surface, with the coastline corresponding to the section and islands representing the portion of the earth's surface rising above the zeroset. We will later see that there is a relationship between the roughness of the boundaries expressed as a fractal dimension, and the roughness of the surface. This provides one of the commonly used ways to measure surface dimension. For the present, we are concerned only with the Korcak relationship.

The quantity analogous to the length along the axis between crossing points is the area of each island. A log plot of the number of islands whose area is greater than a, which can be written as $N(A \geq a)$, vs. a, produces a straight line. The slope of this line for the islands of the earth produces a slope of about 0.6. The fractal dimension of the boundary lines or coastlines is exactly twice this value, or 1.2. For an isotropic self-similar surface (which the earth is probably not due to the anisotropy of erosion) the surface dimension would be greater by exactly 1, or 2.2. This illustrates the means by which Poincaré sections or zerosets can be used to reduce a difficult-to-measure surface dimension to a related quantity in a lower dimension which may be more easily measured.

We will encounter this step repeatedly in the chapters that follow. A word of caution is required. The use of a section plane to reduce the dimension by 1 is only applicable to exactly self-similar objects. Most surfaces of the types of interest in this book are not self-similar but rather self-affine. In brief, this means that the vertical direction is not the same as the lateral directions parallel to the nominal surface, and that the scaling of magnitude with dimension is different vertically as compared to laterally. This might be expected simply from the physics of the situation, since the forces at work to create most surfaces depend on these directions.

It also happens that one of the consequences of the most commonly used measurement techniques for these surfaces is that only a single elevation value z is obtained for each x, y point. If there are undercuts, bridges or caves in the surface, they are not seen when looking down from the top with a light microscope (interferometer or confocal microscope) or a scanned probe (atomic force or scanning tunneling microscope). The result of this limitation is that even if the surface is actually self-similar, the measured array of elevation values will be self-affine.

It is correct to reduce a self-affine surface or image of a surface using a section parallel to the nominal surface orientation. The boundary lines produced are in fact self-similar even if the surface is self-affine, and measurement of the fractal dimension gives the expected $(D_S - 1)$ relationship to the surface. However, using any other orientation for the section plane does not produce this result. In particular, a vertical section plane corresponds to the very common situation of recording a profilometer trace across the surface. Analysis of this profile is possible, but some of the techniques which have been (mis)used (such as a Richardson plot) are not the correct way to do it. In fact, we will see in a later chapter that these methods may produce straight-line graphs but that the slope of these graphs does not have the expected relationship to the fractal dimension of the sectioned surface. Analysis of self-affine vertical section profiles is possible, but requires selection of an appropriate technique.

If the surface is also anisotropic, that is if the x and y directions are not equivalent, it may produce a further restriction on the choice of analysis methods. It happens that most processes which produce surfaces, both natural and man-made, are not perfectly isotropic. Fracture, wear, and deposition all have a clear directionality associated with them. Corrosion may be influenced by crystallographic directions or inhomogeneities of the substrate. And, of course, the erosion of the surface of the earth has both of these problems: the materials themselves have

strong directionalities arising from their geological history, while weather patterns have inherent directionality.

Deposited Surfaces

There are other classes of surfaces which have a fractal character besides those just described. Gouyet's classification calls these type 1 fractals, which are essentially solid volumes whose surface boundary has a fractal geometry. As noted above, type 2 fractals are the boundaries of structures which are fractal, such as clusters, and type 3 fractals involve networks such as pore structures (like the Menger sponge mentioned above). While neither of these is our primary concern here, either may sometimes be important when dealing with the characteristics of surfaces.

For example, many surfaces of commercial importance are produced by the agglomeration of particles onto a substrate. The important variants of this approach are whether the particles arrive by random diffusion or with ballistic energy, and whether the agglomeration is simply a matter of particles sticking together or whether there is a chemical reaction involved which may either increase or reduce the local concentration gradient and sticking probability. It has been known for some time that these types of coatings produce surfaces which have a roughly self-similar appearance. They have even been described as looking like a "cauliflower" because of the presence of irregular mounds of many different sizes, so that a picture (typically from a scanning electron microscope) does not betray the magnification used.

Depending on the method of production, the material under the surface may be dense or porous. If it is dense, it typically shows cones of growth which begin at the surface and enlarge upwards. The largest cones eventually close off the growth path for smaller ones, so that the structure of the surface continually evolves toward a coarser and coarser arrangement of cones, each of which is one of the "cauliflowers" seen in the microscope. The key to the self-similar size distribution is the time distribution of the initiation of growth and the rates of growth and spread of the cones.

Open structures of particles deposited on a surface may have a fractal structure described by the same type of mass fractal as used for classic particle agglomeration in two or three dimensions. This is a very rich and well-studied field. In the simplest example, consider the formation of a soot by the following process: individual particles move randomly due to Brownian motion; if a particle touches a growing cluster, it sticks. Figure 14 shows a simple example structure simulated in two dimensions. It has a similar appearance to the formation of frost crystals on a window, electrical discharges in a dielectric, viscous fingering of fluids in a thin layer, and many other natural phenomena. This is diffusion-limited aggregation, and it produces very open dendritic structures because of the low probability that a particle can random walk down between the growing dendrite arms without touching them. As the size of such a cluster grows (as measured by the distance between farthest points or the average radial distance of particles from the center), it becomes less and less dense.

In effect, we have a Cantor dust whose dimension is less than that of the space which it occupies. A plot on log–log axes of the total number of particles (or mass) vs. radius forms a straight line, whose slope gives the fractal dimension. If the particles were packed together to fill space, their total number would increase as the cube of radius. The agglomerate would have a dimension of 3, the same as the topological dimension of the "embedding space" in which it formed. For a cluster grown with the rules described above, the dimension of the cluster is less than 3, and for some sets of rules can be less than 2. This means that even the projection of such a cluster is not dense, but shows a lacelike pattern with innumerable holes.

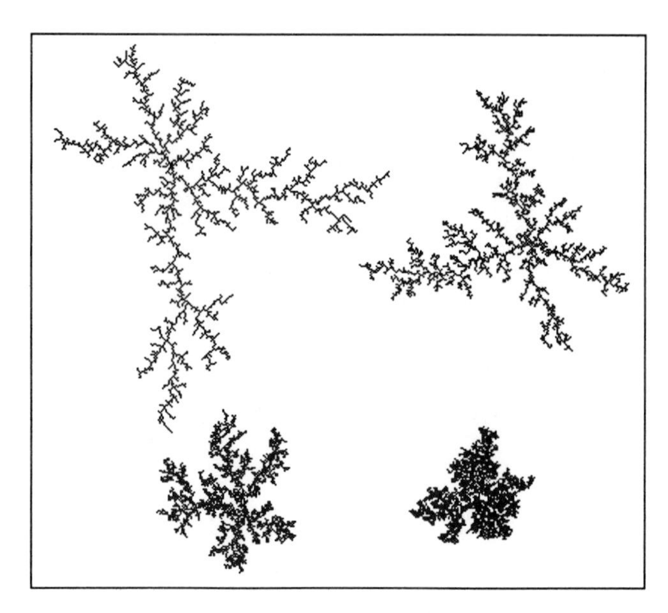

Figure 14. Diffusion-limited aggregation simulations in two dimensions produced by allowing particles to random walk on a square grid and stick whenever they touch. The sticking dimensions are 100%, 25%, 5%, and 1%.

Modeling of these cluster processes can include many variants on the rules. These include changing the probability that particles will stick when they touch, and perhaps varying that probability with direction to simulate crystallographic effects. It is also possible to give the particles a nonrandom component of motion to simulate ballistic deposition, to allow particles to combine together to form separate clusters which may then stick together, and so forth. Simulation of these processes in 2- and 3-dimensional spaces using computers has become a very widespread technique, applied to many topics in chemistry and physics.

It is a simple step to carry out the agglomeration on a substrate. Allowing particles to random walk until they strike the surface or another particle which has already stuck to the surface directly or indirectly, can build up a coating. Biasing the direction of motion or varying the sticking probability of the particles can be used to simulate physical conditions of coating deposition. The fractal dimension of the clustering can be measured by plotting (on log–log axes) the total number of particles as a function of distance from the substrate. Since the particles are all assumed to have the same mass, this is called the mass fractal dimension.

Figure 15 compares an example in which the particles can move only in a vertical plane with the more general case in three dimensions. Notice that in three dimensions, the structure can be so open that only isolated points appear in a section plane. When this kind of deposited agglomerate is viewed from above, only the highest point at each x,y location is visible. The resulting surface is not fractal, although the internal structure of the coating is.

Pore Structures

Porous materials can be produced in many ways in addition to particle agglomeration, of course. Sintering, chemical reaction, corrosive attack, and other processes can all give rise to pore networks. Other, man-made networks such as communications, road, and power grids also share some of the same mathematical characteristics. The important characteristics of any grid are the number of connections or links which meet at each node, and the fraction of such links or nodes which are open or occupied. This can be dealt with generally using the language and tools of percolation theory (Stauffer and Aharony 1991). There have been attempts, as will

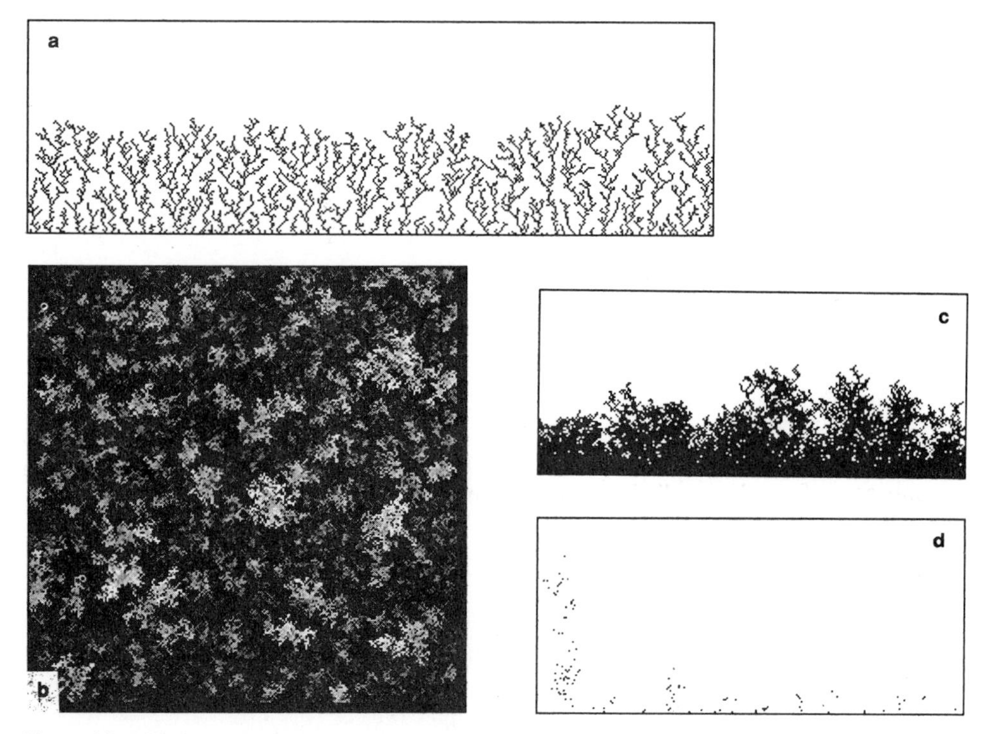

Figure 15. Diffusion-limited deposition on a substrate, comparing the result for motion in a plane with motion in three dimensions: **(a)** diffusion-limited aggregate on a line in two dimensions, with 100% sticking probability, **(b)** normal view of DLA deposition in three dimensions onto a surface with 100% sticking probability (brightness is proportional to elevation), **(c)** projection of 3D DLA in (b), and **(d)** vertical section through 3D DLA in (b).

be discussed later, to use the same approach to investigate fracture of materials as the breaking of "bonds" in such a network. More directly, percolation theory can describe the process of phase transitions in materials, magnetization and the formation of domains, and other processes that do not seem at first glance to have a "network" character.

The connection between networks and clustering is actually rather close. Imagine a network or grid of points in two dimensions, and consider what would happen if the links between the points were randomly selected and broken. How many of the links would have to be broken before the entire network "came apart"? When that happened, what would be the size distribution of the fragments? For the case of a square grid in two dimensions, the answer is that slightly more than 59% of the links must be present for the integrity of the network to remain, and that the size distribution of the isolated islands of network which do not form part of the continuous "backbone" has the power-law form which we associate with fractals.

Percolation networks are fractal. This has important consequences for trying to recover the oil from a reservoir, because it identifies the requirement for the pore network in the rock needed to extract the oil, and the size distribution of oil pockets that will remain unrecovered. The same theory applies to gelation of macromolecules in biochemistry, the spread of epidemics or forest fires, and efficient programming of multiprocessor computer systems or the telephone network. Do you recognize the self-similarity or network clustering in connecting home telephones to exchange offices, of connecting those exchanges to regional offices, and

so on up to the largest distances? What density of telephone calls would cause the network to saturate? It is the universality of these mathematical relationships that interests many of the people who study them.

One particular type of network that seems to occur frequently in natural phenomena is the branching pattern which can be seen in the alveoli of the lungs, in river patterns, and in plants. This is a pattern in which a single stalk branches into two (or more) branches, which then branch again, and so on. The total length of the network increases rapidly as smaller branches are added, but the area of volume occupied by the network remains finite. In most real examples, the number of such branchings is quite finite, unlike mathematical fractals in which the iteration proceeds to infinity. Nevertheless, the behavior over a range of scales obeys the usual self-similar fractal rules.

It might seem that the rules for such branching would have to be quite complex to create the natural structures we routinely encounter, such as an oak tree, and to account for the observed variation in these structures while finding the important differences between the rules which create one structure (the oak tree) and another (a pine tree). Not so. It has been shown that only a very few rules, applied iteratively, are needed to produce branching networks which mimic both the important universal characteristics and the natural variability of real structures. If Nature has encoded the rules for such branching in the genetic code, it may not take very many characters in the code to specify the distinctive shapes of plants, or the efficient branching structure of blood vessels, or of the dendritic branching of neurons.

L-Systems

There are several different but broadly equivalent ways to describe these rules. One is known as a "context-free grammar" or an "L-system" (L for Language or Lindenmayer). It is capable of generating many kinds of fractals, including networks and boundaries. Consider a sequence of characters of the form F–F–F–F. If this is interpreted as commands to a "turtle" which draws lines on a sheet of paper, we might have:

```
F = move forward one unit
- = turn 90 degrees to the left
```

Then this string of characters specifies the drawing of a square. Now suppose we introduce a rule which says that every instance of the character F is to be replaced by the string F–F+F+FF–F–F+F, where the only new character introduced is +, meaning a turn 90 degrees to the right. This corresponds to drawing the first stage in the iterative generation of the Koch island shown back in Figure 2. Iteration consists of re-applying the rule to the string.

Changing the rules can be done in several ways. One is to redefine the meaning of characters, for instance making the turn angle 60 degrees instead of 90 degrees. Another is to introduce a richer set of characters, such as M (move forward without writing) or X (do nothing, but rules may substitute other strings of characters for an X later on), or turn angles in three dimensions instead of in two. A particularly important addition to the language is the use of brackets to indicate saving the current state of the turtle (location and orientation) and returning to it later.

For instance, the rule

```
F becomes F[+F]F[-F]F                                          (1)
```

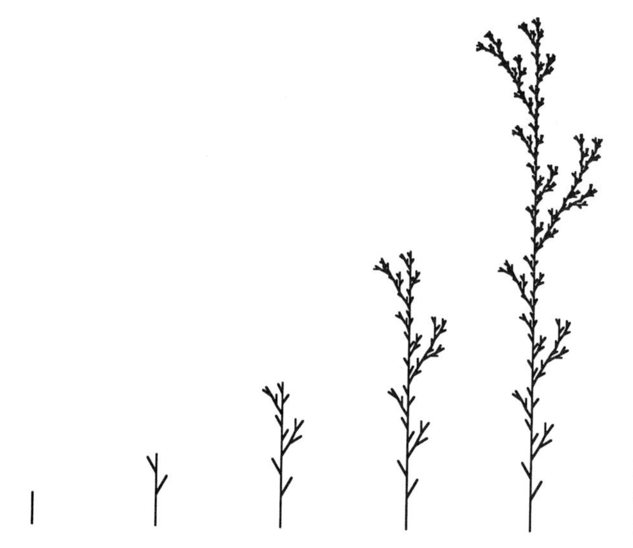

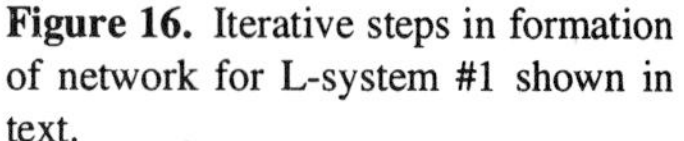

Figure 16. Iterative steps in formation of network for L-system #1 shown in text.

will draw a "weed" as shown in Figure 16 after five iterations. The turn angle for the + and – commands is 30 degrees. Keeping the same angle, but introducing the "dummy" character X, we can write a procedure consisting of a pair of rules:

```
Initial string = X
Rule 1 X = F[+X][-X]FX
Rule 2 F = FF
```
(2)

Iterating this rule produces the network shown in Figure 17. Changing only Rule 1 can produce quite different networks, which have much of the appearance of plants, as shown in Figure 18.

Of course, these networks are as completely deterministic as the Koch islands, and so it is necessary to find ways to add some randomness to them. This is most readily done by having multiple rules which apply to the same character, with probabilities associated with each. Then at each iteration and for each character substitution, a random number is generated and used

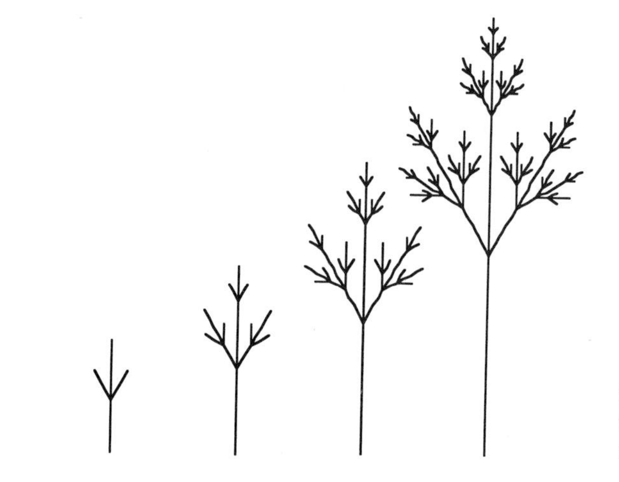

Figure 17. Iterative steps in formation of network for L-system #2 shown in text.

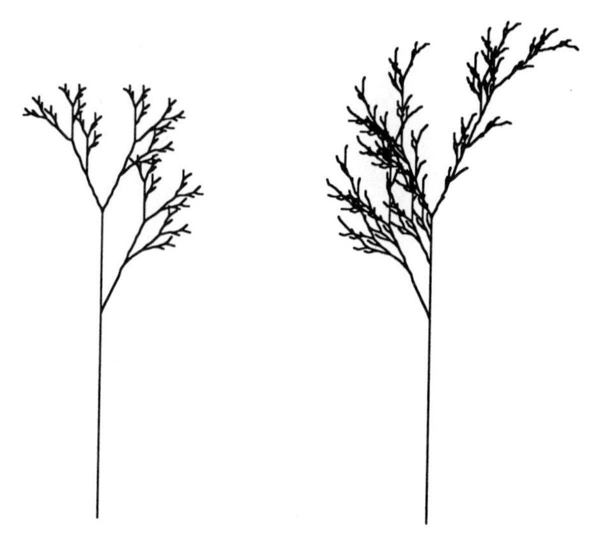

Figure 18. Comparison of networks formed by changing rule 1 in L-system #2 in text: Left: X = F[+ X]F[– X] + X; right: X = F – [[X] + X] + F[+ FX] – X).

to choose one of the multiple rules. For instance, let us change the rule used to generate Figure 17 by offering a choice of three rules:

```
Initial string = F
Rule 1a (p=0.34) F = F[+F]F[-F]F
Rule 1b (p=0.33) F = F[+F]F
Rule 1c (p=0.33) F = F[-F]F
```

(3)

Each time this process is repeated, the result will be slightly different as shown in Figure 19. However, all of the "weeds" grown by this set of rules are recognizably of the same "species." Similarly, Figure 20 shows a set of networks grown with the following rules:

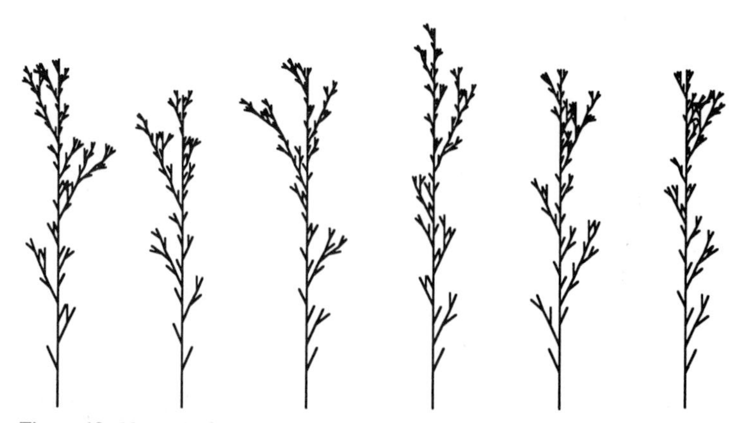

Figure 19. Networks formed by several executions of L-system #3 in text.

Figure 20. Networks formed by several executions of L-system #4 in text.

```
Initial string = F
Rule 1a (p=0.40) F=FF-[-F+F+F]+[+F-F-F]
Rule 1b (p=0.30) F=F[+F]F[-F][F]
Rule 1c (p=0.15) F=F[+F+F]F
Rule 1d (p=0.15) F=F[-F-F]F
```
(4)

By adding appropriate commands for three-dimensional branching, allowing some variability in the branching angle, and adding a few more characters to the alphabet to represent specialized structures such as leaves and flowers, a wide variety of real plants can be mimicked. Using computer graphics to render these stick drawings as plants produces images such as shown in Figure 21. Similar routines have been used to generate plants that do not look familiar, but nonetheless look real. This has enabled movie makers to create exotic landscapes for science fiction movies far more easily and cheaply than the construction of a physical set.

Finding the structural units that correspond to a particular network or class of networks, such as the alveoli in the lungs, electrical discharges in a dielectric, cracking in a high-strength

Figure 21. Three-dimensional flowering plant generated with an L-system.

metal alloy, or the change in dendrite branching in neurons affected by Alzheimer's disease, is a difficult problem that must generally proceed with considerable trial and error. Figure 22 shows an example of a neuron from human brain tissue, stained for visibility in an electron microscope image. That there are such rules and that they are in principle discoverable is known (Barnsley, Ervin et al. 1986; Barnsley 1988; Khadivi 1990; Barnsley and Hurd 1993). The best-known example of this is the generation of a fern by iteratively combining smaller copies of the same basic shape.

The four rules below are able to generate a realistic image of a fern. Each rule corresponds to a rotation, displacement and shrinkage of a subelement of the structure. The rules are applied by starting at any point, selecting one of the rules (with the frequency shown by the probability values, from 1% to 84%), and then moving from the current point to the next point according to the rule. This point is plotted, and the procedure iterated to produce the entire figure. The more points, the better the definition of the result. Figure 23 shows an example. The entire fern with 20,000 points shows the self-similarity of the overall shape. As a portion of the image is blown up for examination, more points are required. Finally, the limit of magnification is set by the numerical precision of the values in the computer, as indicated in the figure.

From a mathematical point of view, the four transformations are basis functions which can be added together in proportion (the *p* or probability values) to produce the overall object. In Fourier analysis, the cosine terms are basis functions and the coefficients specify the amount of each to add to generate the final result. Here, the first problem is to find the correct basis functions for a given image or surface. It is expected that only a few will be needed; Barnsley has shown the ability to compress and reconstruct full-color images of arbitrary scenes with typically only a few terms.

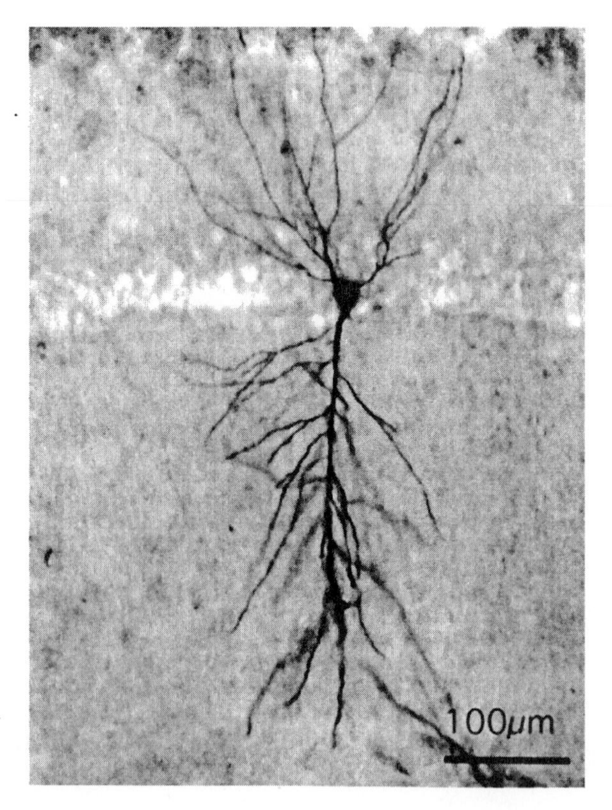

Figure 22. Electron microscope image of a stained neuron showing its branching pattern.

Transformations (Basis Functions) for the Fern Image

```
1 (p=0.840)         x'=+0.821·x + 0.845·y + 0.088
                    y'=+0.030·x - 0.028·y - 0.176
2 (p=0.075)         x'=-0.024·x + 0.074·y + 0.470
                    y'=-0.323·x - 0.356·y - 0.260
3 (p=0.075)         x'=+0.076·x + 0.204·y + 0.494
                    y'=-0.257·x + 0.312·y - 0.133
4 (p=0.010)         x'=+0.000·x + 0.000·y + 0.496
                    y'=+0.000·x + 0.172·y - 0.091
```

The same principle has been applied to the compression of photographic images, where is it known as the Collage Theorem (Barnsley, Ervin et al. 1986; Barnsley 1988; Barnsley and Sloan 1992; Barnsley and Hurd 1993). This is shown schematically in Figure 24. The problem of finding the correct basis functions is, however, far from trivial. Knowing that such functions must exist gives few leads to discovering them. Each consists of a mapping (a combination of translation, scaling, rotation, and warping) and a relative contribution or probability. How these

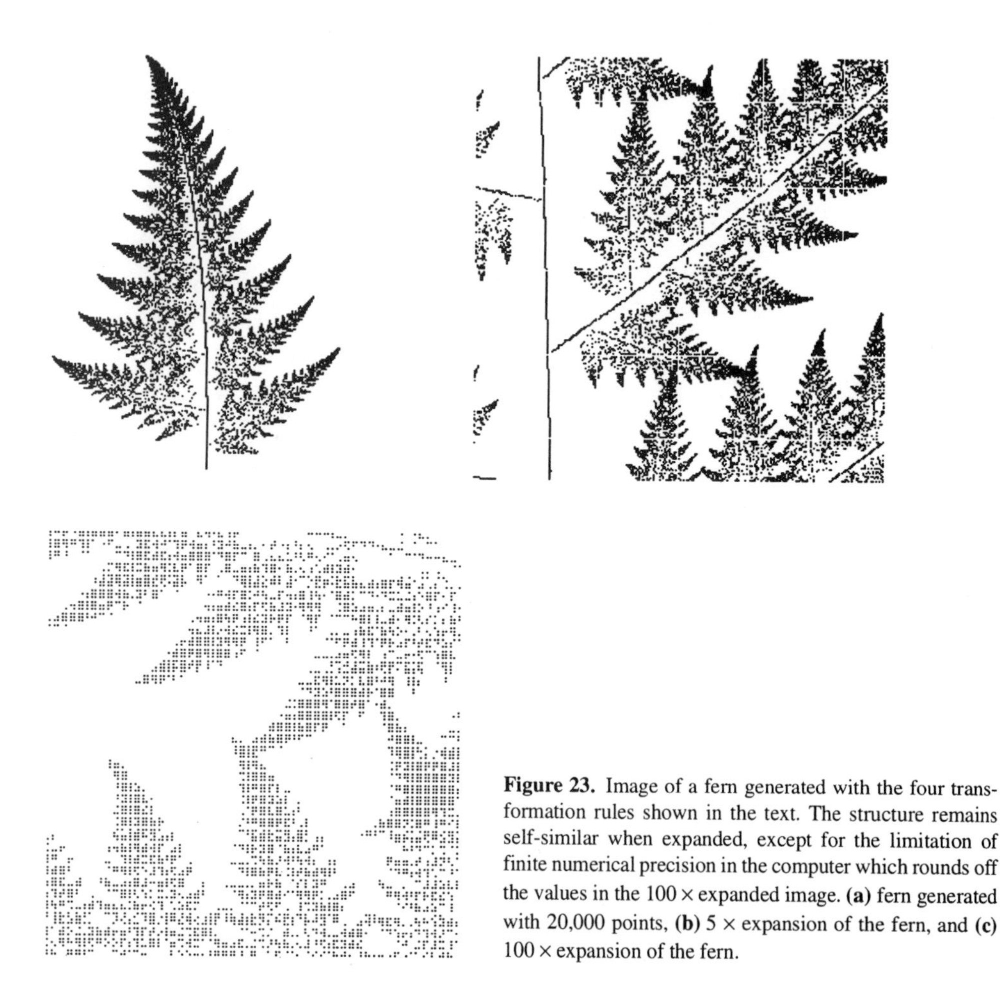

Figure 23. Image of a fern generated with the four transformation rules shown in the text. The structure remains self-similar when expanded, except for the limitation of finite numerical precision in the computer which rounds off the values in the 100 × expanded image. **(a)** fern generated with 20,000 points, **(b)** 5 × expansion of the fern, and **(c)** 100 × expansion of the fern.

can be interpreted for a fractal surface image to supplement the simple fractal dimension is not known. Limited attempts have been made to apply the method to images of fractal surfaces (Wang 1993).

Application of this technique to analyze a profile (Oliver 1992) shows that the number of affine transformations needed to preserve all of the exact detail in the original is equal to the number of points in the profile minus one. Consequently, there is no compression achieved and no compact characteriztion of the fractal. Of course, when the Barnsley method is applied to image compression, some of the terms are set to zero so that the compression becomes "lossy." Reconstruction does not recreate each pixel exactly but replaces them with visually plausible values. The art of deciding which self-affine transformations are important is apparently very hard to automate, and in any case it is not clear how they might be interpreted to describe the structure of a fractal surface.

Whether the rules of the grammar for a network would convey useful information about the origin of or differences between instances of networks is likewise unknown. The relationship of these networks to the surfaces which are the principal interest of this book is tenuous. Some of the measurement methods for surfaces will also sense some or all of the internal porosity in a material, and so the pore network would be important. So-called BET (gas adsorption) measurements of the surface porous materials (Brunauer, Emmett et al. 1938) using

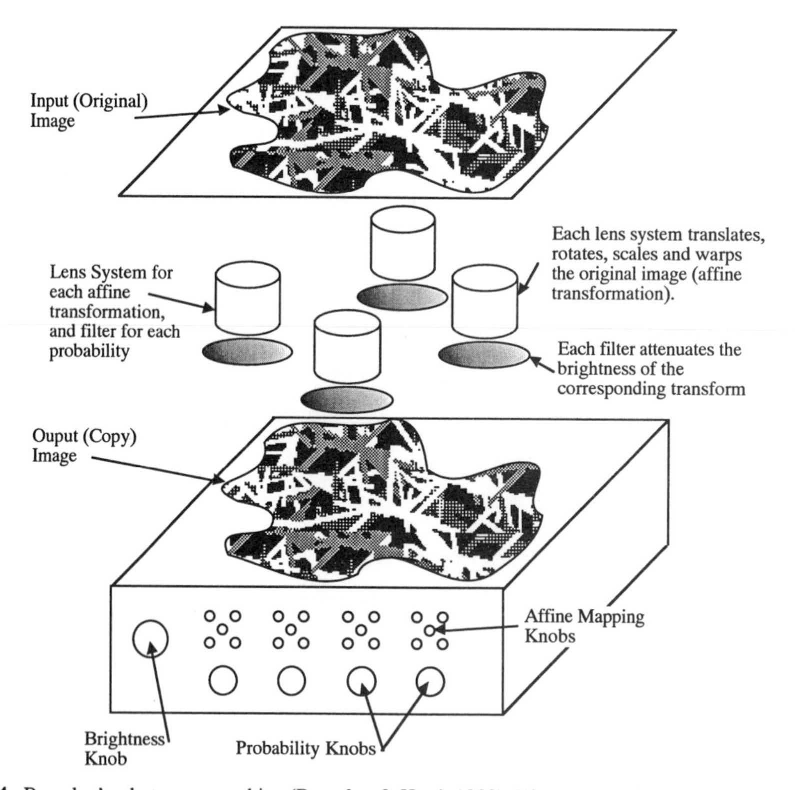

Figure 24. Barnsley's photocopy machine (Barnsley & Hurd, 1993). His collage theorem states that there is a set of mappings and probabilities which produce an output identical to the input to any desired degree of accuracy. These geometrical transforms and probabilities constitute his image compression algorithm (U.S. Patent # 5065447). For a fractal surface image, the mappings may provide information beyond the simple fractal dimension value, which may be related to the history or properties of the surface.

molecules of varying size will sample different amounts of the pore structure depending on the diameter of the pores and the ability of molecules to penetrate the network (Avnir and Farin 1984). Most such measurements on real objects produce "surface" fractal dimensions that approach 2.99... when the pore structure is a richly branched network, indicating that the total surface essentially fills the volume of the object. This tells something important about the object, but does not offer any discrimination about the possible structural variations of the network.

Growing a forest of such branched structures on a surface would produce the functional negative of a pore network, and certainly such a surface could be considered a fractal. However, it seems closer to the type 2 fractal constructed by agglomeration (in this case of branches) than to the type 1 fractal. And indeed, the dimension of a "surface" produced by measuring only the uppermost point at each x,y location in such a forest is not fractal.

2

Measuring the Fractal Dimension of Boundary Lines

Richardson Plots

One of the earliest observations of fractal behavior of boundary lines was carried out by Richardson, although it was published obscurely (Richardson 1961). Driven by an interest in finding relationships between the economic and military conflict between nations and the length of their borders, Richardson tried to measure these lengths from maps. But he quickly discovered that the perimeter lengths varied markedly depending on the scale of the map, regardless of the care he took in performing the measurement.

He then proceeded to measure in an orderly way the increase in the apparent perimeter of boundaries such as the west coast of Britain as a function of the unit of measurement. The simplest technique for this is to set compass dividers to a specific stride length and then "walk" along the boundary on the map, counting the number of steps. This is shown schematically in Figure 1. A small correction is needed for the final partial step at the end. Of course, the exact number of steps and the length of the final partial step will vary slightly for a given boundary depending on the starting point. But if the total number of steps is reasonably large (i.e., if the step is small compared to the total extent of the boundary), this variation is minor. In some cases, several different starting points can be used and the results averaged.

It is also necessary to decide how the step is to be made when the same stride length can intercept the line in more than one location, as shown in Figure 2. The choices are to always

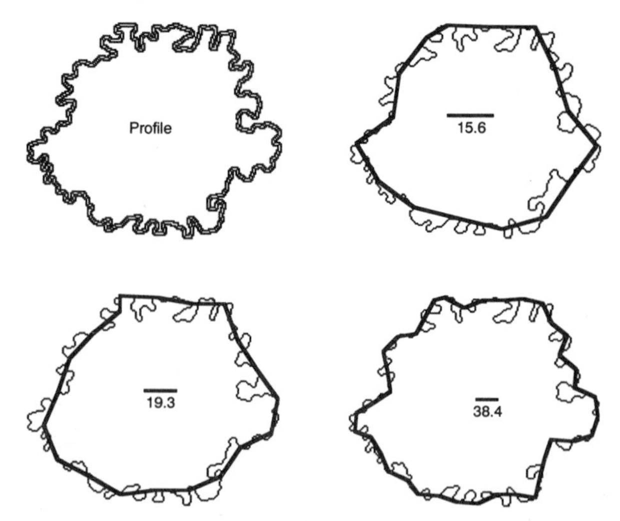

Figure 1. Principle of the Richardson procedure. An irregular outline is converted to a polygonal approximation using different side or step lengths. The total number of sides (including the final partial one) times the length gives an estimate of the total perimeter which increases as the step size becomes smaller. The procedure can be executed by "walking" along the boundary with a pair of dividers set to a fixed stride length.

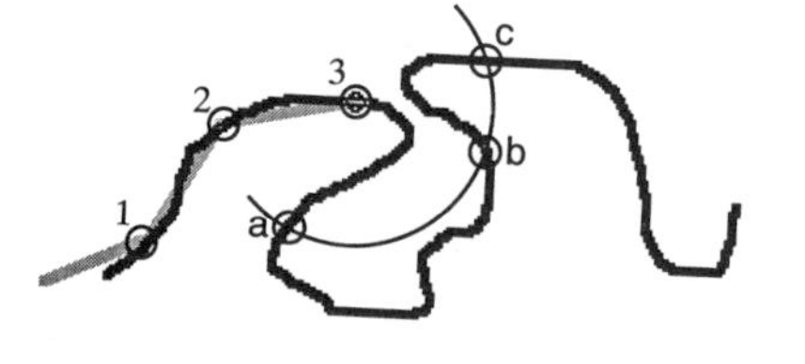

Figure 2. Choosing where to take the next step in a Richardson traverse along a boundary. Coming from the left, the next point selected after 3 could be at a, b, or c.

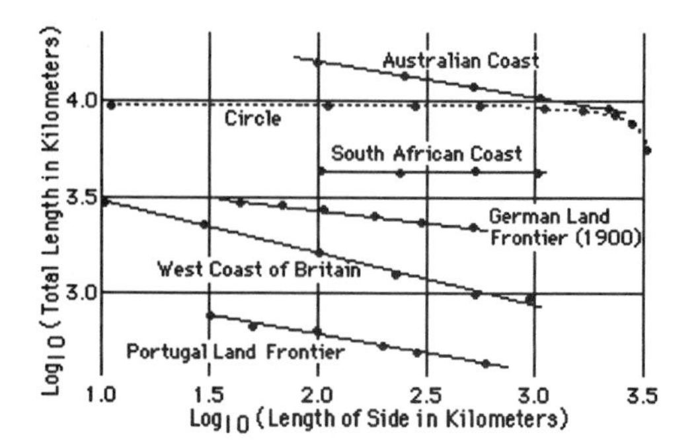

Figure 3. Richardson's plots (1961) of the length of various geographical boundaries versus the distance used for measurement (the length of the side of the polygon used to fit the boundaries).

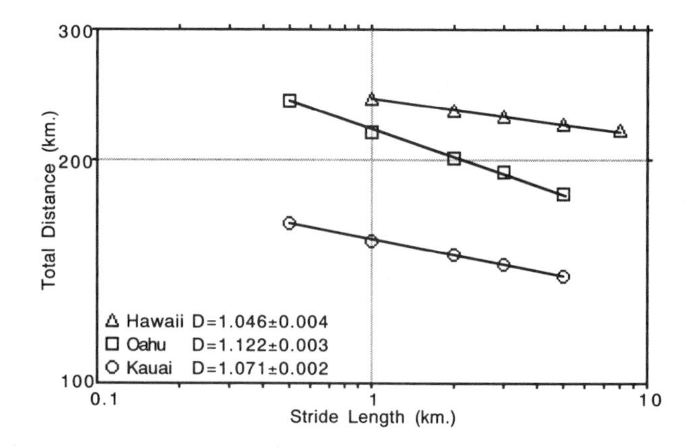

Figure 4. Richardson plots constructed for three of the Hawaiian islands. Hawaii is the youngest and "smoothest" in terms of elevation, Kauai is the oldest, most weathered and most "rugged." The elevation roughness does not correlate with the boundary fractal dimension.

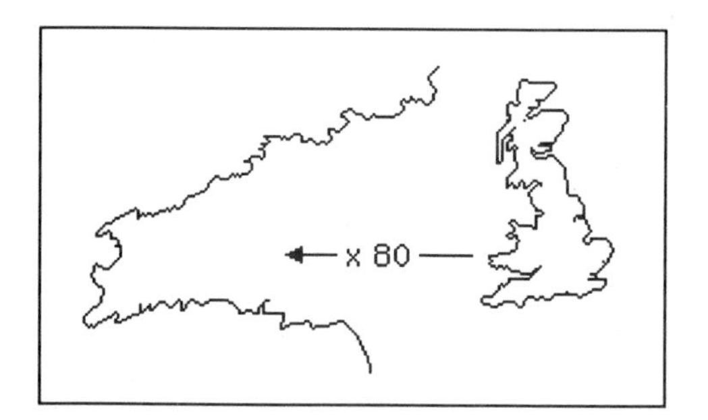

Figure 5. The west coastline of Britain at two magnifications, one at a scale 80 times larger than the other.

take the nearest point or the farthest point, or to swing the dividers alternately from one side of the line or the other. It turns out that these procedures each produce slightly different perimeter values for a given stride length, but do not affect the trend of perimeter as stride length is varied.

Richardson observed that if the measured perimeter was plotted against the value of the stride length using logarithmic axes, the result was a straight line. Furthermore, as shown in Figures 3 and 4, the slope of the line was different for different boundaries. Other researchers have plotted similar graphs for additional boundaries, generally ones of particular interest to them (e.g., the coast of Norway for Jens Feder, Lake Sudbury in Canada for Brian Kaye, etc.). The result is always that the more irregular coastlines as judged by human observers have the higher slope for the line. Perfectly Euclidean political boundaries such as the artificial straight lines which bound the state of Colorado produce a horizontal line on this plot, indicating that the perimeter value does not change with measurement scale, but boundaries that follow natural lines such as rivers or coastlines always have some roughness and produce a plot with some slope.

The increase of perimeter length as the measuring scale is reduced reflects the fact that real boundaries are composed of irregularities at all scales, and as the magnification of examination is increased, more of the roughness is revealed. But the particular result of the Richardson analysis, that the line is straight on a log–log plot, means more than this. The roughness of the boundaries is self-similar, meaning that the amount of increase in observed boundary length is the same at any scale. This means that given an image of the boundary with no scale marking, the visual appearance of the roughness would give no clue to the actual magnification. Figure 5 illustrates this schematically for the boundary of the west coast of Britain, one of the boundaries studied by Richardson. Magnifying the boundary by 80 times produces a line that is visually just as "crooked" as the original. This kind of self-similarity is the hallmark of a fractal, and the fact that it appears so often in natural phenomena is the reason that this geometry has made such a sudden and massive impact on the natural sciences.

The slope is negative, since the largest value for the perimeter length is obtained with the shortest stride length as shown in Figure 6. The slope of the line is between zero (the Euclidean limit) and one. The fractal dimension of the line is simply the sum of the magnitude of the slope and 1.0, the topological dimension of a line. The symbol D_L or D_B will be used for the fractal dimension of the line or boundary, and for a line this can be written as 1.d or (1 + d) to indicate that the fractional part of the dimension is obtained from a measurement such as a Richardson plot. Later on, we will encounter fractal surfaces for which it will likewise be convenient to write 2.d or (2 + d) as the dimension for the surface, abbreviated as D_S.

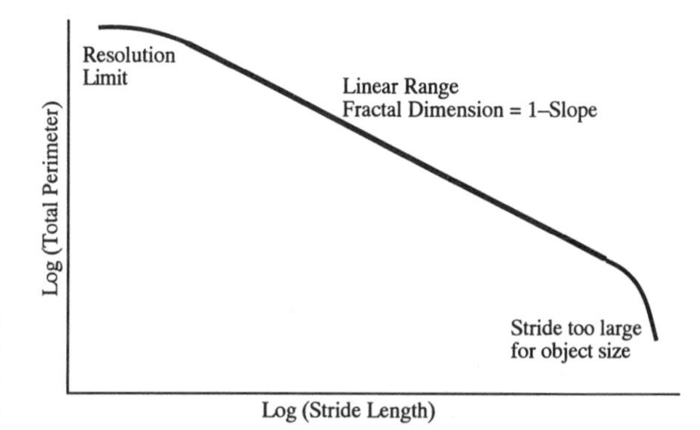

Figure 6. Schematic diagram of a Richardson plot. The straight portion may cover many orders of magnitude on log–log axes, and gives the fractal dimension.

Actually performing a structured walk with dividers on an image or a map is difficult. The small errors due to wobble of the dividers or enlargement of the holes accumulate and may bias the plot. Even more important is the fact that the scale of the image or map limits the range for the stride length to about one order of magnitude. The finite resolution of the image or map, and of unassisted human vision, makes very small stride lengths impractical, while at larger stride lengths the number of steps that can be taken before the boundary is completely traversed is small. The schematic Richardson plot in Figure 6 shows the deviation of the line at both extremes.

If additional images with higher magnification, or maps with a smaller reduction scale can be obtained, measurements may be performed over many decades. For instance, Lovejoy (1982) showed data from clouds measured at various scales (from local images and weather radar to satellite imagery) which show a straight line over six orders of magnitude. Cloud boundaries ranging from small single clouds to entire storm systems share the same self-similar roughness, giving a fractal boundary dimension of about $D = 1.36$. The measurement parameter comes, however, not from a Richardson plot of the boundaries of an individual cloud, but from the variation of measured perimeter and area of many clouds of different sizes, a technique discussed below.

There is a clear upper limit implied by the size of Earth, but it might be interesting to see if the formation of clouds on other planets such as Jupiter could extend the relationship to even larger dimensions, or if the different physical system there (different gases, gravity, winds, etc.) would produce a different slope. At small dimensions, some researchers (Underwood and Banerji 1986; Williford 1988) have taken the apparent deviation of the plot from a straight line as an indication that some other, more complex relationship is present. However, other work on similar samples (metal fractures) suggests that the limitation is an artifact of the finite resolution of the imaging method used and possibly of misapplication of the Richardson method to a self-affine rather than self-similar profile as discussed below. Improving the image resolution from that of a light microscope to a scanning electron microscope extends the straight line farther.

Of course, it is clear that at some dimension, perhaps that of metal grains, crystals within rocks, or even the individual atoms, there must be a change in the physical nature of the boundary. In fact, for some objects there may be other characteristic dimensions at which the nature of the boundary changes for physical reasons, such as the forces which dominate its formation. When atomic forces in the form of crystal structure or surface tension forces are strong, the boundary may become Euclidean.

In other cases, the fractal dimension may be different in one range of dimensions as compared to another. For instance, the roughness of the sand on the beach, which arises from the microscopic wear and fracture of the grains and may have taken place in river environments very different from the beach which is the final resting place for the sand, will not necessarily lead to the same fractal dimension as the boundary of the coast on a scale of meters or miles. Yet both may be fractal. Kaye calls such situations in which the Richardson plot can be conveniently divided into two (or more) straight line regions "multifractals," but because that word is also used with a very different meaning in some areas of physical and mathematical interpretation, we will use the terminology "mixed fractals" here. The difficulties in this interpretation, and other effects which may produce similar, confounding effects in the Richardson plot, are discussed in Chapter 6.

Notwithstanding these limitations, the fact remains that in many situations a single number, the fractal dimension, summarizes concisely and (apparently) meaningfully the "roughness" of a boundary. Figure 7 shows several feature profiles. They happen to have come from a microscopic observation of particulates, but they could equally well be corn flakes or islands. The roughness does not reveal any intrinsic scale. The number shown for each outline

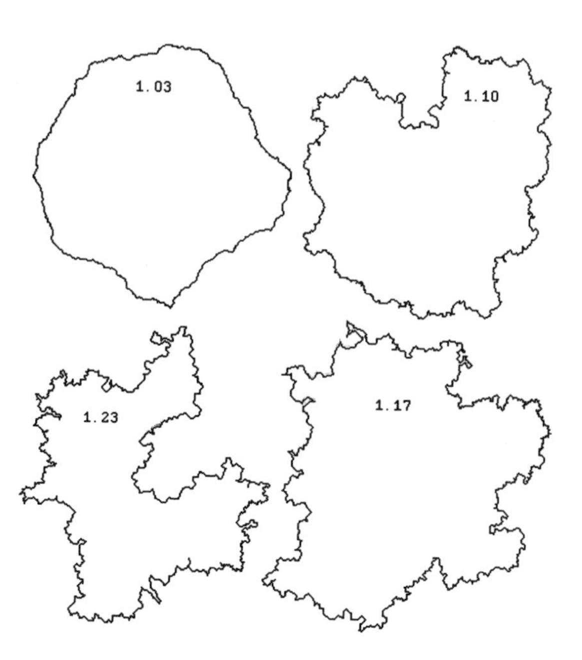

Figure 7. Several fractal outlines with varying fractal dimensions.

is the fractal dimension. The observed visual roughness or irregularity of the outlines increases with the fractal dimension. In many tests with human observers it appears that ranking of the apparent roughness of surfaces, outlines, and other structures produces the same order as measurement of the fractal dimension.

This is interesting, because human vision is easily tricked in attempts to estimate other shape descriptors such as circularity. Even comparing size parameters such as length can be confused by differences in color, orientation, or juxtaposition. This sensitivity is exploited in many visual illusions and magic tricks. Finding that human vision can apparently respond to fractal dimension is a further indirect evidence that this is an important physical property that helps to describe the natural world.

Using a Computer

A computer system can be used to facilitate the measurement of perimeter lengths and the construction of a Richardson plot. The simplest method is to place the image or map onto a digitizing tablet and trace the boundary line. Digitizing tablets use a variety of signals to locate the position of the pen or crosshairs. Some use radio frequency signals embedded in the tablet and a detector coil in the pen, some use capacitance or electrical resistance between conducting surfaces within the tablet and respond to stylus pressure, and some use ultra-high-frequency sound waves which are triangulated to locate the pen, to give just a few examples. But all of these devices are generally capable of locating the stylus tip or crosshairs to better than 0.1 mm and often to better than 0.01 mm in x and y dimensions (unaided human vision can typically resolve points about 0.1 mm apart at normal viewing distances). The tablets transmit these coordinates to the computer at high speed so that as the boundary line is traced a series of closely spaced point coordinates is recorded. If the tracing speed is reasonably constant, the point spacing is also fairly uniform.

Since the list of stored x, y coordinates in the computer represents a polygonal approximation to the boundary, it is possible to sum up the total length of the line or closed boundary

by adding the distances from point to point (Schwarz and Exner 1980). From the Pythagorean rule, this is simply

$$\text{Length} = \sum_{i=1}^{n} \sqrt{(x_i - x_{i-1})^2 + (y_i - y_{i-1})^2}$$

where the index i counts through the list of points. This perimeter value is the maximum value obtainable from the given image or map, with the points spaced closely together to produce a very small stride length. The average stride length can in fact be determined by taking the mean of the individual steps.

If every second point in the list is used, the stride length is approximately doubled, and the perimeter is reduced. The calculation is the same as shown above except that the index i is restricted, for example, to even values and the $(i-1)$ subscript becomes $(i-2)$. If the procedure is repeated using every third point, every fourth point, etc., the stride length will continue to increase, the total length will continue to decrease, and a Richardson plot can be constructed to determine the fractal dimension.

When every nth point is used for the calculation of the perimeter around a closed path, it is possible to choose any of n different starting points. Each will give a slightly different total length. The best way to deal with this (since it is the computer and not a human who is doing all of the arithmetic) is to perform the calculation for every possible starting point and average the results. This improves the precision of the resulting fit, and takes insignificant additional time, since the greatest labor is in the manual line tracing.

This is also the largest potential source of error, since most human operators find it difficult to trace accurately along an irregular line. It is in fact quite common to trace consistently on one side or the other (usually the outside of the line depending on which side is considered to be the feature being measured), and this will usually result in an underestimate of the roughness. This potential source of error, as well as the labor involved in the tracing, has encouraged finding ways to use automatic image acquisition and processing to bring the image into the computer and delineate the boundary.

Digitized Images

Video cameras produce an image which is convenient for digitization in small computers. The original signal consists of an analog voltage which varies along each raster line across the image. The digitization process consists of subdividing each line into a number of samples called pixels (picture elements) by a high-speed clock, and then measuring the voltage in that time segment to a fixed resolution. This same approach is used for many other raster scan imaging devices such as flat bed image scanners for macroscopic photographs, the scanning electron microscope, scanning confocal light microscope, scanning atomic force microscope, etc. Most of these produce images which initially have from 500 to a few thousand pixels across each line and a similar number of lines. Each pixel may have grey values which range from black to white over a range of 256 (2^8), 4096 (2^{12}), or even more steps. For a full-color image, the total space required to store such an image may be several million bytes of computer or disk memory.

The most typical images used for computer-assisted image analysis consist of an array of square pixels. These are separated into foreground and background pixels based on some classification, such as color, brightness, local texture, etc. Most of the segmentation methods have a very difficult time in automatically selecting the "correct" threshold setting which will define the boundaries. Since pixels have a finite size, they average the values from both sides

of the boundaries they straddle. But the act of segmenting the image requires assigning each pixel to either the features or the surroundings. This will cause some error in any case. In addition, human selection of thresholds is frequently biased as people tend to select values which make the boundaries as smooth as possible.

Figure 8 shows an example in which the "best" threshold clearly lies somewhere in the range of brightness values between the two peaks in the image histogram which correspond to the two phases present in the ceramic. This histogram is a plot of the number of pixels in the original image having each of the possible shades of grey. The peaks in the histogram correspond to the two different phases present in the structure, and a common segmentation technique, called thresholding, is to select ranges of brightness values that correspond to each of the phases (Russ 1990a).

In this image, which is fairly typical of many situations, the peaks are separated by a broad valley, and the threshold setting can be placed anywhere in this range of values. Any of the settings in this range may produce a foreground–background separation that appears plausible, but clearly no more than one can be correct. The different settings produce quite different estimates of the volume fraction of each phase present, as well as different boundary length and roughness. There is no universal solution to this problem, although in particular situations finding the setting that minimizes the change in feature area or perimeter length can be used as an automatic method.

There are some additional problems which may be present in this kind of classification. First, it is possible that the image has a gradually varying relationship between brightness (or color, etc.) due to nonuniform illumination, curvature, etc. There are ways to correct for this but they require some assumptions and additional work. Furthermore, if the measured signal does not vary linearly from one phase to another, for example when a logarithmic response video camera is used, the "midpoint" between the two extremes is not the arithmetic average. Finally, detector imperfections can bias the result. In the case of a video camera, for instance, the bright areas in the image tend to "bloom" or spread out into neighboring dark areas, so that bright features will appear too large (and dark ones too small). This will also tend to artificially smooth rough boundaries.

When image processing must be used to prepare an image for thresholding, the problem of accurately locating the boundary and preserving the details of its shape may be exacerbated (Russ 1992a). Some image processing operations shift boundaries, and many of them broaden the boundary so that it is more difficult to ascertain its true location. Many techniques artificially smooth boundaries, and some rely on finding the smoothest boundary to automatically establish a threshold. While this is a good strategy for images in which the boundaries really are smooth, for instance if they arise from physical processes such as surface tension or crystallography that may be expected to produce Euclidean rather than fractal outlines, obviously the results would be biased for a boundary that is really fractal.

On the other hand, the presence of electronic noise in the image, typically appearing as a random or Gaussian random variation in the brightness of pixels, can artificially increase the fractal dimension. This is because pixels near the boundary generally have brightness values near the threshold limit, and the addition of the noise can move them above or below the threshold, creating a "froth" of additional points along the boundary that increases the apparent length of the perimeter. There are a variety of effects that produce similar results from other kinds of imaging devices, such as the greater depth of penetration of the electron beam in the scanning electron microscope in regions of low density or atomic number, and the effect of internally generated electrical fields shifting the boundaries observed with the scanning tunneling microscope. Even if these effects are known, which is not often the case, they can be difficult to correct.

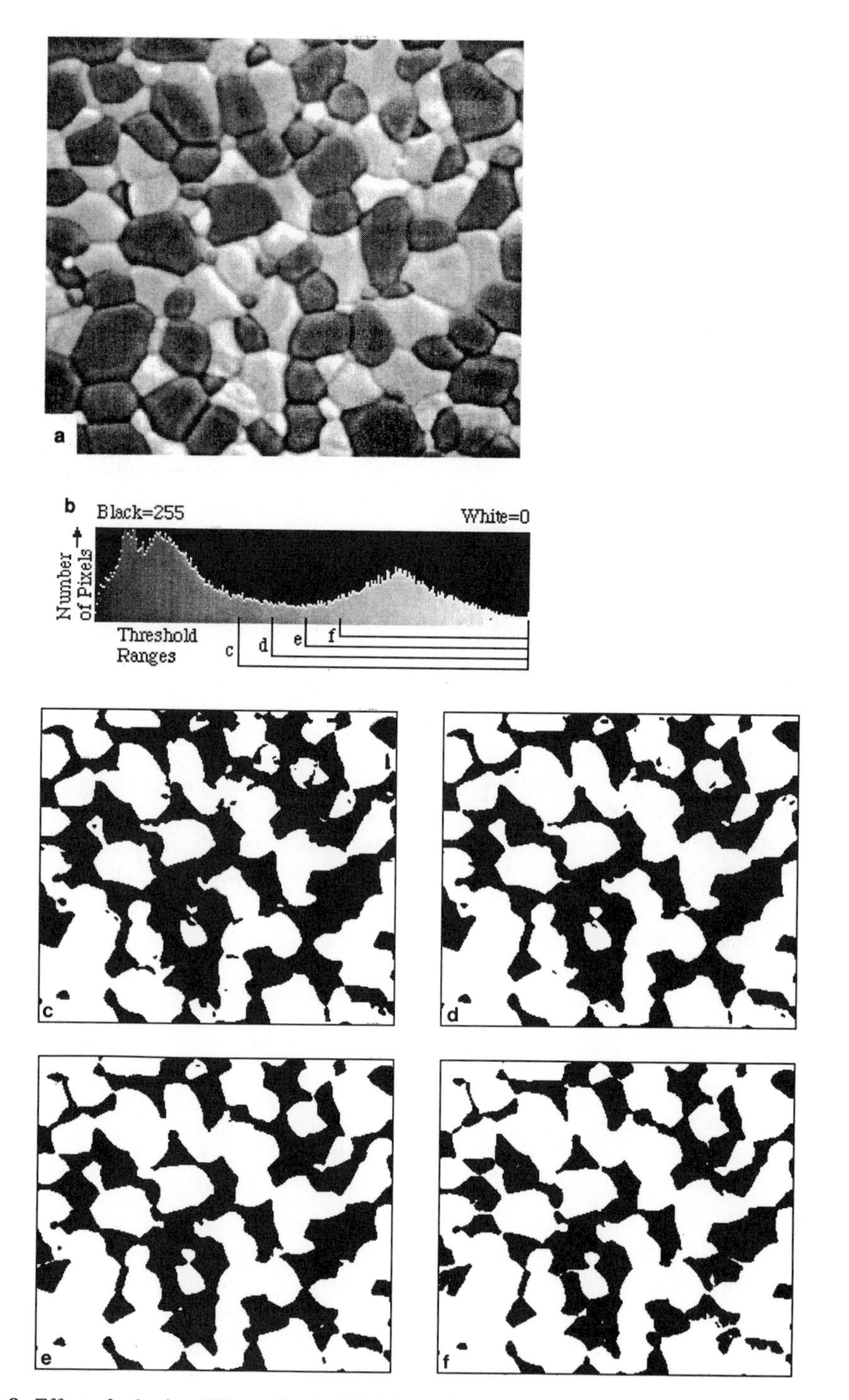

Figure 8. Effect of selecting different threshold brightness levels to discriminate the two phases (alumina and zirconia) in an SEM image of a ceramic. The "best" setting, which assigns the pixels to the phases with the least error, lies somewhere between the two peaks, but not necessarily midway or at the lowest point. The area fraction and boundary morphology of the phases vary with the different settings.

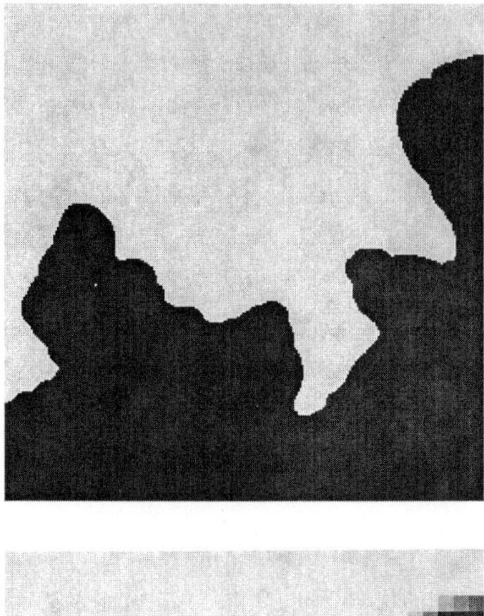

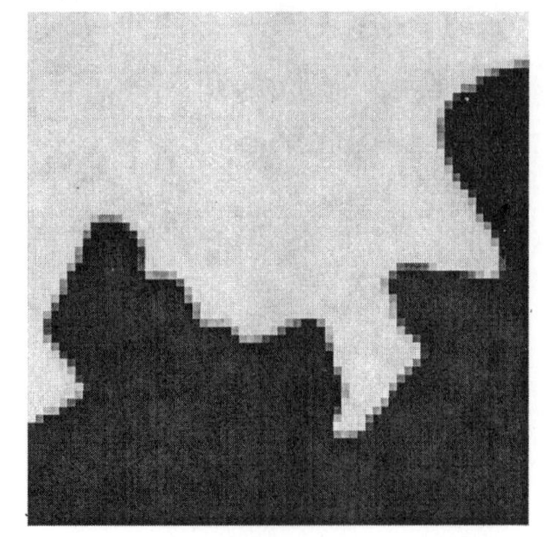

Figure 9. Diagram showing the effect of finite pixel size on boundary definition. Pixels along the boundary between the dark and light phases have intermediate brightness values which are averaged over the finite area of each pixel, depending on the relative size of the pixels.

The alternative to the usual "pixel-based" representation is boundary representation where lines are drawn on the image defined by the transition between pixels of color, brightness, or whatever the defining property is for the features. Figure 9 shows an illustration in which the pixels which straddle the boundary have an intermediate grey value according to where they lie with respect to the boundary. Interpolating between the two values for the feature and surroundings locates points which can be connected into a polygonal boundary representation. Using any reasonable number of points produces somewhat simplified and hence smoother boundaries. In any case, it cannot be known whether boundary irregularities are present which are smaller than the pixel dimension. For a truly fractal boundary, it is presumed that these small scale deviations are present, but the larger-scale irregularities can be used to obtain a characterization of the fractal dimension. Figure 10 shows an example in which the boundary irregularities are partially obscured by the finite pixel dimension.

Because the pixels are originally arrayed on a square grid, converting boundary points to a series of locations which are uniformly distributed along the boundary is rarely possible. Even for the case of a high resolution digitizing tablet used to trace a boundary, the coordinates are quantized in the x and y dimensions. This means that the sides of the polygon are not exactly of equal length, and this will bias the slope of a Richardson plot for very small scale stride

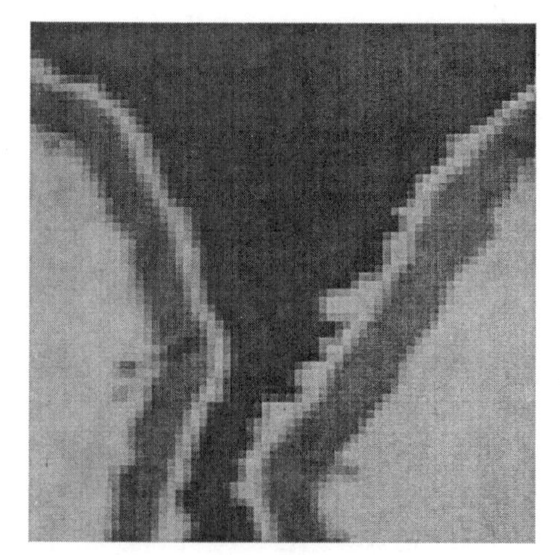

Figure 10. Metallographic image of a cross section through a coating, enlarged to show the individual pixels in the digitized image. Intermediate grey scale values for the pixels indicate averaging of the values from the different phases, but the pixel size obscures some of the fine boundary details.

lengths. A partial improvement can be obtained by starting the polygon or structured walk at every possible point and averaging the result. However, unlike the use of multiple starting points discussed above for a polygon with equal side lengths, it is not clear that averaging will produce an unbiased result for the case in which the vertices are points fixed on a square pixel grid.

Regardless of how the boundary is defined, the use of a pixel image of features and background imposes several limitations on the measurement of the fractal dimension of the boundary. One is the size of the pixels, which sets the lower limit of dimension for any kind of plot. If the pixels are smaller than the resolving power of the original optics, the use of the pixel size as the lower limit will show a leveling off in the log–log plot that is artificial. Another limitation comes from the various forms of bias mentioned above which generally produce smoother boundaries than may actually be present, or noise which generally produces a rougher apparent boundary.

In addition, the measurement techniques which work best on a pixel array are not the same as those most appropriate when the boundary lines are available directly, such as the Richardson plot or structured walk. We will examine each of these methods in turn. It should be kept in mind, however, that these different techniques actually measure slightly different fractal dimensions than the similarity or Hausdorf dimension produced by the Richardson plot (and in many instances different from each other). For an ideal boundary line, self-similar (not self-affine) and consisting of continuous data which can be measured at any scale, the numeric values of the different dimensions should all approach the same value as the measurement scale becomes smaller. For digitized pixel data, even without the various sources of bias and error mentioned above, this is not true.

One way to convert pixel information to boundary points which form a polygon is to represent the pixels as a series of lines. This run-length encoding is simply a list of the starting and ending points, or the starting points and lengths, of each line segment which passes through the features. Figure 11 shows a schematic diagram comparing run-length encoding and boundary representation. For raster-scan data, it is possible to create this list as the data are acquired without ever storing the individual pixel values. However, most modern image analysis systems do store the entire image so that it can be processed in various ways to correct image defects such as shading, extract texture, or other information which may be useful in segmenting the image, or simply to allow the appropriate threshold value to be selected interactively after the data have been acquired.

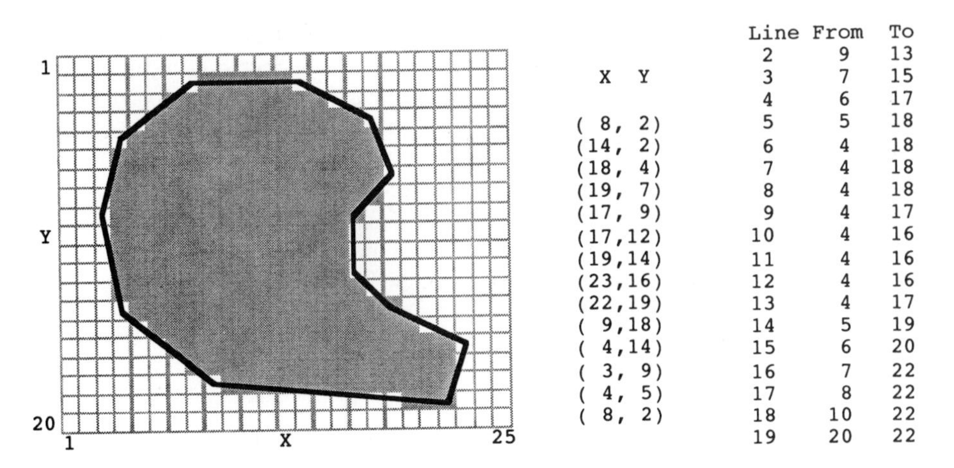

X	Y
8	2
14	2
18	4
19	7
17	9
17	12
19	14
23	16
22	19
9	18
4	14
3	9
4	5
8	2

Line	From	To
2	9	13
3	7	15
4	6	17
5	5	18
6	4	18
7	4	18
8	4	18
9	4	17
10	4	16
11	4	16
12	4	16
13	4	17
14	5	19
15	6	20
16	7	22
17	8	22
18	10	22
19	20	22

Figure 11. Comparison of boundary representation and run-length encoding to compactly describe the size and morphology of a feature.

When run-length data are available, the perimeter length can be approximated (at the resolution scale of the pixels) by connecting the end points of successive chords in the table to form a polygon, and determining the total length with the Pythagorean theorem. A series of such polygons can be constructed using every scan line, every second scan line, every third line, and so forth as indicated in Figure 12. As for the varying stride length approach, these polygons will not have the same perimeter; the coarser ones will bridge over the smaller irregularities, and so a log–log plot of total perimeter vs. the spacing of the raster lines can be constructed whose slope gives a fractal dimension.

These polygons do not have very uniform side lengths, as the horizontal segments are as long as the chords, and the segments connecting the ends vary widely depending on the angle. Whereas in the ideal Richardson plot, all of the polygon sides have the same length, in this case there are some short and many longer ones. For a perfectly isotropic boundary with the same number of polygon sides at all angles, the result would be to displace the log–log curve

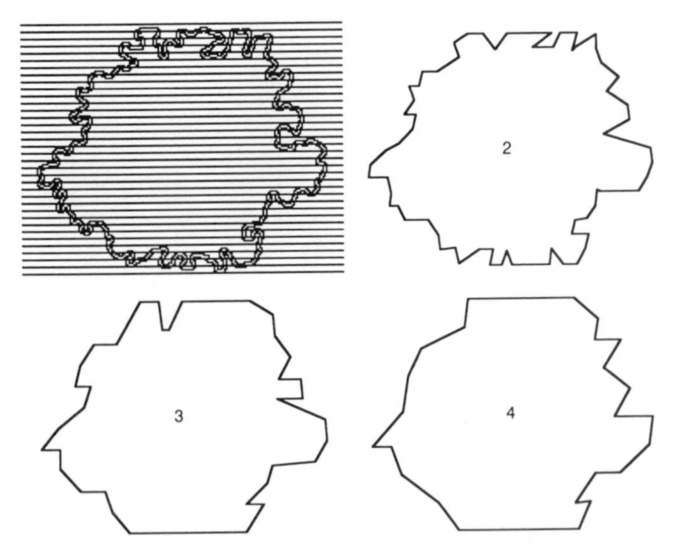

Figure 12. Creating progressively coarser polygons from a feature outline by skipping scan lines.

sideways but not to alter the slope. If the feature boundary is not isotropic, the result is less predictable and depends on the orientation of the feature with respect to the raster lines.

A modification to the basic Richardson method has been proposed (Tricot 1989a; Normant and Tricot 1991) which constructs convex hulls around portions of the fractal curve between the points selected by walking along the profile with a selected stride length. The convex hull is a polygon that minimally encloses that portion of the profile, at least with respect to the finite precision with which points on the profile are known. The total length of the convex hulls is then plotted as a function of the spacing of the points, or the stride length. It is claimed that this provides better precision than the basic method, but it requires a very large number of points to define the line (the authors suggest 10^4), involves considerably more work, and does not overcome the basic objections to the Richardson method discussed above.

Mosaic Amalgamation and the Kolmogorov Dimension

Another technique with less sensitivity to orientation is mosaic amalgamation (Kaye 1978a; Kaye, Clark et al. 1986; Russ 1986; Kaye 1989b). Like the method just described, this relies on progressively coarsening the image representation. The original image gives the finest scale approximation to the boundary length, and larger scale approximations which do not make use of all of the available image resolution are used to construct log–log plots of total boundary length vs. measurement scale.

In mosaic amalgamation, the pixels from the original image are averaged together to form a new image in which the pixels are larger in area. Figure 13 shows this schematically. The larger pixels ignore the fine details of the boundary, and so the measured perimeter will decrease. For a fractal object, it is expected that a log–log plot of perimeter against the size of the pixels used will show the usual straight line, whose slope gives the fractal dimension. Figure 14 shows an example in which this is done by averaging together the grey scale values of the original image to produce a series of coarser representations, each of which is then thresholded.

Coarsening is performed more rapidly using the binary (black and white) image present after the original image has been thresholded. Matching patterns of pixels to a preset list replaces each 2×2 block of pixels with a single larger pixel. As shown in Figure 15, blocks which are predominantly foreground or background take that identity. When half of the pixels are of each class, there is an equal probability, and a results table, which may either be structured or random, is used to select the outcome. Of course, this procedure is repeated and the image progressively coarsened into 2×2, 3×3, etc. blocks to produce a log–log plot of total perimeter vs. the size of the pixel grouping.

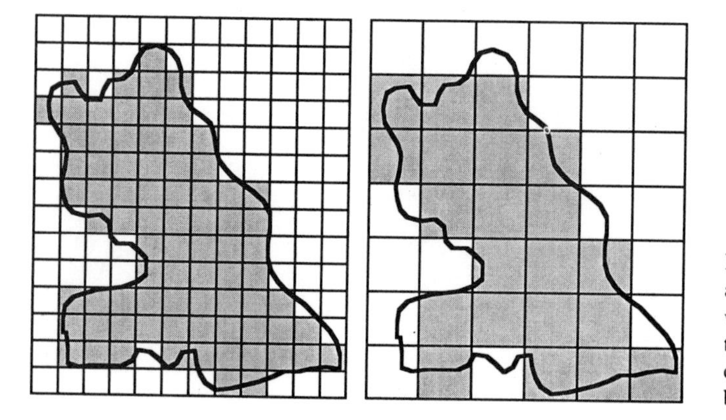

Figure 13. Diagram of mosaic amalgamation, in which pixels with twice the width (four times the area) produce a coarser representation of the boundary.

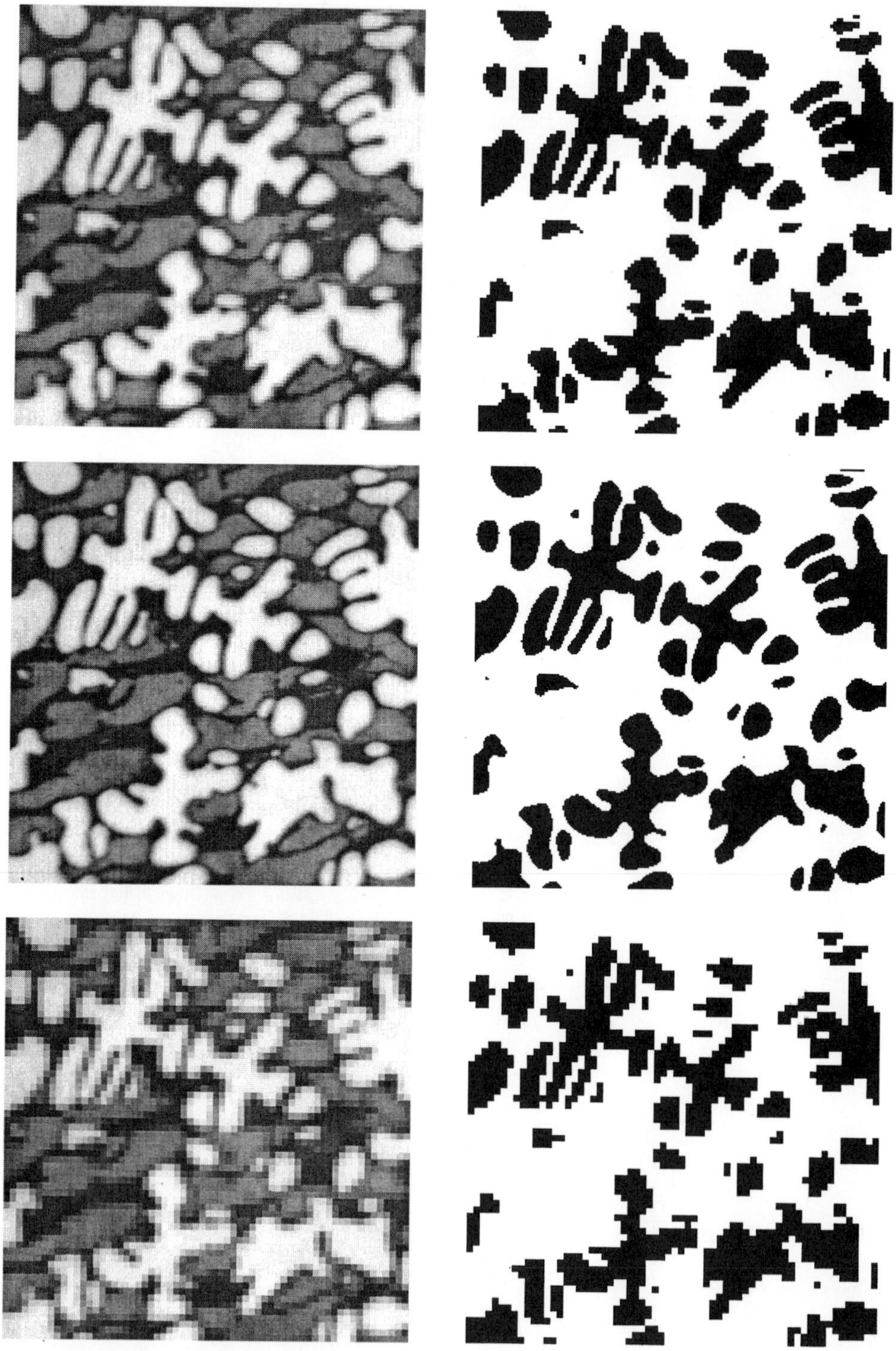

Figure 14. Progressive coarsening of a grey scale image by averaging the grey scale values into larger pixels (two and four times the original pixel size), followed by thresholding.

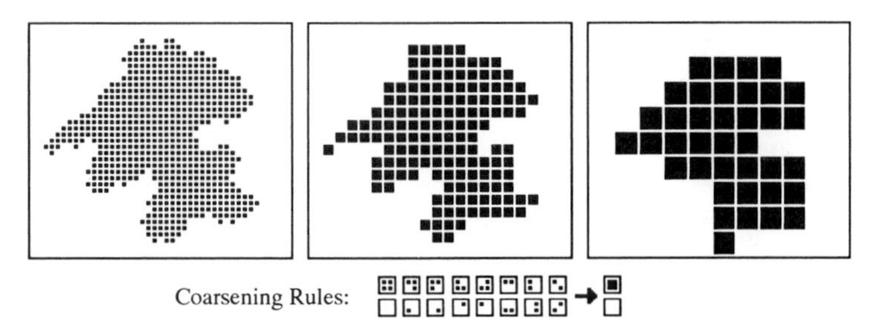

Figure 15. A black-and-white (binary) feature showing the original pixels and successive coarsening steps by mosaic amalgamation with the rules shown.

The dimension produced by mosaic amalgamation is technically not the same as the Hausdorf or similarity dimension produced by the Richardson technique. Instead, it is a Kolmogorov dimension. As noted above, for the ideal case of isotropy and continuous data, the numeric values approach the same limit. There is a simpler way to determine the Kolmogorov dimension manually, called box counting (and so the result is sometimes called the box-counting dimension). In this method, a grid of squares is placed over the boundary or line profile and the number of squares through which any part of the line passes is counted. This process is repeated with different grids having different size squares, and the number of squares "visited" by the profile plotted vs. the length of the side of the square, on the usual log–log scale.

This should properly be done with random placement of the grid on the profile, which is recommended for most manual counting and measurement procedures. However, when the same algorithm is used in a computer with digitized data consisting of discrete x, y coordinates, it becomes convenient to double the size of the squares progressively by dividing the coordinates by two and truncating the integer result. This can cause some bias in the results. An additional problem arises from the rather narrow range of dimensions over which it is practical to extend this operation. It is not possible to make the squares smaller than the resolution of the data, nor so large that they approach the dimension of the entire array. If there are only about 100 data points, then a box size larger than about 16 points wide (2^4) would give very few counts, and as shown in Figure 16, some boxes might "fall off" the ends of the data. Consequently there would be only 4 points in the entire plot! Profiles with 1000 points or more produce results with significantly greater precision. The Kolmogorov dimension generally has a relatively poor precision as compared to most of the other measurement methods.

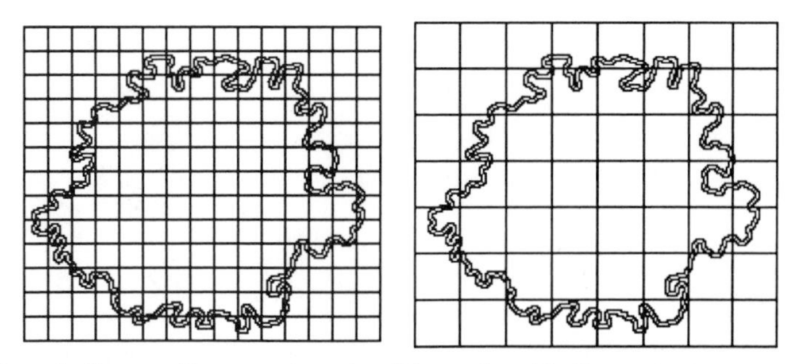

Figure 16. Schematic diagram of box-counting using different size grids. Some grid sizes may not exactly fit the image dimensions, producing edge effects.

The mosaic amalgamation and box-counting methods are only suitable for self-similar profiles, not for the more general case of self-affine ones. This is because the horizontal and vertical dimensions of the pixels in the mosaic, or the squares in the grid, are the same. It would be necessary to use rectangles instead of squares, and to vary the aspect ratio of the rectangles with dimension, to accommodate a self-affine curve. Knowing the fractal dimension of the curve and the scaling properties in the two directions would make this possible, but of course unnecessary.

The Minkowski Sausage

Minkowski was another of the turn-of-the-century mathematicians who considered ways to describe "monster" curves which were continuous but had undefined lengths and derivatives (Minkowski 1901). The Minkowski dimension is different from the Hausdorf or Kolmogorov dimensions (Bouligand 1929). From a mathematical point of view, the Hausdorf value is a true dimension while the others are not. Falconer shows (Falconer 1990) that ideally the Minkowski dimension is greater than or equal to the Hausdorf dimension, and there are many cases in which the inequality holds. The discretization of the data points for the boundary generally ensures that the two will not be equal and may make either value larger. While the Minkowski dimension is easier to obtain and may be useful for comparing two lines, there is no fundamental reason to assume that the $D_S - 1$ relationship to the surface holds, even in the absence of anisotropy or other effects. The Kolmogorov dimension is also not a true dimension, and in general will provide an estimate of the upper limit of the Minkowski dimension, which as noted may already be greater than the similarity dimension.

Fortunately, the exact numeric value of most of these measured dimensions is not as important in most applications as characterizing the differences in values associated with surfaces whose history or properties are different. In this case, the precision is more important than the accuracy and as long as there is enough data, and it is measured using the same method, comparisons may be valid and useful. That reasoning provides the justification for using these different definitions and procedures for obtaining estimates of the fractal dimension of profiles and of surfaces.

The Minkowski method for measuring the fractal dimension of a boundary is illustrated in Figure 17. A circle is swept continuously along the line and the area which is covered, called the Minkowski "sausage," is determined. This value is then plotted as a function of the circle diameter, and the slope (on the usual log–log plot) gives the dimension. The important difference between this and the Richardson method is that the circle is moved so that its center lies on every point of the line. The Richardson technique is equivalent to placing a series of circles along the line which touch at their edges, and counting their number as a function of diameter.

Manual implementation of the Minkowski approach is impractical, but it lends itself well to computerization. Flook (1978) used a series of erosions and dilations on pixel-based images to perform the measurement. These are neighborhood operations which add or remove pixels along the feature boundary. Figure 18 illustrates the method. Dilation adds any background pixel which touches the feature, while erosion removes any feature pixel which touches the background. Performing each of these operations N times and counting the pixels which are affected as a function of N produces a plot (Figure 19) whose slope (on log–log axes) gives the fractal dimension.

Unfortunately, erosion and dilation on a square pixel array are not isotropic operations. The neighborhood pixels in 45 degree directions (corner touching) are farther from the central pixel than those which share an edge. If the boundaries are not isotropic, the results will be

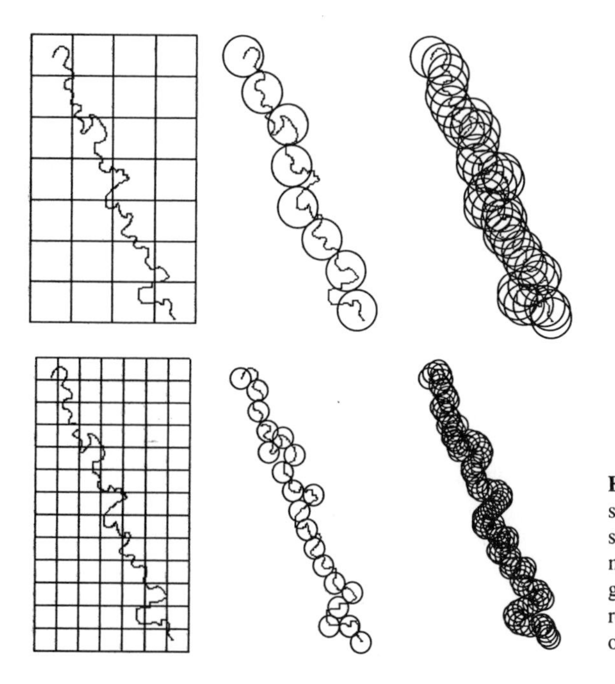

Figure 17. The principle of the Minkowski sausage (right) as compared to the structured walk (center—Richardson dimension) and grid counting (left—Kolmogorov dimension), for two different resolution values (grid size, stride length, or dilation distance).

biased. Figure 20 shows this effect. In image processing operations, more complicated erosion and dilation rules are used which alternate between testing all eight neighbor positions and just the four edge-touching neighbors, or which count the number of neighbors (Russ 1992a). These can reduce (but not eliminate) the anisotropy inherent in square pixel arrays, but only work for larger erosion distances. The anisotropy of the erosion/dilation approach with discrete pixel data produces a result which is more closely related to the box counting than to the Minkowski method. Another difficulty with iterative erosion and dilation is the length of time required for a large number of steps.

Implementation with the Euclidean Distance Transform

A more efficient and isotropic method of performing erosion and dilation uses the Euclidean distance map or distance transform. This is an image processing operation that is applied to black and white images to produce a grey-scale result in which each pixel has a brightness value equal to its distance from the nearest background point. The same procedure can assign values to the background based on their distance from the nearest point on a feature. There is a very efficient procedure which generates the Euclidean distance map without iteration (Danielsson 1980). Thresholding the distance map for either the features or the background produces uniform erosion and dilation to any desired distance from the original boundary, without iteration. Figure 21 shows the results in comparison to the classical erosion and dilation illustrated in Figure 20.

Figures 22–25 show an example in which an image of a paint pigment particle has been processed with this method. The Euclidean distance map for the particles and the background together is shown in Figure 23. Once the distance map has been generated, it is possible to threshold the image at various brightness levels to select pixels within various distances from the boundary. These are the Minkowski sausages, as illustrated in Figure 24. Because the distance map is isotropic, the width of the sausages is uniform regardless of whether the feature

boundary is isotropic or not. It is not really necessary to perform the thresholding operations. Counting the pixels as a function of their brightness (distance) values produces a plot (Figure 25) that directly gives the Minkowski dimension (Russ and Russ 1989; Russ 1992a). Combined with the efficiency of constructing the Euclidean distance map, this method for determining the fractal dimension is very fast.

Although it is more isotropic, more efficient, and generally more precise than the box-counting method, and easy to implement on a computer, the Minkowski method described

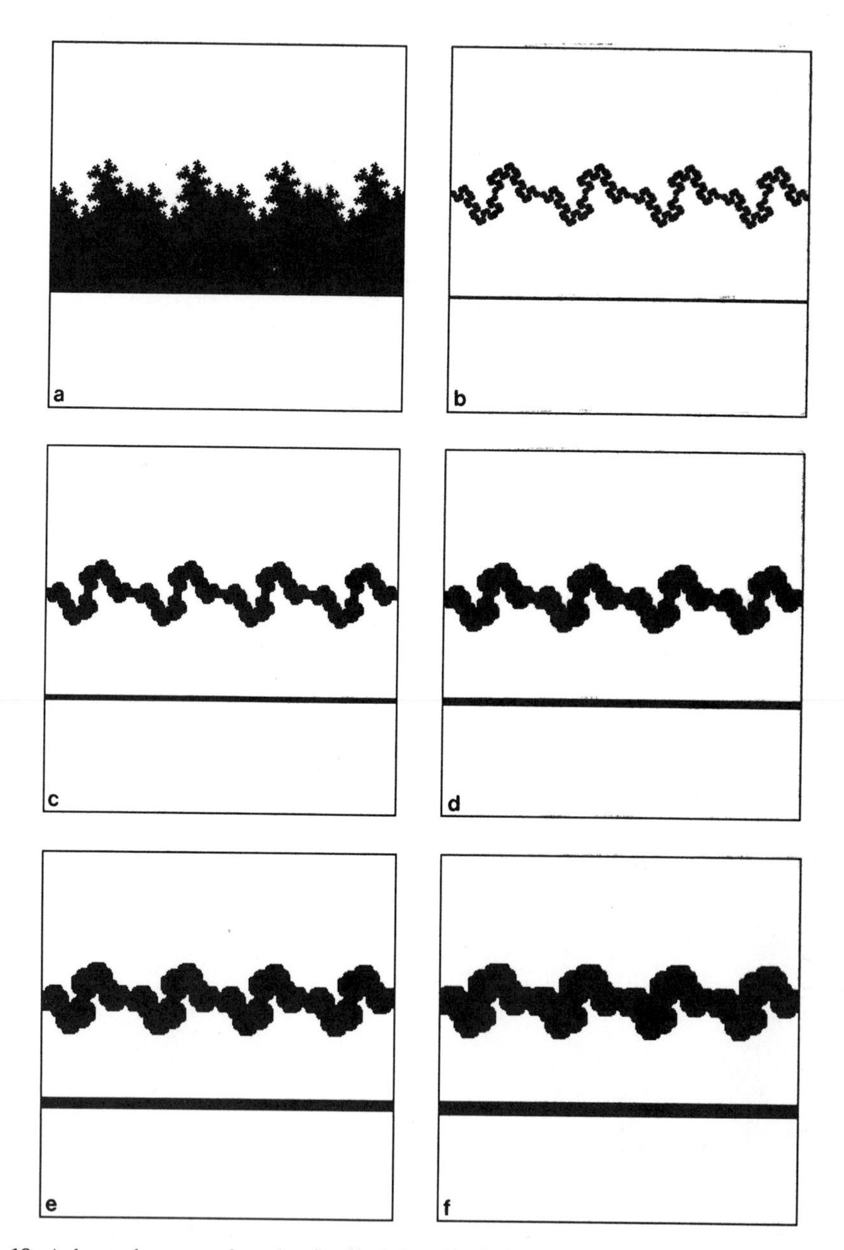

Figure 18. A shape whose upper boundary is a Koch fractal and whose lower boundary is a Euclidean line, with successive dilations of the boundaries to construct the Minkowski sausages.

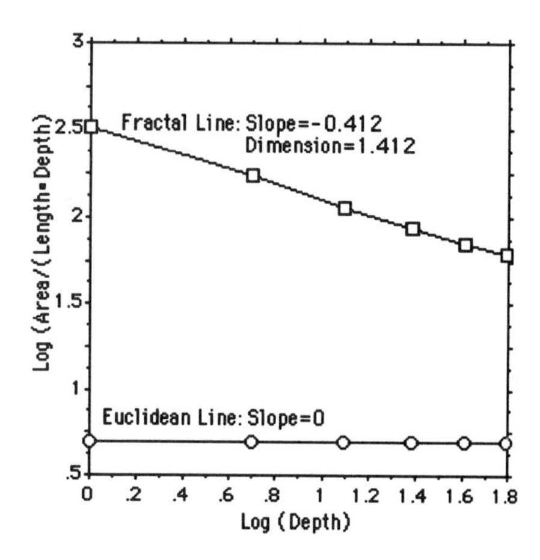

Figure 19. Determination of the fractal dimensions of the boundary lines in Figure 18. The "depth" is the number of dilations of the boundary used. The Koch curve has a dimension of 1.412 while the straight line dimension is 1.00.

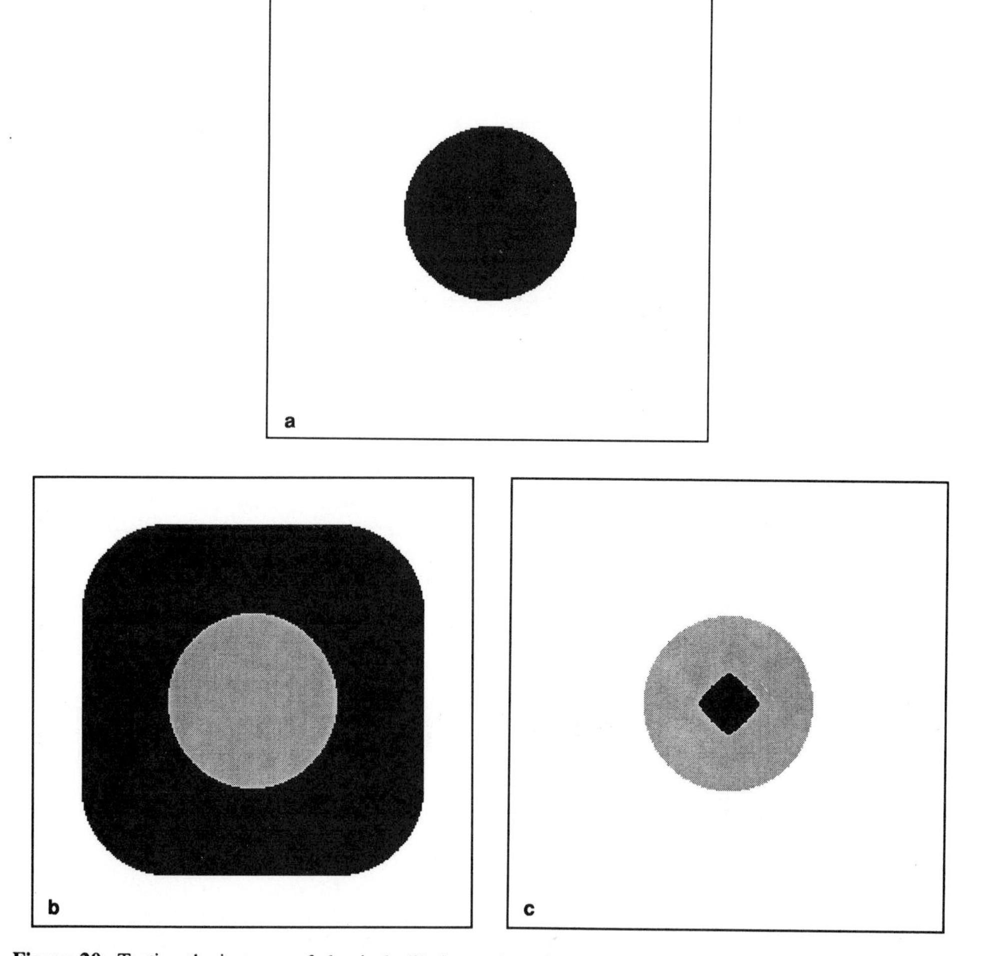

Figure 20. Testing the isotropy of classical dilation and erosion: **(a)** original circle; **(b)** after 50 repetitions of dilation; **(c)** after 25 repetitions of erosion.

above suffers from one limitation which is the same as for box-counting and other methods discussed. It is only applicable to self-similar and not to self-affine boundaries. In Chapter 1, it was mentioned that sections through a fractal surface which are parallel to the nominal surface plane produce islands and coastlines which may be self-similar, but vertical sections do not. Many instrumental techniques for characterizing surfaces produce elevation profiles, which are self-affine. These cannot be measured using the Minkowski method as described.

However, there is a modification of the Minkowski approach that can be correctly used for self-affine profiles. Consider the vertical elevation profiles shown in Chapter 1. For any vertical elevation profile, the curve can only be self-affine and not self-similar. Consequently, the vertical and horizontal dimensions do not scale in the same way, and so the structuring element cannot be a circle. Using an ellipse whose aspect ratio varied with size according to the (unknown) scaling relationship between horizontal and vertical dimensions would make it possible to construct the correct Minkowski plot, but of course this is not practical for the measurement of observed profiles.

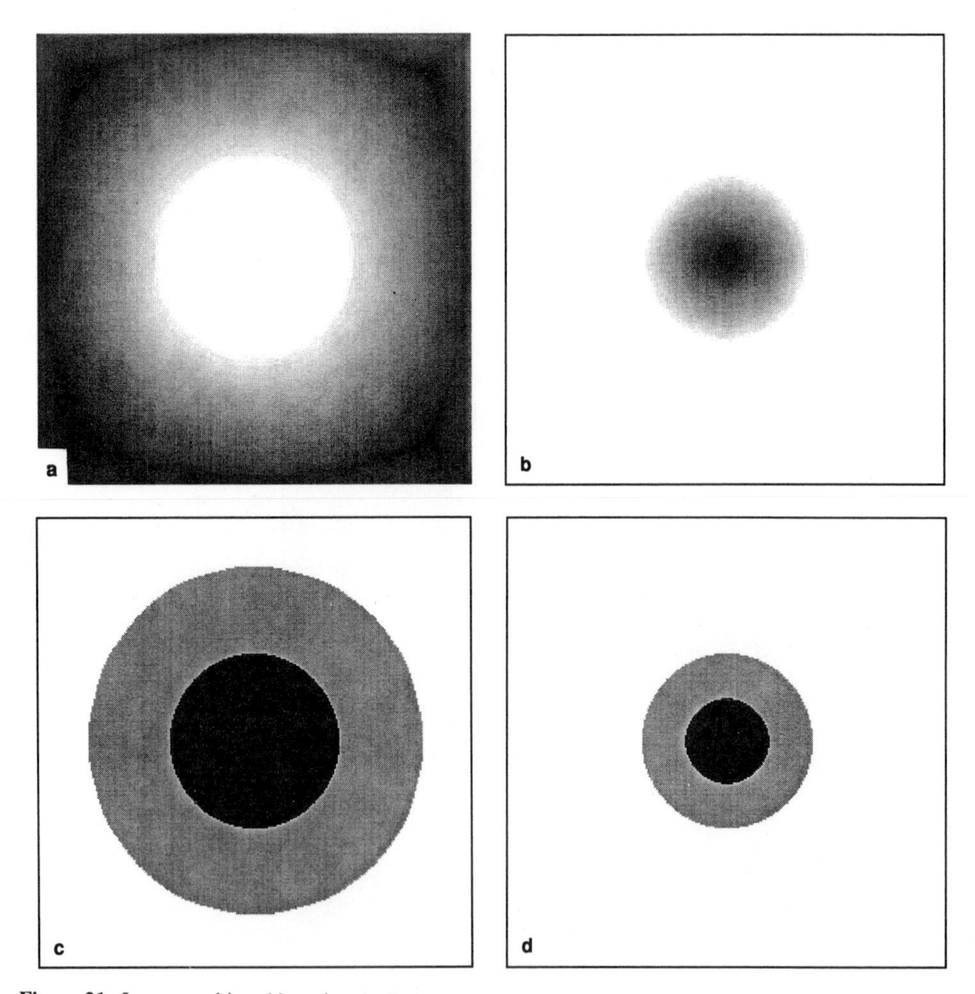

Figure 21. Isotropy achieved by using the Euclidean distance map for dilation and erosion (compare to Figure 20): **(a)** the EDM of the background around the circle; **(b)** the EDM of the circle; **(c)** dilation achieved by thresholding the background EDM at a value of 50; **(d)** erosion achieved by thresholding the circle EDM at a value of 25.

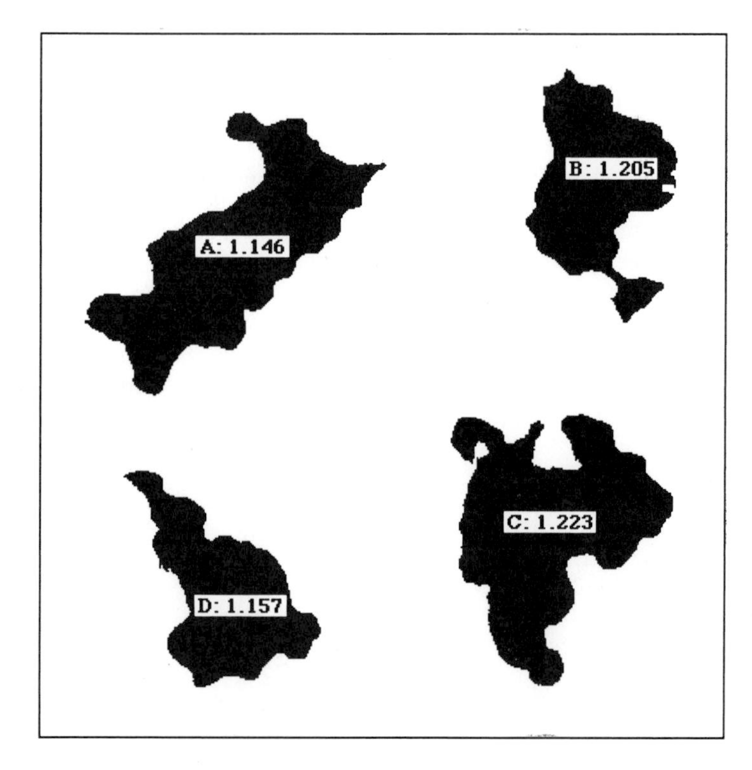

Figure 22. Binary image of four paint pigment particles, each one labelled with its fractal dimension.

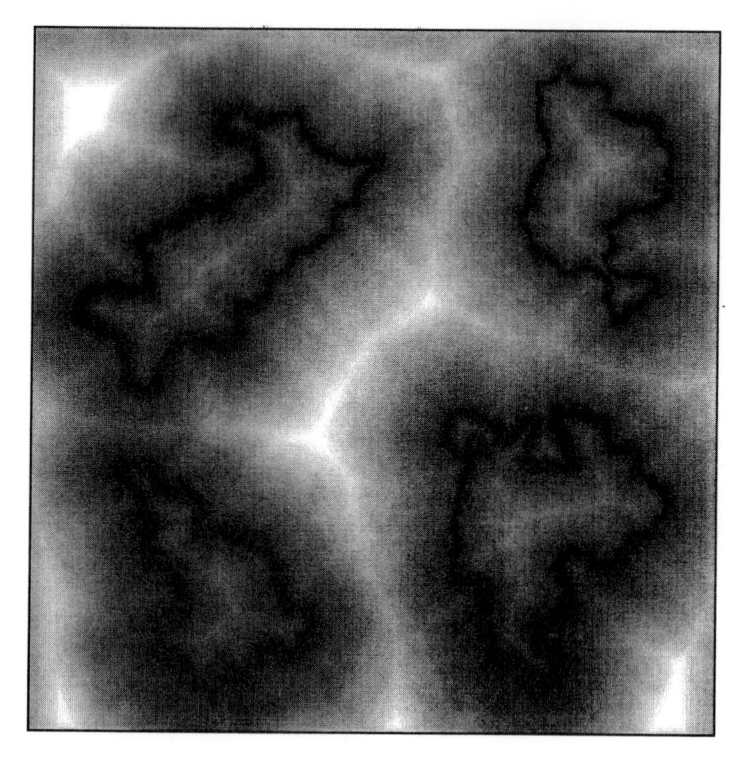

Figure 23. Euclidean distance map of both the features and the background in Figure 22.

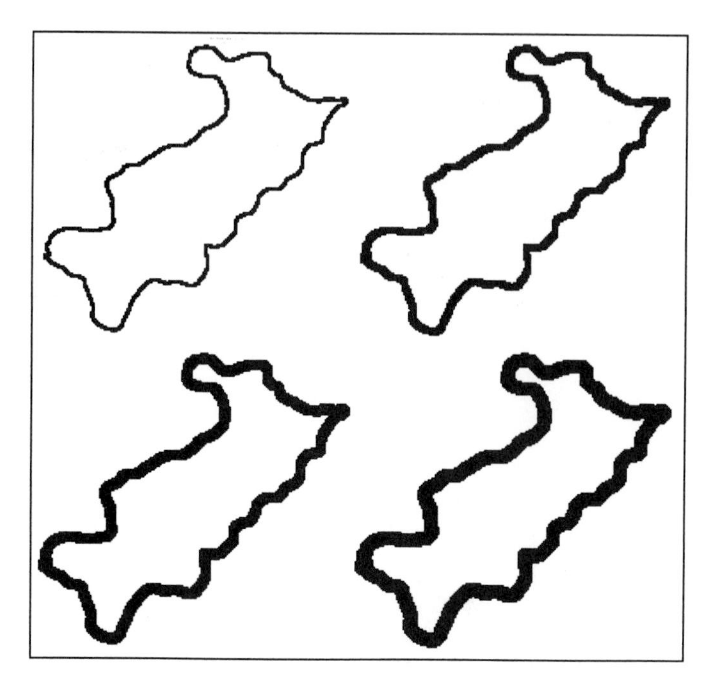

Figure 24. The results of progressively thresholding the EDM image of feature A in Figure 22 at different pixel values, to obtain the pixels within varying distances of the boundary. These are Minkowski sausages.

If the important scaling dimension is along the horizontal direction, then a structuring element consisting of a line segment parallel to the horizontal axis can be used. A series of line segments of different lengths are moved continuously along the profile and the area swept out plotted as a function of line length (using log–log axes). The slope of this line gives the fractal dimension. This is equivalent to dilation of the curve using a horizontal structuring element, as shown in Figure 26.

A more efficient implementation of the technique when the profile is stored in the computer as an array of elevations at equally spaced points along the horizontal axis can be

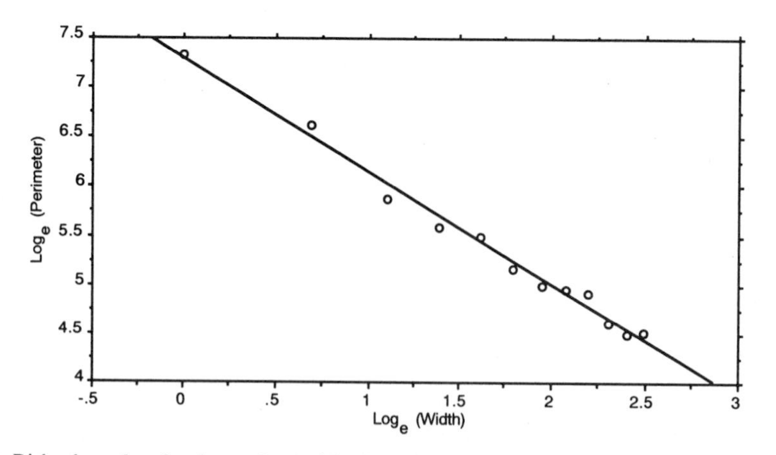

Figure 25. Richardson plot of perimeter (area of the thresholded band divided by width) vs. width for feature A in Figure 22. The slope of the plot is –0.146, giving a fractal dimension of 1.146.

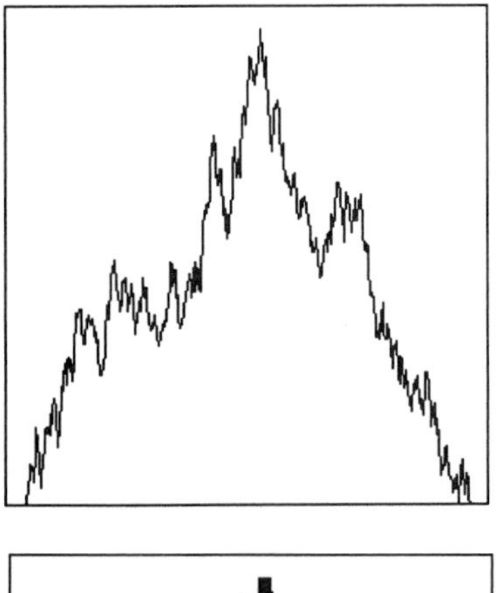

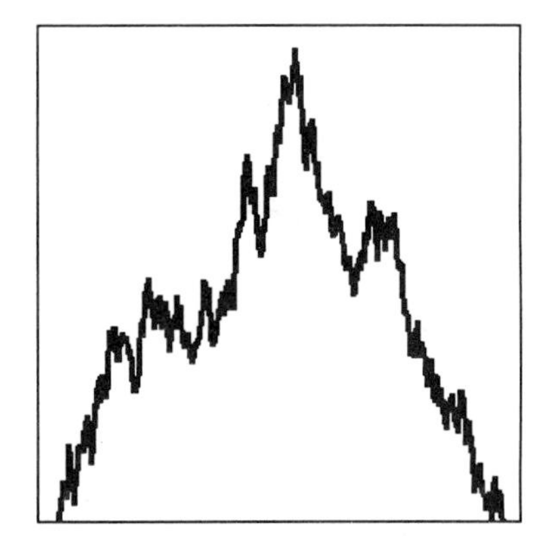

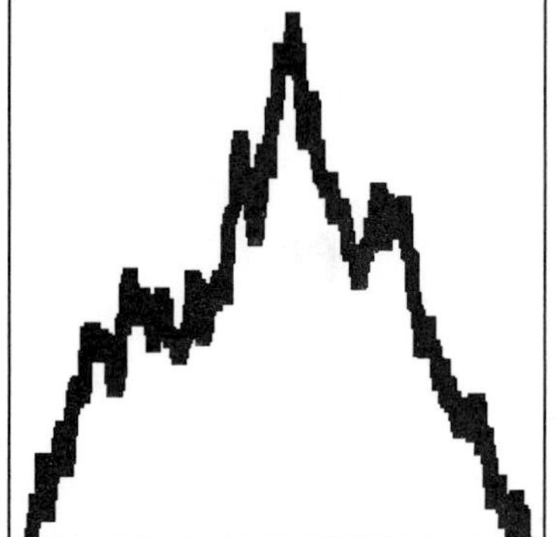

Figure 26. Dilation of a self-affine vertical elevation profile is performed in the horizontal direction, to produce the appropriate Minkowski sausage.

accomplished by finding the minimum and maximum values in various intervals. When this method of performing the calculation is used, the method is sometimes called the variational technique (Dubuc, Roques-Carmes et al. 1987; Tricot, Champigny et al. 1987; Tricot, Quiniou et al. 1988; Dubuc, Quiniou et al. 1989). Another fast way to implement the Minkowski measurement for vertical elevation profiles is to perform the one-dimensional equivalent of the Euclidean Distance Map, producing a grey scale array which can be counted to produce the log–log plot. This is illustrated in Figures 27 and 28. All of these approaches give identical results, differing only in the speed of implementation.

Fitting Lines to Data

When the log–log plot has been constructed for the Minkowski fractal, or for that matter for any of the other methods (Richardson, box-counting, etc.), it is necessary to fit a straight line to the data to determine the slope and hence the fractal dimension. Since the points are not uniformly spaced, and the regression fit should not necessarily weight all of the points the same, this is not an entirely trivial operation (Draper and Smith 1981). For the Richardson and box-counting methods, there may be more than one point corresponding to the same stride length or grid size, if multiple starting points for the walk or placements of the grid are used. This also complicates the fitting process.

For images based on an array of pixels, or vertical elevation profiles measured at equally spaced horizontal points, the number of different points in the fit is limited by the number of points on the profile. In most cases it is not possible to use step lengths, grid sizes or dilation

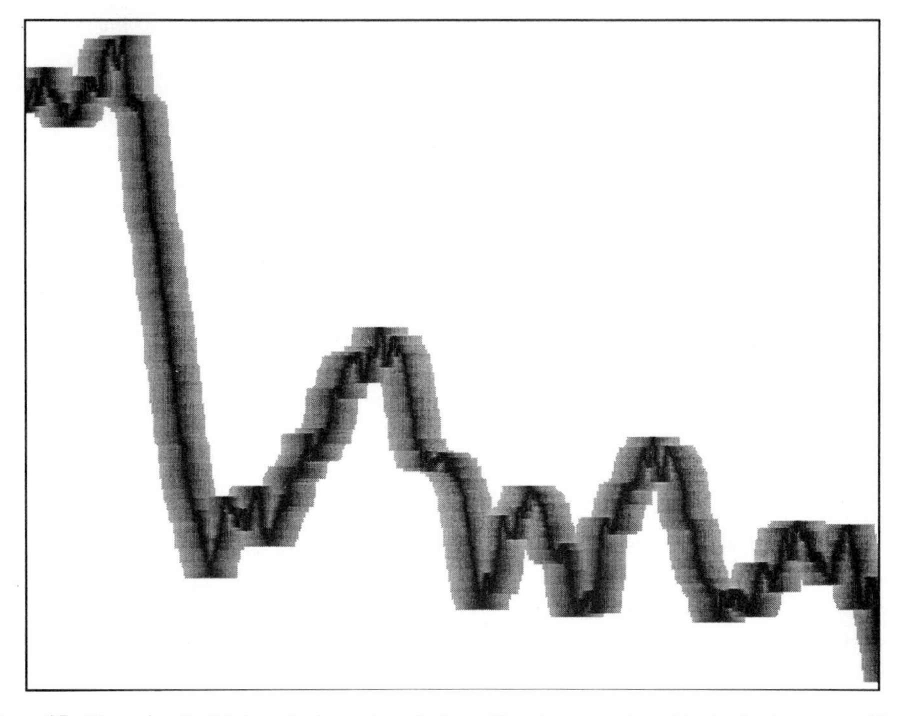

Figure 27. Measuring the Minkowski dimension of a fractal line (generated by midpoint displacement with α = 0.7) using a horizontal Euclidean distance map. The point shading gives the horizontal distance of each pixel from the line.

distances greater than about a factor of ten less than the total number of points. For a profile with only 100 points, the plot will contain only about 10 points and the precision of the slope value will be limited. Profiles with thousands of points are needed to obtain slope values with three meaningful decimal digits of precision, assuming that the points really do fit a straight line. Generally, the Minkowski method gives the best precision for a given number of profile points, and the box-counting method the worst. The Hurst and Fourier methods discussed in Chapter 4 are often better and worse, respectively.

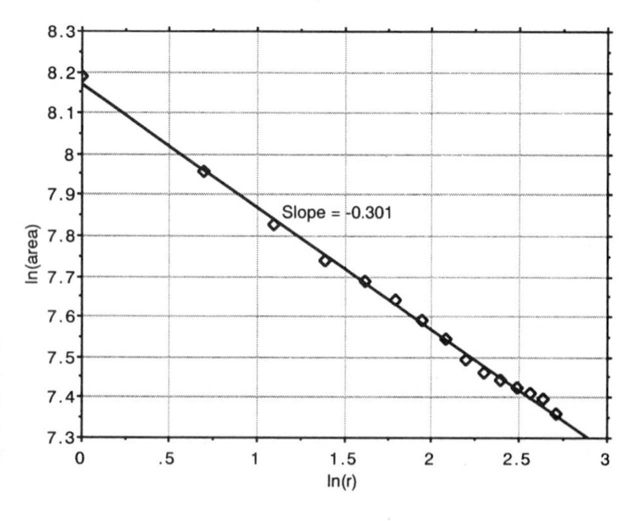

Figure 28. Minkowski plot produced from the brightness histogram of Figure 27. The fractal dimension of the self-affine line is measured as 1.301, in good agreement with the value of α used in generating the line.

If only short profiles are available for measurement, for instance in measuring the profiles of many small particles, the usual approach is to measure each and determine an average. There is no reason to expect this to give the "right" answer. If the profile contains regions with different local dimensions, or has a scaling dimension which is not constant, the log–log plot may not be straight. The interpretation of these plots is discussed in the chapter on mixed fractals.

Fitting a straight line to a series of data points is one of the most commonly performed procedures in data analysis. Given a series of points x_i, y_i, the least-squares fit line is written $y = mx + b$, where m is the slope and b the intercept. For N points, the best fit values of m and b are calculated from summations of the x_i and y_i values as

$$m = \frac{N \sum x_i \cdot y_i - \sum x_i \sum y_i}{N \sum x_i^2 - \left(\sum x_1\right)^2}$$

$$b = \frac{\sum y_i - m \cdot \sum x_i}{N}$$

This calculation, commonly performed by software programs and even by some pocket calculators, assumes that the x_i values are the independent variable and y is the dependent variable, and that the values of x have no error; all of the misfit of the line to the points is in the vertical or y direction. It also assumes that the quality of all of the data points is the same, and that there is no reason to expect the errors for some of the data to be greater than for others.

For the application of this method to log–log plots used to evaluate fractal dimension, none of these assumptions is particularly good. There is certainly some error in the x values. The precision of determining the stride length, box dimension, etc., is imperfect. The vertical dimension is often determined by counting, which should have very high precision; the standard deviation in the number of things counted is the square root of the number. Most important of all, the plotted values for both axes are logarithms of the measured distances. Whatever the precision of the actual measured values, it is more likely to be a fixed linear measure than a fixed percentage of the value. Consequently, the conventional linear least-squares fit in effect weights the points at one end of the line more than at the other end, which can bias the slope if the data are not actually straight.

Finally, even assuming that the use of a linear least squares method is suitable for the data, how does one judge the results? The expected errors in the slope and intercept values are computed as

$$\sigma_m = \left[\frac{N \cdot S}{N \sum x_i^2 - \left(\sum x_i\right)^2}\right]^{1/2}$$

$$\sigma_b = \left[\frac{S \cdot \sum x_i^2}{N \sum x_i^2 - \left(\sum x_i\right)^2}\right]^{1/2}$$

$$S = \frac{\sum y_i^2 + N \cdot b^2 + m^2 \sum x_i^2 - 2 \cdot \left(b \sum y_i - m \cdot b \sum x_i + m \sum x_i y_i\right)}{N-1}$$

These give the one-standard-deviation uncertainties in the values of m and b in the equation, but they do not indicate whether a straight line fit was actually an appropriate model for the data. One rather straightforward way to evaluate this is to also perform a fit to a higher

polynomial and then compare the results of the two fits. The higher-degree polynomial will always, of course, be able to fit the data better. But it uses up one more degree of freedom in the process, and the improvement in the fit may not be that great. A comparison of the reduced chi-squared of the linear fit results to that for a quadratic fit can be used to see if the quadratic fit is statistically superior. The test is based on the ratio of reduced chi-squared values (which will have an F distribution). In many of the examples shown in this text, this test is performed using a critical value of F corresponding to ($p = 0.25$), and the symbol § is shown after the standard deviation value as a warning if the quadratic fit is significantly superior to the linear fit. This probability means that one time in four the warning will be given in error, and that the data called into question are in fact adequately described by a straight line relationship in accordance with fractal theory.

Other Methods

There are other measurement procedures that are sometimes used to determine the fractal dimension of a profile. In general, these methods produce dimensions that are numerically close, but not identical to the Hausdorf (Richardson), Minkowski, or Kolmogorov techniques. Two of these, the Fourier and Hurst transforms, use the difference between the vertical elevation of points as a function of the horizontal distance between them to construct plots which describe the self-similarity of the data. Both the Fourier and Hurst methods can be extended to measure surfaces and to deal with the anisotropy of surfaces. They are discussed in detail in Chapter 4.

The mass fractal, discussed in Chapter 1 in the context of clusters and networks, can also be applied to profiles and to surfaces. A plot of the number of points as a function of elevation will exhibit the usual straight line plot on log–log axes. However, this plot is difficult to construct because the points must be sampled uniformly along the profile, not along the horizontal axis. Since a fractal profile is inherently rough at all scales, and the derivative is not continuous, there is no direct way to find the points.

The Korcak fractal dimension of a profile was also discussed in Chapter 1. It is constructed by making a cumulative plot of the number of horizontal intercepts as a function of length, for any arbitrarily selected elevation value. The points where the profile crosses the horizontal line comprise a zeroset, and the distribution of horizontal spacings between points is fractal. A log–log plot of the number of distances which exceed a value l as a function of l has a slope that is exactly one less than the fractal dimension of the profile. Because of the horizontal sectioning, the method can be correctly applied to a self-affine profile. Figure 29 shows two elevation profiles, with different fractal dimensions, and their Korcak plots.

However, for practical purposes, this method is not very efficient. For profiles which are digitized as a series of elevation values at uniformly spaced horizontal points, the point spacing defines the shortest horizontal distance that can be measured. Of course, the actual profile has an increasing number of ever-shorter horizontal intercepts. But the ones shorter than the point spacing cannot be detected or measured. At the other extreme, there must be a few very long chords, but the probability of encountering them is restricted if the total span or number of points in the digitized profile is not large. In a typical plot of the number of intercepts as a function of distance, the number of long intercepts is artificially reduced by the finite length of the total span. Consequently, fitting a line to determine the slope of the plot can only be performed for the shorter lengths, beyond which the plot curves downwards. This was illustrated in Chapter 1. Selecting the region for fitting is essentially arbitrary.

Another particularly simple method (at least conceptually) that can be applied to a vertical elevation profile or to an entire surface is to calculate the standard deviation of the elevation

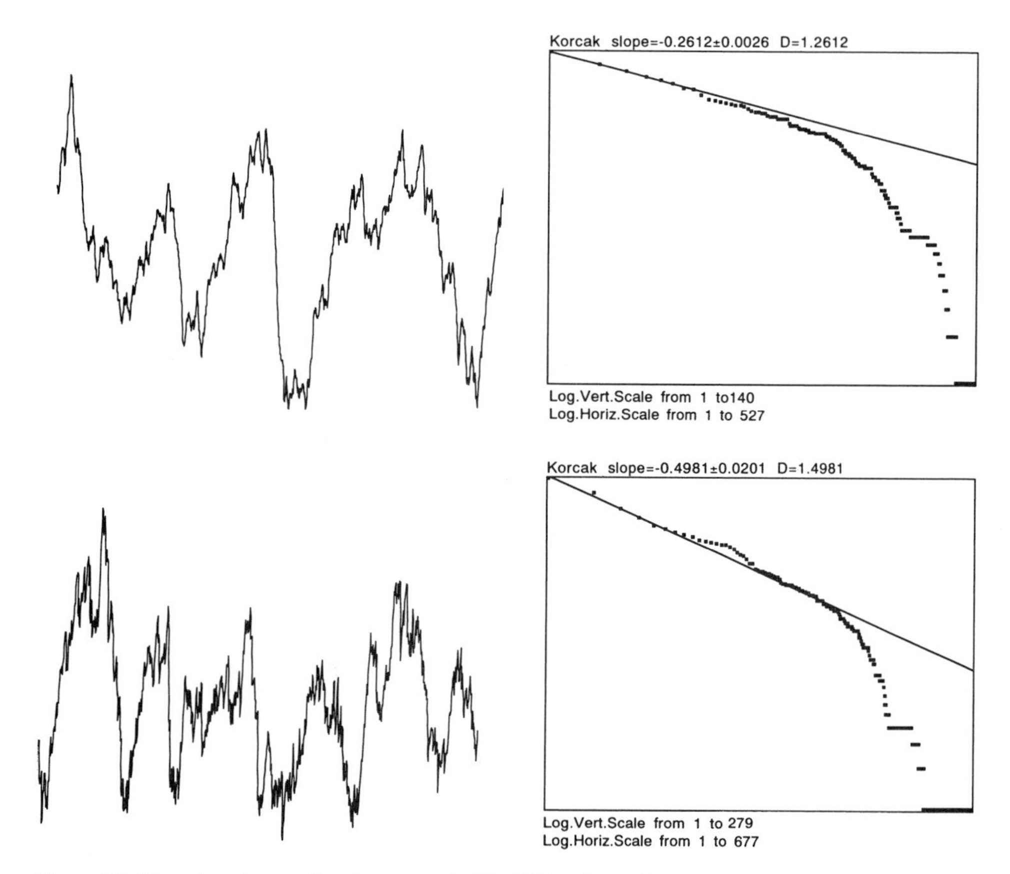

Figure 29. Two elevation profiles (constructed with 1024 points using an inverse Fourier method to have different fractal dimensions) and their corresponding Korcak plots. The drop-off in the plots for long intercepts is due to their systematic undersampling in the finite extent of the profiles.

values within a window of width w. This is best done by placing many windows of the same width (or area, for a surface) on the data, either systematically or randomly, and calculating the average value of the standard deviation for the multiple windows. The procedure is then repeated with windows of different sizes, and the standard deviation plotted against the window size. The slope of the log–log plot gives the fractal dimension, albeit with fairly poor precision or efficiency. This method comes directly from a consideration of the familiar behavior of Brownian motion, which in one dimension produces a fractal plot of distance from an initial point vs. time whose dimension is 1.5 and for which the standard deviation increases as the square root of the time. For profiles with other dimensions, sometimes described as "fractal Brownian" profiles, the principle is the same but the rate with which the standard deviation increases with time (or whatever the horizontal axis represents) changes (Mandelbrot and Van Ness 1968).

One implementation of this measurement method (Malinverno 1990) defines a window width along the x axis starting with a large fraction of the entire scale. This is then reduced in steps by multiplying by, e.g., 0.9, down to a minimum of 10 points. For each window, the points within the window are "detrended" by subtracting a best fit straight line, or just the line

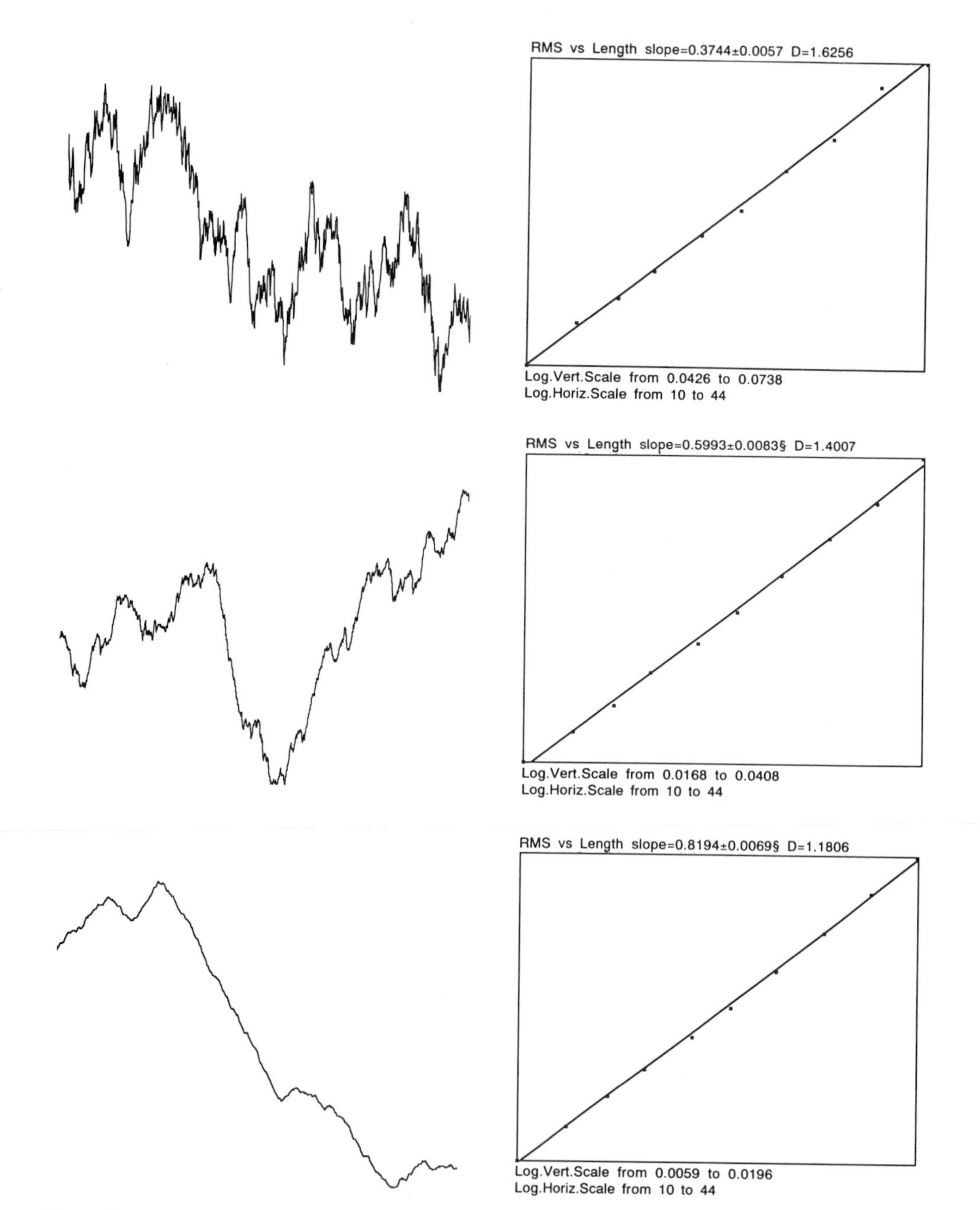

Figure 30. Plot of the increase in the variance of the elevation values in three profiles as a function of the horizontal window used for the calculation. The profiles were constructed using an iterated displacement method to have different fractal dimensions.

connecting the end points. Then the remaining RMS (root-mean-square) deviation or "roughness" is calculated as

$$\mathrm{RMS}(w) = \frac{1}{n_w} \sum_{i=1}^{n_w} \sqrt{\frac{1}{m_i - 2} \sum_{j \in w_i} (z_j - \bar{z})^2}$$

where n_w is the total number of windows of length w in the data, m_i is the number of points (the −2 corrects for the two degrees of freedom used to get slope and intercept of the detrending line), z_j are the residuals from the line, and $\bar{z}$ is the mean residual in the ith window w_i (which will be zero for the least-squares line fit but not for the simple end-point fit). The number of windows is high since windows of each size overlap by 50% along the data axis. The slope of the log (RMS roughness) vs. log (window length) plot gives the fractal dimension. This is illustrated in Figure 30 for three elevation profiles generated to have widely varying fractal dimensions. This method is similar in some respects to the Hurst R/S method described in Chapter 4. It does not require that the data points be uniformly spaced, as required for many other methods.

Malinverno (1990) showed that for a self-affine data series, such as a profile, the numerical result from this technique agreed with Fourier transform (FT) power spectrum analysis for $D \leq 1.7$. Schepers, van Beek et al. (1992) call this method "relative dispersion" analysis, and compare it to the Hurst and Fourier methods described in Chapter 4. Using profiles generated by different methods, they conclude that the Fourier method provides the highest accuracy, while the use of the variance produces estimates that are biased towards a profile dimension of 1.5 (Brownian motion) and the Hurst method produces values that tend to be too high.

A closely related technique, also very inefficient, is the "variogram" method (Mark and Aronson 1984), which plots the number of randomly selected point pairs classified by log of the difference in elevation between them. A similar method has been used (Roy, Gravel et al. 1987) for points on surfaces.

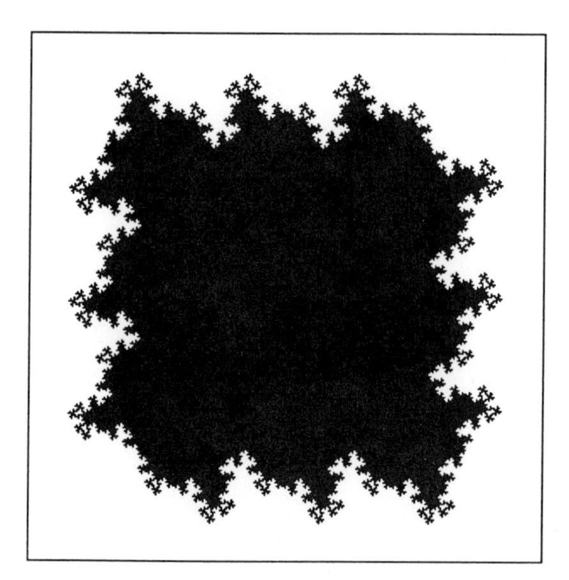

Figure 31. A fractal shape constructed iteratively by elaboration of a Koch island. Each iteration increases the perimeter length by a constant factor so that the similarity or Hausdorf dimension is log(8)/log(4) = 1.50

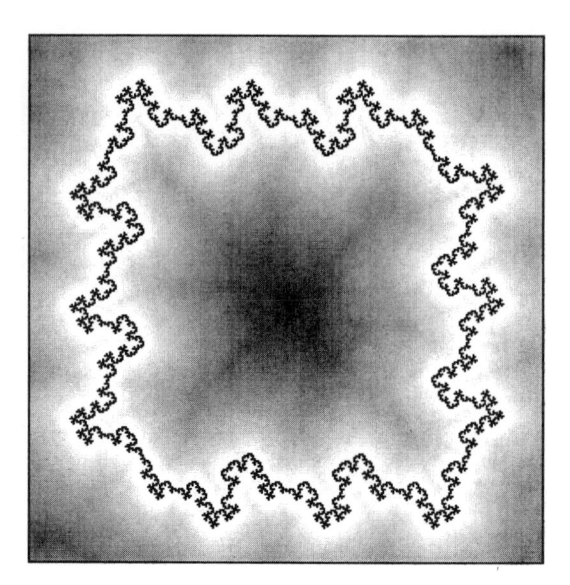

Figure 32. Euclidian distance map constructed around the shape in Figure 31. Brightness values represent the distance of each pixel from the nearest point on the boundary.

Comparison of Dimensions

The various dimensions discussed here, and in other chapters where they are extended to surfaces, are not identical. Only the Hausdorf dimension which results from Richardson plots, or from mass fractal plots, is technically a "dimension" according to formal definition. The Minkowski dimension may be equal to the Hausdorf dimension for a "well behaved" curve. Le Méhauté (1991) describes a set as "very regular" if the Hausdorf dimension is identical to the Minkowski dimension, but in general the Minkowski dimension is greater than or equal to the Hausdorf dimension. This also applies to the Korcak value. Further, the Kolmogorov dimension estimates the upper limit of the Minkowski dimension (Falconer 1990).

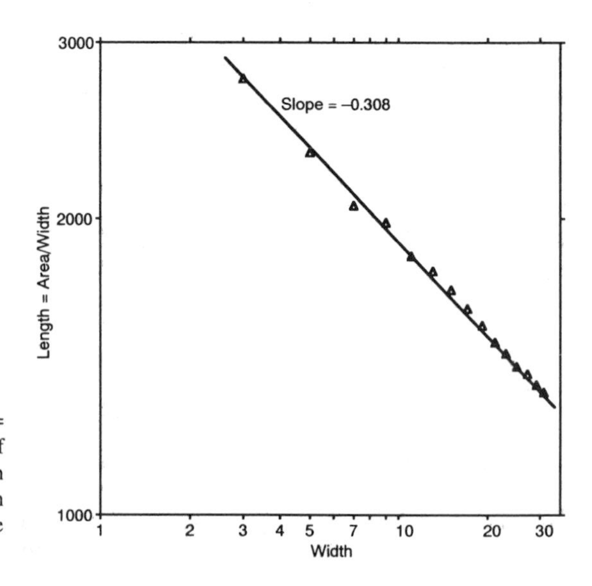

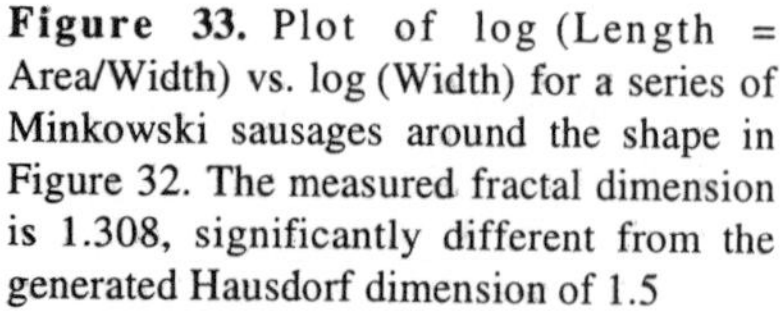

Figure 33. Plot of log (Length = Area/Width) vs. log (Width) for a series of Minkowski sausages around the shape in Figure 32. The measured fractal dimension is 1.308, significantly different from the generated Hausdorf dimension of 1.5

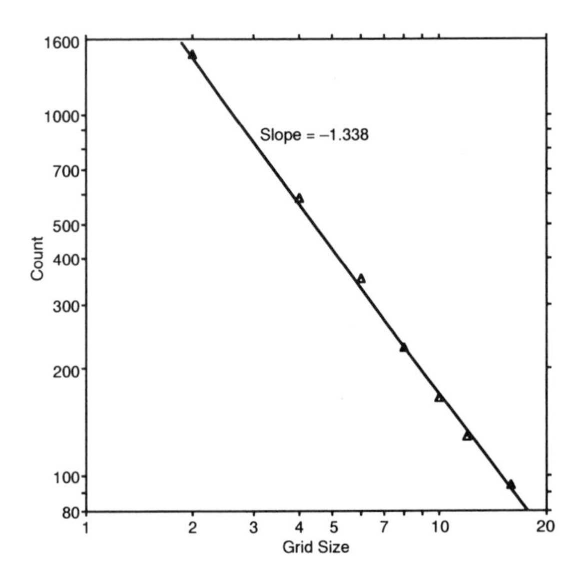

Figure 34. Plot of log (Count) vs. log (Grid Size) for box-counting results on the shape of Figure 31. The measured fractal dimension is 1.338, different both from the Minkowski result in Figure 33 and the generated Hausdorf dimension of 1.5.

This order of magnitude of values is further complicated for real data by the fact that the values (both elevation and horizontal position) are not really continuous but are digitized with finite precision. The regular sampling of points in the horizontal direction, or for the outlines of islands in both the x and y directions, affects the numerical values of the various measurement procedures in different and not completely predictable ways.

It is often assumed that the numerical values obtained from the various measurements are the same, and for some generated mathematical fractals carried out to a fixed level of detail they may be, but this is a dangerous assumption for real data. In most cases, it is only possible to compare the numerical values from different profiles or surfaces using one measurement procedure. As an illustration of the kinds of differences that may be observed between different measures, consider the Koch island shown in Figure 31. This is a structured fractal constructed with a Hausdorf dimension of 1.5. The Euclidean distance map of the island and surrounding background is shown in Figure 32. The resulting Minkowski plot in Figure 33 gives a dimension of 1.308. Box counting (Figure 34) gives a Kolmogorov dimension of 1.338. Neither is close to the theoretical value. Indeed, both of the values are less than the Hausdorf dimension, rather than greater than it as predicted. This is due to the finite size of the pixels in the boundary, and the discrete sampling of the data.

Selection of the procedure to be used may be based on the form in which the data are acquired, the required precision, and the relative efficiency of performing the measurement. For data in the form of elevation profiles with uniform point spacing in the horizontal direction, the Minkowski method based on a horizontal structuring element, performed using the variational approach, is one of the more efficient methods, with reasonable precision.

The precision of all these methods depends on the number of points in the profile. None of the methods is good to more than about 1 digit (to the right of the decimal point) for short profiles with only 100 points or so. With 1000 points or more, precision of 2–3 digits is possible. However, it is important to distinguish between the precision or reproducibility of the measurement, and the absolute accuracy. The discussion above of the effects of finite point spacing and sampling, and the differences between the various dimension measurements, should give pause to any consideration of the accuracy of these measurements.

An even more basic consideration is that most measurements cover only a relatively short range of dimensions. Unlike mathematically generated fractals, real data cannot be ideally fractal over all scales. Even over limited ranges, real data often show minor deviations from the ideal straight-line behavior that has been illustrated in some of the examples. Fractal characterization of such data may still be very useful as a way to compare data sets, but interpretation of the numeric values as though the data were from an ideal mathematical fractal should be avoided.

3

The Relationship between Boundary Lines and Surfaces

Analogies

A fractal surface can be understood by direct analogy to the fractal boundary lines or profiles already discussed. It is possible to construct a regular fractal surface using the same approach as the Koch island, as shown in Figure 1. In this case, the surface is initially a Euclidean plane. Adding and removing blocks in a regular pattern produces an increase in surface area without changing the enclosed volume. Repeating this operation with ever smaller blocks allows the surface area to increase without limit. The fractal dimension describes the rate at which the surface area increases. For the particular structuring element shown in Figure 1, this is 2.792.

A random fractal which appears more "natural" can be produced by a similar iterative method in which the placement and magnitude of the structuring elements is randomized. This is directly analogous to the random midpoint displacement technique described before. The full details and an algorithm for this method will be presented in Chapter 6. Figure 2 shows such a surface at each stage of the iterative procedure. Notice that as the iteration proceeds and the step size becomes smaller, the details of the surface emerge.

Figure 1. Iterative steps in the creation of a fractal surface using structured operations analogous to a Koch island.

In this example, it is especially clear that the profiles along the x and y axes are also fractals, since they are produced in the same way as the random midpoint displacement technique already discussed. It is also true that any profile across the surface will be fractal, whether this lies along the x or y axes, or in some other direction. This will become the basis for some of the measurement techniques to be discussed below. Intersections of the surface with planes oriented in other directions will be discussed in detail shortly. There are also a variety of other models, some iterative and some not, for producing fractal surfaces with different characteristics. These are the subject of Chapter 6. For the moment, our purpose is to understand some of the characteristics of natural fractal surfaces, and the ways in which they can be measured.

It should be possible to perform a measurement of the dimension of a fractal surface in a way exactly analogous to the Richardson method for a profile. The measurement element for a surface is a triangle, just as a line segment is used for a boundary line. Placing triangular tiles of a particular size on a surface so that it is exactly and completely covered is always possible provided that the triangle corners must lie on the surface and any intersection of the surface with the interior of the triangle is ignored. From a mathematical point of view, this presents no difficulties. Of course, as a physical measurement procedure for a real surface, this could present some problems and in fact this direct measurement method is not used because the tiling procedure is impossible to carry out for discrete elevation data. The use of identical-sized triangles means that the locations of the corners produce a tessellation of the surface that requires being able to locate any point on the surface, which is by definition continuous but not differentiable.

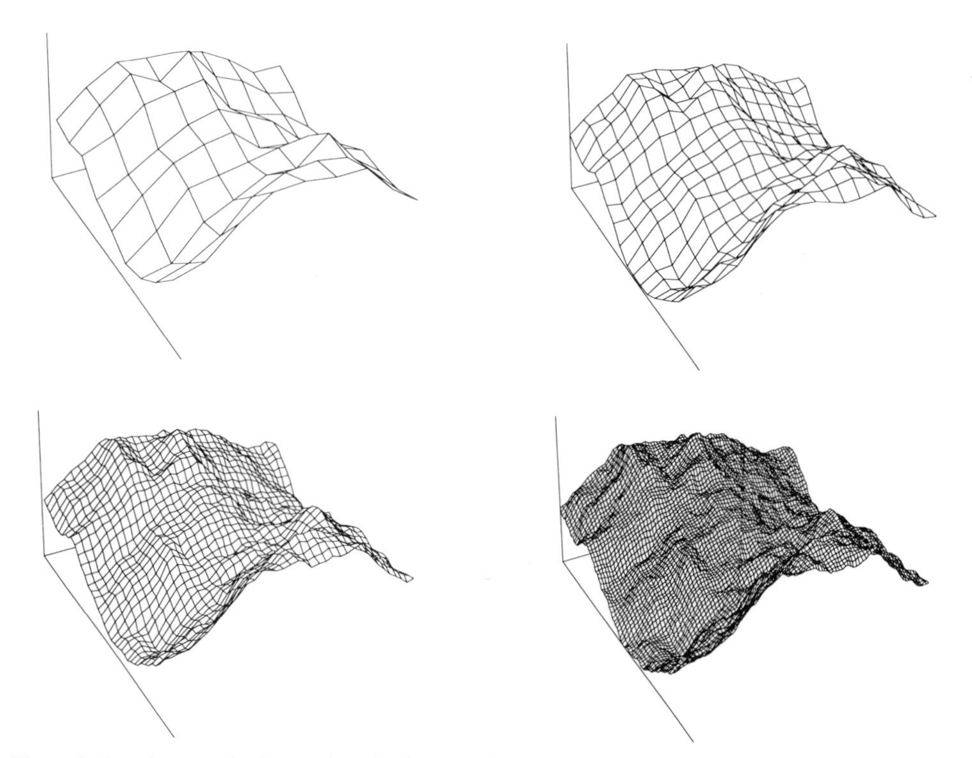

Figure 2. Iterative steps in the creation of a fractal surface by random midpoint displacement. The process is analogous to the creation of a profile by the same procedure.

Figure 3. Tiling of a fractal surface with equilateral triangles, illustrated for two different triangle sizes. A log–log Richardson plot of the area estimated by the triangles vs. the triangle edge dimension gives the fractal dimension.

Conceptually, as shown in Figure 3, it is possible to use tiles of various dimensions to tile a surface, to count the number of tiles and plot their total area as a function of side length (on log–log axes) and produce a Richardson plot for a surface. The slope of this plot is, as for the boundary profile, negative. The magnitude of the slope is the fractional part of the fractal dimension. Adding two (the topological dimension of a surface) produces a fractal dimension $D = 2.d$ with a value between 2 (the Euclidean plane) and 2.99… (a very rough surface which extends outward into the third, or volume dimension).

In practice, it is very difficult to actually perform this kind of tiling. The placement of the triangular facets can only be accurate if the surface values are continuously recorded. In the most common methods of surface measurement, the elevation values are recorded on a grid of x, y values which are uniformly spaced along a horizontal projection plane. These points are not uniformly spaced along the actual surface, and interpolation to determine the elevation or z values between points is both time consuming and misleading, since by definition a fractal surface has detail which extends down to all scales and displaces the points from the values obtained by interpolation. Also, the placement of the tiles on the surface cannot proceed in any simple pattern such as a raster following the underlying x, y grid, because the placement of the tiles on the surface will proceed more quickly in the nearly horizontal regions and more slowly elsewhere.

Direct measurement of surface area using triangles of different sizes has been described (Clarke 1986) and applied both to topographic maps of the earth and to scanning electron microscope (SEM) images. In both cases, stereoscopy was used to obtain the elevation of selected points (Thompson 1987). It is a very difficult task to construct the tessellation, and the range of dimensions that can be covered is small. Consequently, the precision and accuracy are both poor.

Direct Methods

One of the physical methods used to measure the fractal dimension of a surface does actually perform a kind of Richardson technique. Surface adsorption of gas molecules on catalysts and similar surfaces has long been used to measure the available surface area. This so-called BET technique, named for Brunauer, Emmett, Teller (1938), can be performed with gas molecules of different sizes. Of course, when this is done the measured surface area changes. The use of simple gases like N_2 has been extended to complex organic molecules, so that a total range of sizes of several orders of magnitude can be covered. It is necessary to take

into account the shape as well as the size of the larger molecules. The plot of total number of molecules (a measure of the surface area) vs. an effective size for the molecules produces a straight line plot on log–log axes, whose slope gives the surface dimension (Pfeifer, Avnir et al. 1983; Pfeifer and Avnir 1983a; Pfeifer and Avnir 1983b; Pfeifer and Avnir 1983c; Avnir and Farin 1984; Pfeifer, Avnir et al. 1984a; Pfeifer, Avnir et al. 1984b; Farin, Peleg et al. 1985; Avnir 1989; Daoud and Martin 1989; Elber 1989; Fripiat 1989; Le Méhauté 1989; Pfeifer and Obert 1989; Farin and Avnir 1989a; Farin and Avnir 1989b; Avnir and Farin 1990; Pfeifer and Cole 1990; Pfeifer, Obert et al. 1990; LaBrecque 1992; Piscitelle and Segars 1992; Ludlow and Hoberg 1990).

There are three ways that gas adsorption data are used to measure a fractal dimension:

1. A series of monolayer covers of the sample are made with different gases whose cross-sectional dimension σ varies. The number of molecules is observed to vary as $n_m \propto \sigma^{(-D/2)}$, where D is the surface fractal dimension.
2. A single adsorbate molecule is applied to particles of different size. The surface area per unit mass S is related to the surface fractal dimension D and particle radius R by $S \propto R^{(D-3)}$.
3. The pore volume distribution from gas adsorption or mercury intrusion data is used to determine the surface fractal dimension D from the change of cumulative pore volume V with respect to pore radius ρ. Then $-dV/d\rho \propto \rho^{(2-D)}$.

It is the first of these methods that provides the most direct measure of the fractal dimension for an intact surface, as discussed in this book. The method is similar to a Richardson approach, but the packing of adsorbate molecules is "around" the surface rather than strictly on it, as indicated in Figure 4, and this may modify the results somewhat. However, the method is still strictly only applicable to self-similar as opposed to self-affine surfaces.

One strength of all three techniques is that they can access all of the "hidden" portions of the surface, undercuts, pores, etc., which may not be seen by techniques that record a single elevation value at each x, y coordinate. Also, the method covers the surface continuously without regard to actually measuring any elevation points, and is capable of working at very small dimensions (down to the size of monatomic gases). It is in this range of dimensions that the surface properties of many of these catalyst surfaces are important. The results are only appropriate for self-similar and isotropic surfaces, but it appears that many catalyst surfaces (and processes) may fit into this category (Sernetz, Gelléri et al. 1985; Sernetz, Bittner et al. 1989; Sernetz, Willems et al. 1989)

The traditional model for the distribution of active sites on a catalytic surface predicts that the activity is proportional to $r^{-\nu}$, where ν is 1, 2, or 3 depending on whether the active sites are faces, edges or corner sites. However, experiment often reveals non-integral exponents. Pfeifer (Pfeifer, Avnir et al. 1983; Pfeifer and Avnir 1983a; Pfeifer and Avnir 1983b; Pfeifer

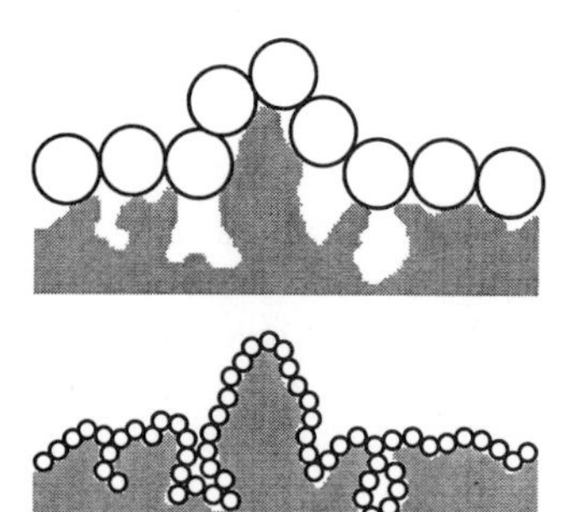

Figure 4. Principle of the BET measurement of surface area by adsorption of a single layer of molecules of different sizes. Larger molecules ignore the fine details of surface morphology.

and Avnir 1983c) shows that this is because the surface is fractal, and $v = D - 3$, where D is the surface fractal dimension. Diffusion or chemical reaction on surfaces is also dependent on the roughness (Rammal 1984; Blumen and Köhler 1990; Cook 1988; Evesque 1989; Havlin 1989; Kopelman 1989). Diffusion depends both on D and the connectivity. For a simply connected surface (no bridges or loops) the time needed to move a distance X is proportional to $X^{1/D}$. For a planar surface, $D = 2$ and this reduces to the usual $X^{1/2}$ case.

There are a number of other ways to perform a measurement of the fractal dimension of the surface directly, and we shall return to them shortly. First, it is interesting to consider the relationships between line and surface fractals and the ways that these may be useful for measurement.

One of Mandelbrot's "speculations," ideas which he has put forward without a complete derivation or proof, is that intersecting a fractal surface of dimension 2.d with a plane will produce an intersection boundary whose dimension is exactly 1.d. This is actually a special case of the intersection of a fractal surface with another surface. We will see in later chapters that the ($D_S - 1$) rule is only true if some restrictions are placed on the surface having to do with its randomness, uniformity, and isotropy. It is not at all clear that most real fractal surfaces meet these conditions. In addition, Preuss (1990) has criticized the accuracy of adding 1.0 to the boundary dimension to get a surface dimension because of the small number of points used to determine the profile dimension across a typical image. However, this idea provides a good place to start.

Figure 5 shows the basic idea. A fractal surface has been intersected with a horizontal plane, producing a "shoreline" where the surface ("sea level") intersects the land. Measuring the dimension of the shoreline requires exactly the same methods previously discussed for fractal lines, for example a Richardson plot. In fact, since the Richardson method was originally applied to physical shorelines, it might be expected that the rougher the land surface the more irregular the shoreline. This provides a qualitative justification for measuring the fractal dimension of the boundary line and adding 1.0 to obtain the surface fractal.

While it seems reasonable to expect a very rugged terrain to produce a tortuous shoreline (for instance the fjords of Norway), and conversely a very smooth terrain to produce a smooth shoreline (for instance, the beaches of Florida), it does not seem that the surface of the earth is actually a very isotropic or uniform fractal, and this correlation does not hold in quantitative detail. Figure 4 of Chapter 2 shows results measured for three of the Hawaiian islands (Russ 1991a). The measurements were made with stride lengths covering about one order of magnitude, on maps with a 1:225000 scale. Kauai, the oldest, has by far the most rugged terrain due to erosion and weathering, while Oahu is somewhat younger and Hawaii (the "Big Island") is very young and still growing due to volcanism. Hawaii is effectively a smooth shield volcano whose surface is quite smooth. One might expect the shoreline fractal dimensions to follow the same trend, but they do not. The factors of weathering and erosion which sculpt the surface (rain and wind) are not the same as those which influence the shoreline (waves).

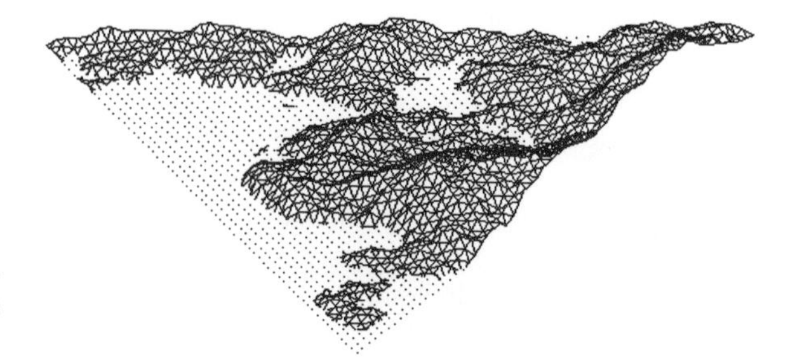

Figure 5. A fractal surface cut by a plane. The outline is also fractal, with a dimension that in some cases is 1 less than the surface.

Another example of anisotropy comes from the example of clouds cited in Chapter 1. The fractal dimension of cloud boundaries viewed from either above or below is consistent (D = 1.36) over many orders of magnitude, but it is quite different from the dimension of clouds viewed horizontally. Different physical forces, including wind patterns, vertical changes in temperature and pressure, etc., produce these different patterns of roughness, although both boundaries are fractal.

A fractal surface for which the scaling in the vertical direction is different from that in the lateral directions is self-affine rather than self-similar. It was mentioned earlier that for a self-affine profile, the Richardson technique is inappropriate and produces an erroneous value for the dimension, although it may show á straight line plot on a log–log scale. Mandelbrot has shown that in the limit as the measurement length becomes small, the slope of this line may approach $1/H$, where H is $(2 - D)$. This means that the magnitude of the slope of the line and the apparent fractal dimension can exceed 2.0, which has no meaning for a line. The same thing is true for self-affine surfaces. The tiling method, which is analogous to the Richardson method, produces a log–log plot of measured area vs. dimension which reveals the fractal nature of the surface, but the slope does not provide a measure of the dimension, and may lie outside the meaningful range from 2.0 to 2.99...

Zerosets

Passing a plane through the surface in a vertical direction produces an elevation profile which is self-affine rather than self-similar, and must be measured using the same techniques discussed in Chapter 2 for self-affine profiles. However, passing a horizontal plane through the surface (as shown schematically in Figure 5) produces an intersection profile which does obey the $D - 1$ relationship, provided that the surface is isotropic. This is a consequence of the fact that all of the points on the profile have the same value of z, which is the direction in which the scaling factor is different.

This is called the zeroset or Poincaré section of the fractal, and has an analogy for a self-affine line profile. Intersecting such a profile with a straight horizontal line produces a series of intersection points whose spacing is fractal. In fact, this set of points is just a Cantor dust as described before. Amongst other natural phenomena, the spacing of these crossing points models the temporal pattern of noise occurring on transmission lines. A plot of the frequency of occurrence of various distances between successive intersection points vs. the distance or interval, on a log–log scale, produces a Korcak plot whose slope gives the fractal dimension. This was shown in Chapter 2. Of course, this is a very inefficient way to actually measure the dimension of a profile.

The direct analogy of this approach for the intersection of a horizontal plane with a fractal surface is to measure the areas of the lakes which lie below the surface. Figure 6 shows an image of a fractal surface that is thresholded at one brightness level. The thresholding produces lakes, shown in the rendered surface, which can be measured. The number of boundaries of lakes or islands from this single elevation level is small, but the thresholding can be repeated at many different elevations. The regions that touch the edges of the image cannot be used, because their full area cannot be determined.

A log–log plot of the number of islands or lakes whose area exceeds value A as a function of A produces a straight line plot as shown in Figure 7, whose slope gives the fractal dimension of the surface (Mandelbrot 1975; Kaye 1988; Isichenko and Kalda 1991). This is the Korcak method for measuring the surface dimension, and it is also a relatively inefficient method for obtaining the surface dimension. However, it has been applied to cases ranging from metal fractures to the surface of the earth, confirming the expected relationships.

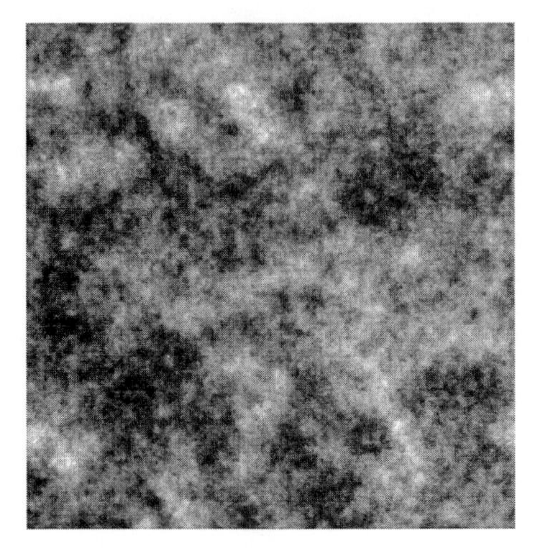

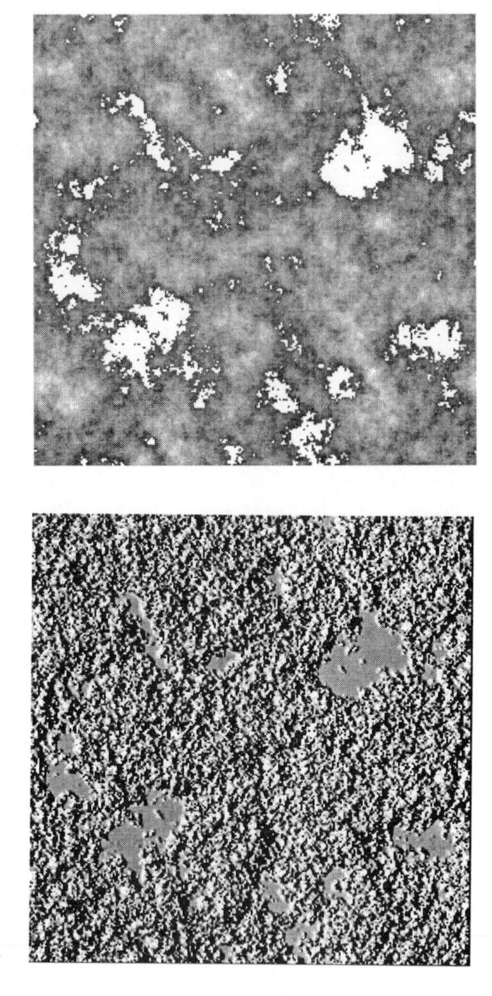

Figure 6. A fractal surface (generated with inverse Fourier transform), thresholded at an arbitrary elevation level and the islands shown as white regions and as lakes in a rendered view of the surface.

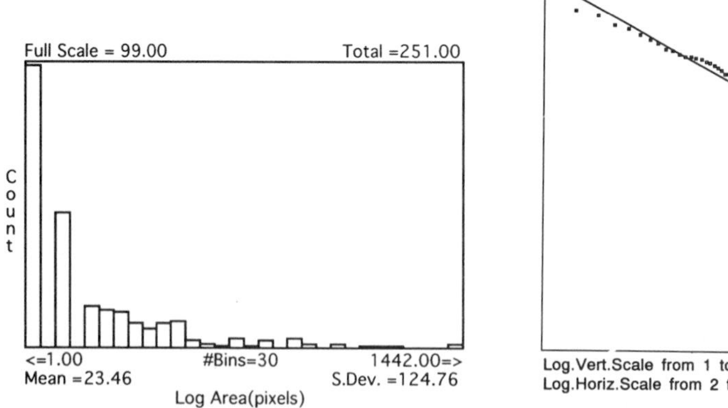

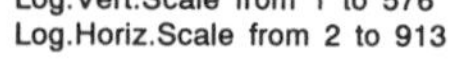

Figure 7. Data from Figure 6. The area plot for the islands in the thresholded image shows many small and fewer large sizes. Repeating the thresholding at many levels and constructing a Korcak plot of the cumulative number of islands exceeding each area value provides a measure of the fractal dimension.

Mandelbrot reports that from a size distribution of the islands of the earth, the surface dimension is about 2.3, which is in approximate agreement with the result from the boundary profiles. It should be realized that assigning a single dimension to the entire surface of the planet is almost certainly not appropriate, since very different terrains are produced according to the presence of different kinds of rock, different climates and weathering patterns, etc.

Dimensional Analysis

Another use of the islands produced by the intersection of a fractal surface with a section plane uses both the area and perimeter values of each island. If the islands were Euclidean objects, the relationship between the area and a linear dimension such as perimeter would have an exponent of 2 (i.e., Area $\propto$ Perimeter^2). However, it is noted in nature that many such relationships expected on the basis of Euclidean dimensions do not always hold. For instance, the surface area of salamanders does increase in proportion to the 2/3 power of body mass, as would be expected from the Euclidean geometric relationship between area and volume. Likewise, the diameter of tree trunks increases as the 3/2 power of their height, as required by mechanical constraints to prevent buckling. But plotting the brain weight of mammals vs. body weight shows a slope of 0.62 when plotted on log–log axes (McMahon and Bonner 1983). It is not known whether this is related to the increase in the convolutions of the brain cortex in the "higher" mammals, which has been suggested as a fractal. A log–log plot of the brain volume vs. surface area has a noninteger slope that corresponds to a surface fractal dimension of 2.76. These noninteger power law relationships are described as allotropic, and the construction of such plots to determine the dimension is called dimensional analysis. It can also be applied to these fractal islands in a zeroset sea.

A log–log plot of perimeter vs. area for many islands produced by this method has a slope equal to half of the boundary fractal dimension, which is in turn 1.0 less than the surface dimension. Figure 8 shows an example of the sectioning of a surface, and Figure 9 presents several examples of log P vs. log A plots for data from a series of such sections through the surface at different elevation levels. The plots show separately the bright islands and dark "lakes" which are above and below the surface of the intersecting planes, respectively. Since

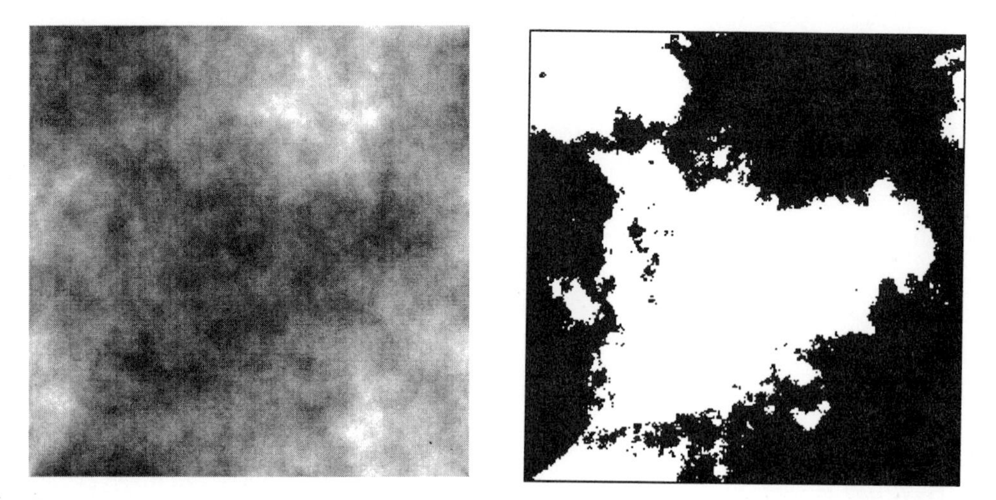

Figure 8. Generated isotropic fractal surface ($D = 2.30$) shown as a range image (grey scale equals elevation), and a binary image of islands produced by thresholding.

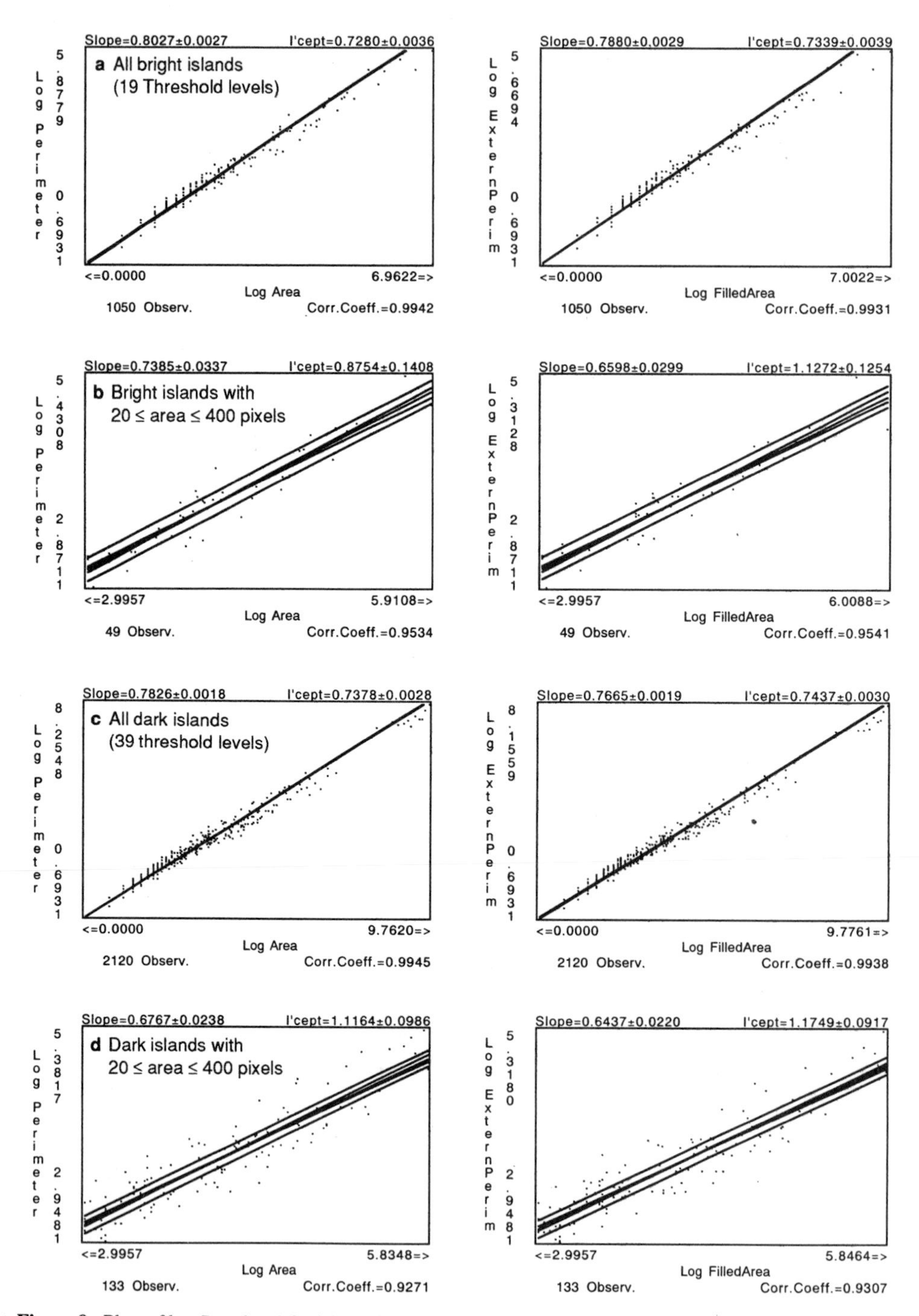

Figure 9. Plots of log P vs. log A for islands from Figure 8. Both dark and light islands are shown, for all of the data and for just those islands with areas between 20 and 400 pixels. Plots are shown both for the net area and perimeter and for the external perimeter and filled area.

there is no inherent up or down direction, both sets of regions will be referred to as islands. The two sets of values should produce identical plots, if the surface is truly isotropic and fractal (either self-similar or self-affine). Also, the slope of the plot should not vary as the elevation value of the thresholding is varied (Figure 10). Some published data (Aguilar, Anguiano et al. 1992; Pancorbo, Anguiano et al. 1993; Russ 1993b) have shown such variations, possibly due to the presence of noise in the original elevation data or anisotropy of the surface (as discussed in Chapter 7).

There are also two sets of plots, one for the net area of islands without any internal "lakes" and including the total perimeter including that around such internal regions, and the other in which the area includes the lakes and the perimeter is only the outer or external boundary. Of course, the slopes are slightly different for these cases since the filled areas are greater than the net areas, and the external perimeter is less than the total perimeter. The filled area/external perimeter values seem to agree slightly better with the expected fractal dimension for the surface. Mandelbrot has used this method for fracture surfaces (Mandelbrot, Passoja et al. 1984), commenting that lakes within islands should be included and islands within lakes should be skipped.

The plots also show separately the fit to all of the data points and just those which lie in the range from 20 to 400 pixels in area. It is recommended (Aguilar, Anguiano et al. 1992; Gomez-Rodriguez, Asenjo et al. 1992) that the smallest islands not be used for the fitting, since they have so few pixels that accurate measurement of the perimeter and even the area is difficult. In addition, large islands may result from merging together of several smaller islands and can bias the plot somewhat. Also, as noted before for the Korcak analysis of island areas, the larger islands are systematically undersampled because of the finite image area.

A practical difficulty with the log *P* vs. log *A* method for determining the fractal dimension is the greater effort needed to measure the perimeters, as opposed to simply counting the islands. The area of islands is relatively easy to determine (by counting pixels), and can be used to construct Korcak plots as shown in Figure 11 for the same islands. In addition, however, the log *P* vs. log *A* method may introduce a bias (Mu et al. 1993) because the smaller perimeter values are measured less accurately and with a stride length (the size of a pixel) which is larger with respect to the total length than it is for the islands with larger perimeters. This of course is a consequence of the finite pixel size of the image, and of the fractal nature of the boundaries which means that the perimeter changes with the size of the measuring element or stride length.

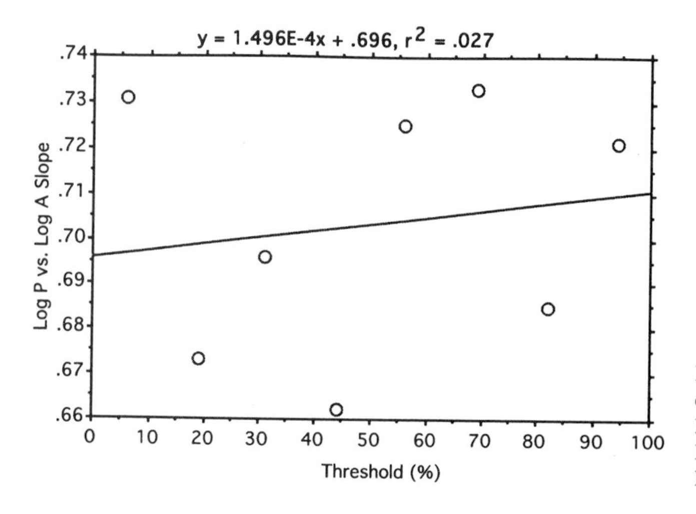

Figure 10. No significant trend of fractal dimension (slope of log *P* vs. log *A*) with threshold level was found in the data from Figure 9.

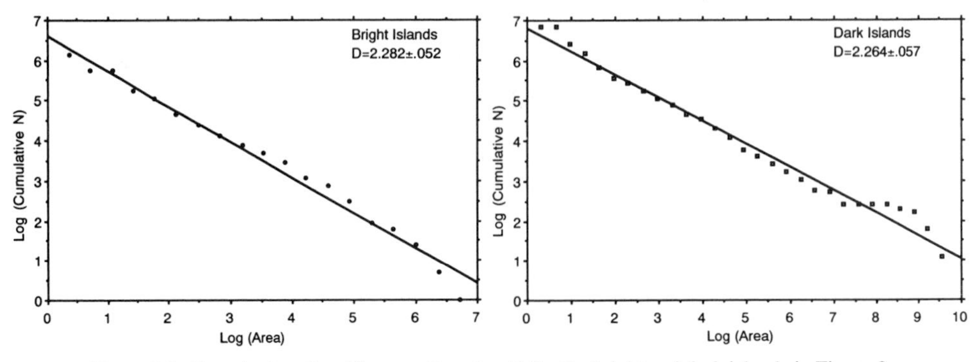

Figure 11. Korcak plots (log N(area > A) vs. log A) for the bright and dark islands in Figure 8.

Imre (1992) also suggests that there may be systematic variations in the shape of small and large islands, which could bias the plots.

There is another difficulty with the perimeter measurement as it is often performed for this technique. A naive measurement of perimeter can be obtained by counting the edges of the pixels, or the number of pixels in the feature which touch any part of the background. This overestimates the actual perimeter, but that is not the problem since it could just displace the plot of log P vs. log A parallel to itself, to higher perimeter values. But in fact the amount of the overestimate is a function of how irregular the boundary is (i.e., the fractal dimension), and also depends on the orientation of the feature with respect to the alignment of the rows of pixels. This is shown in Figure 12. A more accurate measurement of the perimeter, which is used in image analysis systems and is less sensitive to the feature orientation, is the chain code length. Chain code is discussed and illustrated in Chapter 4. It substitutes diagonal lines for the steps in the pixel boundary, and uses the distance from center to center of the pixels rather than the edge length. All of the data shown in log P vs. log A plots in this book were measured by the chain code perimeter. However, the program in the appendix performs dimensional analysis by counting the number of pixels which touch the background, either along a corner or an edge.

The intersection of the horizontal plane with the self-affine fractal surface produces boundary lines which are self-similar rather than merely self-affine, and so it is permissible to use the measurement methods discussed before for self-similar lines to determine the dimen-

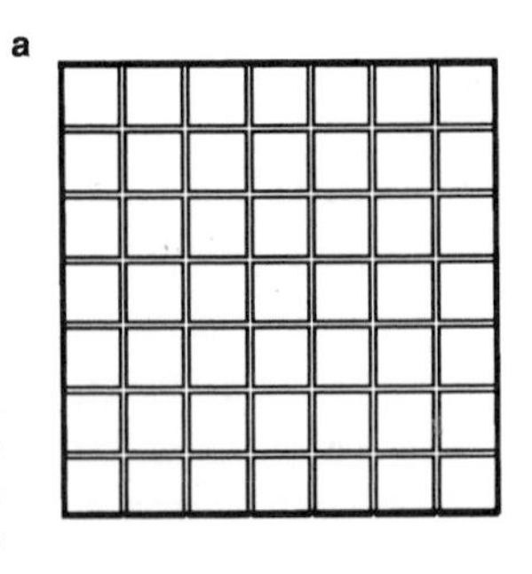

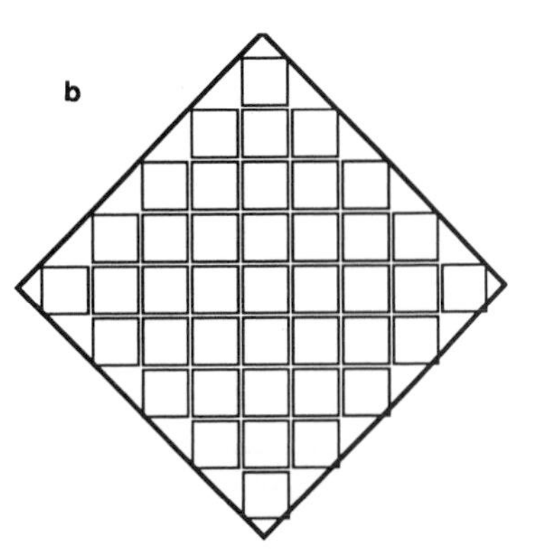

Figure 12. Comparison of perimeter estimation by chain code vs. summation of pixel edge lengths, as a function of orientation: **(a)** sum of edges = 28 units, chain code perimeter = 28.0 units; **(b)** sum of edges = 36 units, chain code perimeter = $20\sqrt{2}$ = 28.28 units.

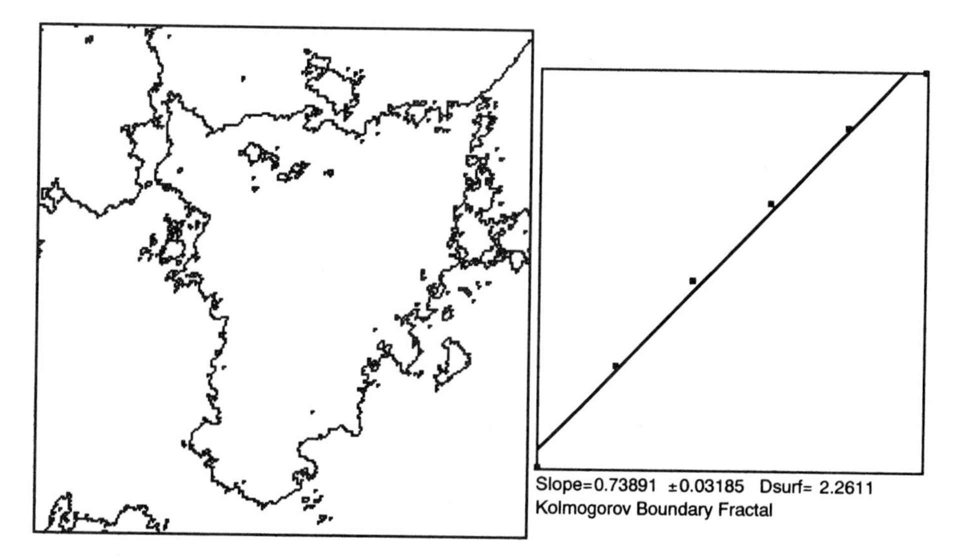

Figure 13. Kolmogorov (box count) dimension for the boundary produced by thresholding the range image in Figure 8 at 50% elevation.

sion. This includes the Richardson method, and so it is appropriate (if usually difficult or impractical) to apply this technique to measure natural shorelines.

Other methods which can be applied properly to self-similar data can be used for these outlines as well. For instance, the Kolmogorov or box-counting method can be used to determine the scaling of the boundary length produced by this cutting or thresholding operation. This can be done by placing a grid on the section image and counting the number of grid squares through which the boundary passes, as a function of the size of the grid. A plot of these values on log–log axes provides a direct measure of the fractal dimension (although with rather limited

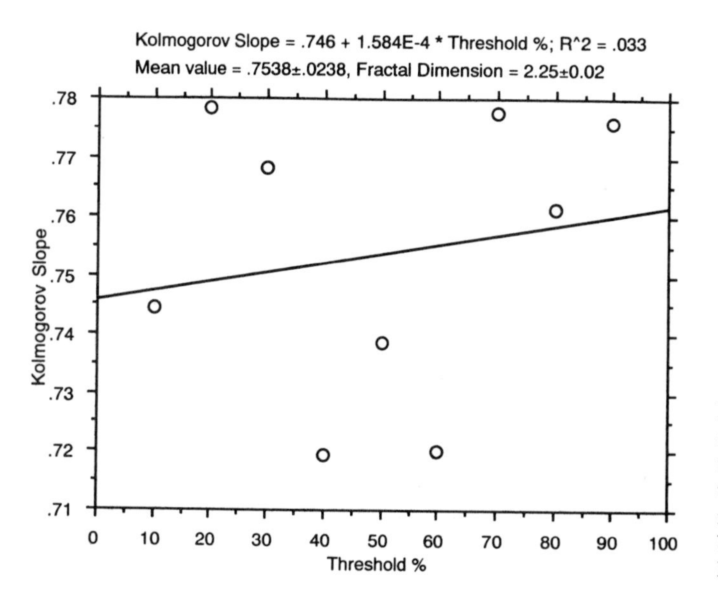

Figure 14. There is considerable scatter but no significant trend of measured slope with threshold level, using a Kolmogorov or box count on the boundaries of thresholded islands from Figure 8.

precision because of the usual problem with box counting of being limited to only a few grid sizes). Figure 13 shows the application of the box-counting method to the boundaries produced by sectioning. Figure 14 shows the considerable scatter of results obtained from varying the elevation used for thresholding. There is no trend with elevation, which is as expected.

There is still an important underlying assumption in the use of the zeroset or horizontal section, which is that the scaling in x and y directions is the same. There are certainly some important cases in which this seems an unwarranted assumption, such as wear surfaces in which a definite directional anisotropy is present. Some other surfaces, such as deposited particulates, may be isotropic and thus permit this approach. For the surface of the earth, the directionality of erosion due to weather patterns raises important questions. For instance, Kaye (1989a) has shown that the fractal dimension of the shoreline along the north side of Manitoulin Island is different from that along the south side, either due to prevailing winds, different geology, or other physical effects. This means that the local land surface is probably self-affine in all three directions, and it is not appropriate to measure the shoreline dimension with a Richardson method. Instead, the methods appropriate for self-affine data must be used.

The Minkowski Comforter

The Minkowski method can be applied to the measurement of self-similar or self-affine surfaces. For a strictly self-similar surface, a sphere is moved to every point on the surface to sweep out a volume (sometimes called the Minkowski comforter, just as the Minkowski sausage is swept out along a line). This structuring element has the same dimension in the x, y, and z directions and so is appropriate only for cases in which the three dimensions scale in the same way.

Modification of the structuring element allows the method to be used for self-affine surfaces as well. By analogy to the case for a self-affine profile, in which the structuring element becomes a line parallel to the horizontal axis, we can use a disk parallel to the horizontal plane when the z direction differs from x and y, but there is no anisotropy in the x, y plane. The disk is moved to every point on the surface and the volume which is swept out is measured and plotted as a function of disk radius.

In practice, the original data often consist of an array of elevation values on a regular grid which can be considered as an image. In this case, the Minkowski measurement can be performed by many image analysis systems using local neighborhood procedures sometimes called grey-scale dilation or rank operators. As indicated in Figure 15, these operators examine a region of pixels around each pixel in the original image, find the brightest or darkest value (highest or lowest elevation) in that neighborhood, and create a new image in which the central pixel gets either the extreme value, or the difference between the two. If the difference is not

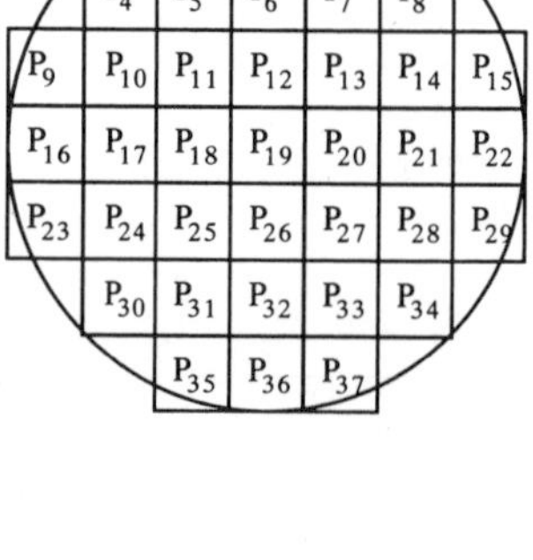

Figure 15. Schematic diagram of pixels in a neighborhood used for ranking. The pixels used approximate a circle. The elevation (brightness) values of the 37 pixels are ranked in order. In a grey-scale erosion or dilation, a new image would be constructed in which the central pixel was set to either the brightest or darkest value in the neighborhood. To construct the Minkowski comforter, the difference between the highest and lowest values would be used. The procedure is repeated for every pixel in the original image.

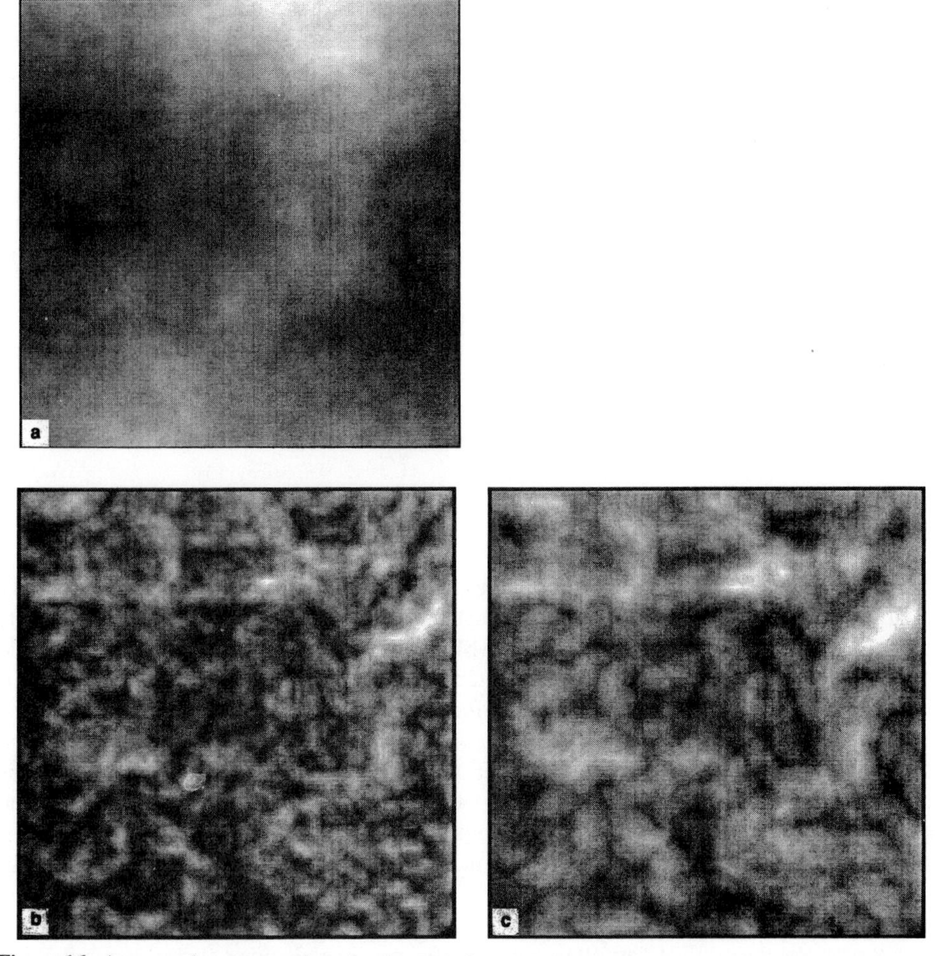

Figure 16. A grey scale representation of an isotropic fractal surface (generated by midpoint displacement with $D = 2.18$) **(a)**. Two of the Minkowski cover images are shown (difference between dilation and erosion using disks of diameter 9 **(b)** and 13 **(c)** pixels).

directly available, saving the separate minimum and maximum images and subsequent subtraction will give the same range information.

If the neighborhood is made circular, or as close as possible to a circle considering the square grid, then the integrated total brightness value of the difference or range image is the volume of the Minkowski comforter for one disk size, corresponding to the size of the neighborhood. Repeating the operation for many different neighborhood sizes will provide the data for the log–log plot from which the surface fractal dimension is obtained. Figure 16 shows an example of this procedure for a fractal surface, and Figure 17 shows the resulting plot.

Of course, the procedure just described is rather slow since the elevation or brightness values of pixels around each location in the image must be sorted, the sorting of large numbers of values (as the neighborhood size grows) is an inherently time-consuming operation, and quite a bit of storage is needed to keep the additional image(s). The same algorithm can be performed more efficiently by carrying out all of the various sorting operations on a single pass through the image, taking advantage of merging partially sorted lists. Adding the

difference values to the total comforter volume for each diameter eliminates the need for storing images. The implementation of the method for maximum efficiency can speed the execution by several orders of magnitude, but the basic algorithm remains the same. This is a direct extension of the "variational" method for determining a Minkowski dimension of a line to the case of a surface (Dubuc, Zucker et al. 1989).

It is worth noting that if the surface is also anisotropic in the x, y directions, as will be shown in Chapter 7, the Minkowski dimension reported by any of these methods is suspect. However, the method can be further modified to measure the dimensions in each direction. In this case, the circular disk is replaced by a line segment parallel to the axis of interest. The neighborhood comparison of pixel values is reduced to a line of pixels instead of a circular region. It is possible to perform the sorting and comparisons for both x and y directions during a single pass through the image, and so the two directional fractal dimensions for a fully affine surface can be evaluated. This method has an advantage over measuring the dimension of a selected elevation profile across the surface parallel to the x and/or y directions because it uses the data from the entire surface and is thus more precise.

Data Formats

Nevertheless, it is the use of elevation profiles across surfaces which is most routinely used to report a fractal dimension in those measurement instruments which report such values. Even in the case in which the surface is truly self-similar in all three dimensions, the elevation profile as measured by techniques such as profilometry, atomic force microscopy, interference microscopy, or at the very different scale of topographic mapping by side-scan radar, stereo imaging, sonar depth finding, etc., is not self-similar but self-affine. This is fundamental to the process of making the elevation profile, which is single valued (only one elevation value is recorded at each x, y location). Any undercuts in the surface are not seen, and only small scale surface deviations which are steeper than the larger scale ones can be observed. The distribution of slope elements is biased, and for the case of discrete elevation data which are uniformly spaced in the horizontal direction, this effect is magnified.

This means that even if the surface is self-similar (and as noted before, many are not), the elevation profile is self-affine and must be measured accordingly. The Richardson method is inappropriate for this, in addition to giving difficulties when the data are uniformly spaced in the horizontal direction rather than along the surface. The Minkowski method with an appropriate structuring element is one method which can be applied correctly, and we will shortly discuss others. It should be noted that in spite of these fundamental problems, the

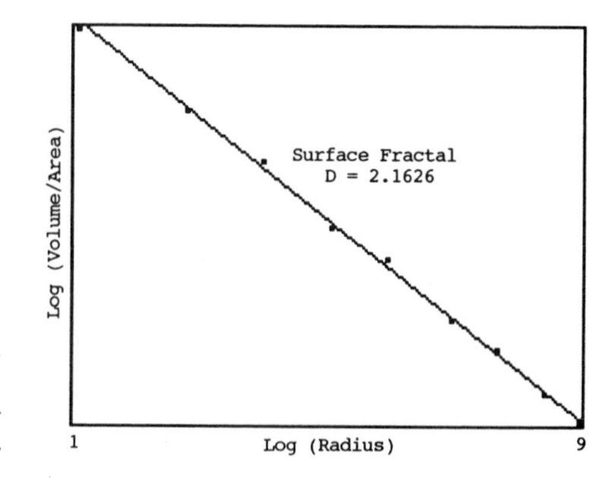

Figure 17. Data from Figure 16, showing the measurement of the surface fractal dimension as $D = 2.16$.

Richardson method has (apparently because of its operational and conceptual simplicity) been misapplied to vertical sections in many cases. As discussed in Chapter 2, the result of this effort is often a straight line on a log–log plot, which confirms the fractal nature of the profile and hence of the surface, but the numerical value cannot be properly interpreted. Of course, as pointed out before, the boundary profile produced by intersecting the surface with a horizontal plane does produce a self-similar boundary which can be correctly measured using the Richardson approach, although it is not necessarily correct to add 1.0 to this value to describe the surface.

Some of the various surface measurement techniques which provide a topographic map of z (elevation) vs. x, y position will be discussed further in the final chapter. Regardless of the scale, ranging from atomic to geographic, these fall into three main categories. First are methods which record elevation along a linear traverse, and perhaps perform many such traverses in a raster pattern to completely cover the surface. An example of this is the atomic force or scanning tunneling microscope, or towed-buoy sonar scans. The second category includes methods such as interference microscopy and side-scanned radar which measure a two-dimensional array of elevation values. These have the advantage that the measurements have the same spacing and resolution in x and y, and instrument artifacts along or across the raster orientation are not present, although they may be introduced in the process of digitization and storage. Both the first and second methods generally store the data as an array of elevation data as a function of x, y on a complete grid of square pixels. The pixel dimensions typically correspond to the best lateral resolution of the technique. This type of image is often called a range image, and can be processed or displayed in much the same way as many other kinds of image processing and analysis.

The third category acquires data as a series of discrete x, y, z coordinates which do not, in general, constitute a complete array of values. Measurements from stereo pair images, whether at the resolution of the scanning electron microscope or aerial photography, fall into this category. The conversion from discrete points to a full grid of values requires a separate interpolation step, and this hides any fractal character to the surface at scales approaching the pixel values. This may bias measurement methods which cover scales down to the pixel dimensions.

There is a fourth category of miscellaneous measurement techniques, including the gas adsorption method described above, as well as photon scattering which is discussed in a subsequent chapter. These do not produce a map of elevation values, but instead integrate information from across the surface and report a few values which are interpreted to yield a surface dimension.

When a map of elevation values is stored, the uniform spacing of the pixels in the x, y plane may create bias or other errors in the evaluation of the fractal dimension because the distance between points along the surface varies with local slope. This problem also arises for fractal line profiles which are digitized. When contour lines are digitized by tracing boundaries, for instance with a graphics tablet connected to a computer, the x, y coordinates have a finite resolution but this is generally much smaller than the dimensions being traced and so the points may be regarded as a set of discrete coordinates which are connected by small straight line segments to produce a polygonal approximation to the boundary. While the sides of the polygon are not all exactly equal, which as noted before may create some minor bias for a Richardson analysis, the dimensions are generally small compared to the important dimensions recorded in the tracing.

These data are quite different, however, than the grid of x, y points which are produced as an iso-elevation contour line by "slicing" the stored pixel array of a range image. Figure 18 shows an example of a fractal surface recorded as a grey-scale range image, with several contour lines. These lines can be constructed in two different ways. The fast method is to locate

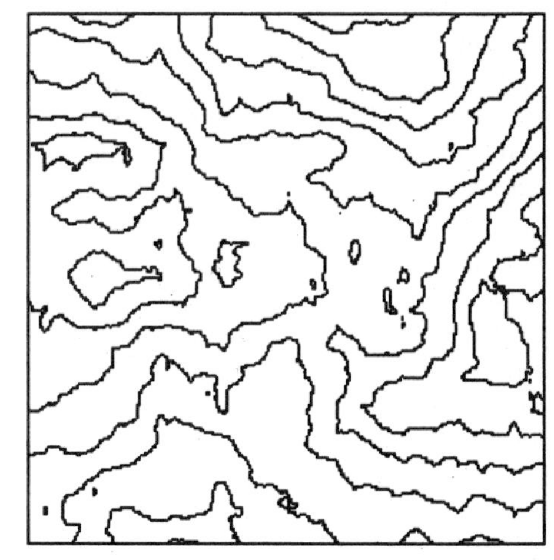

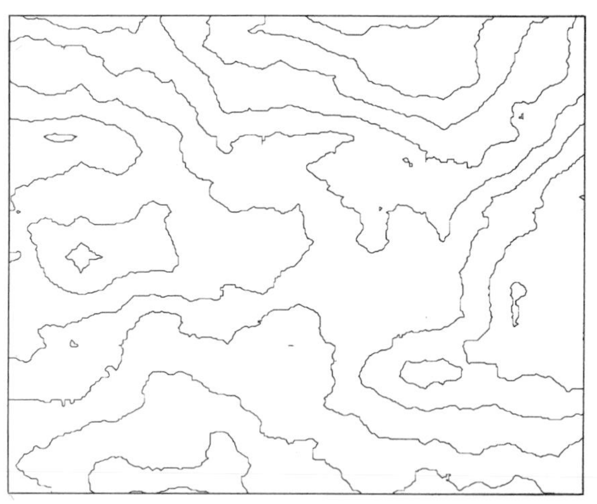

Figure 18. Grey scale (range) image with contour lines drawn by two different methods.

those pixels which lie above a selected elevation but have neighbors which do not. A more accurate contour line in the sense of a topographic map would pass near but not exactly through these pixels, and must be constructed by interpolation. While the appearance of these slice images appears to correspond to the sectioning of the surface by a plane, the bias in pixel locations may either over- or underestimate the length of the actual shoreline, and bias the resulting measurement of the fractal dimension.

Effects of Noise

All real measurement techniques for elevation profiles or surface range images have finite precision. The errors which are made are very technique- and instrument-dependent. Many methods have quite different measurement precision in the lateral direction as compared to the vertical direction. For instance, the interference light microscope has a lateral resolution limited by the wavelength of visible light to a fraction of a micrometer, while the vertical resolution is well under a nanometer. Some other instruments have responses that vary both vertically and with lateral direction; for instance, the scanning tunneling microscope (STM) has lateral

resolution that depends on the shape of the tip, and "noise" which is different in the fast and slow scan directions of the raster.

It is usual to describe the response of the instrument in terms of a power spectrum, discussed in more detail in the next chapter in connection with the Fourier analysis of profiles and images, but for the present it is adequate to note that a classic "$1/f$" noise model can be used as a model for the instrument errors, to investigate the dependence of the varying measurement techniques on the fractal dimension value which is obtained. Adding such instrument noise to a profile or to an image can be readily simulated (Lam and Sander 1992), whereas it can be very difficult to determine the actual noise spectrum for a particular instrument. In principle this might be measured by scanning a perfectly flat surface and recording the signal, but this is impractical here because no such perfect surface exists. Remember that fractal surfaces contain detail at all length scales, and so there is no cutoff limit below which noise can be ignored. Some of the techniques used to estimate real instrument noise contributions are shown in the final chapter. Aguilar, Anguiano et al. (1992) and Pancorbo, Aguilar et al. (1991) measured the instrument noise along scan lines in an STM as $1/f$ (exponent equal to -1.0 ± 0.1), independent of scan rate and other instrument parameters, by recording an image without any sample present. Our own observations of images of very flat surfaces with several instruments (STM, interferometer, etc.) shows that the measured dimension approaches $1/f$ at high magnification, but it is difficult to determine how much of this is due to the microscope and how much to the surface itself.

Noise of the "$1/f$" type can be simulated simply as a fractal profile or image, with a dimension of 1.5 (for a line) or 2.5 (for a surface) as discussed in the next chapter. Adding this contribution to a fractal profile or line in varying proportions to simulate varying amounts of instrument noise allows us to determine the effect of such noise on the measured value for the fractal dimension of the line. Figure 19 shows an example. Adding the "signal" and "noise" profiles in various proportions and measuring the fractal dimension of the resulting self-affine profile with different algorithms can provide some insight into the variation of the observed dimension.

As noted above, several measurement methods may be applied to an elevation profile. A Korcak fractal dimension constructed as a cumulative plot of the number of horizontal intercept lengths that exceed λ as a function of λ can be used. So can a Minkowski or variational method in which the area enclosed between upper and lower envelopes of neighborhood minima and maxima is plotted as a function of neighbor width (Dubuc, Quiniou et al. 1989). And, as discussed in the next chapter, performing a Fourier transform and plotting log (Magnitude^2) vs. log (Frequency) produces a slope from which the dimension can be obtained. Figure 20

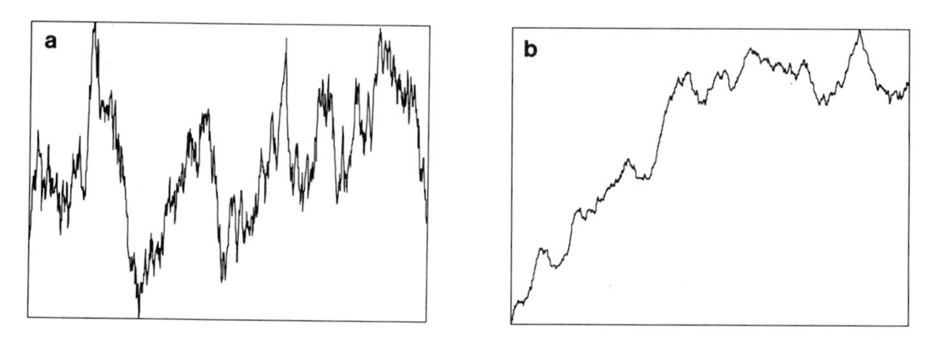

Figure 19. Line profiles used to measure the effect of adding noise on the fractal dimension: **(a)** $1/f$ noise generated by inverse Fourier transform, D approximately 1.5; **(b)** fractal profile generated by iterated midpoint displacement, D approximately 1.2.

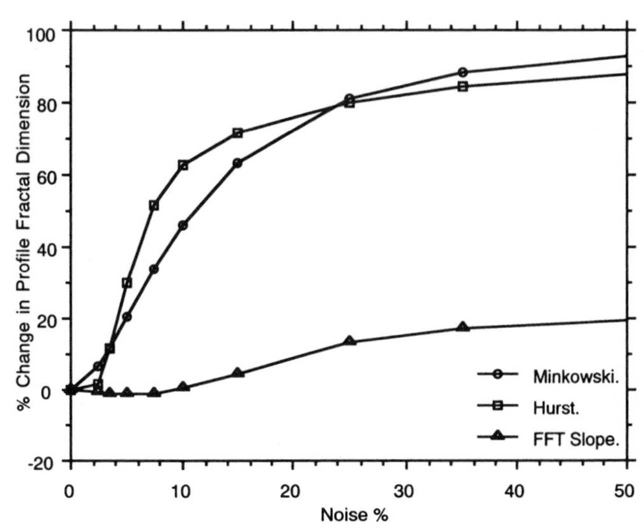

Figure 20. Variation of profile D measured by Minkowski, Hurst, and FT as a function of the amount of added noise.

shows the results from these methods, normalized to show the variation between the two extreme values for the pure noise ($D = 1.5$) and original profile ($D = 1.2$).

Figure 20 shows that the Korcak and Minkowski dimensions are very sensitive to the presence of even a small amount of noise. This observation may be interpreted by noting that these techniques emphasize the minimum and maximum values in each small neighborhood, which are spread farther apart by the high frequencies in the noise. Adding only 5% noise (19:1 signa-to-noise ratio) produces about a 20% change in the measured fractal dimension. Adding about 25% noise (a 3:1 signal-to-noise ratio) produces a measured fractal dimension that reflects only the noise, and the value hardly changes as more noise is added.

The fast Fourier transform (FFT) method, on the other hand, is virtually insensitive to the noise up to about 10% (10:1 signal-to-noise ratio) and does not deviate significantly until the noise reaches 25% (3:1 signal-to-noise ratio). This is because the least-squares linear fit to the log (Magnitude^2) vs. log (Frequency) plot has many data points (512 in the examples shown, which are 1024 point profiles) and the presence of the noise causes only a minor shift in the overall slope as points at the end of the plot are shifted. This suggests that for elevation profiles, the influence of instrument noise which is an order of magnitude smaller than the signal can in fact be safely ignored, within the assumption of $1/f$ noise in this model.

This advantage may be more important than the difficulty of applying Fourier analysis to elevation profiles, which is that the numerical precision of the fit is poor when the number of points in the profile is small. For two-dimensional range images with many more measured points, even this limitation may not be severe.

Simulated Noise

In order to examine these effects, the next step is to examine the effect of adding noise to a range image of a fractal surface. This was done by generating an isotropic $1/f$ noise image (by inverse FT as discussed in Chapter 6) and adding it to the isotropic fractal from Figure 8. This is shown in Figure 21. Table 1 summarizes the results.

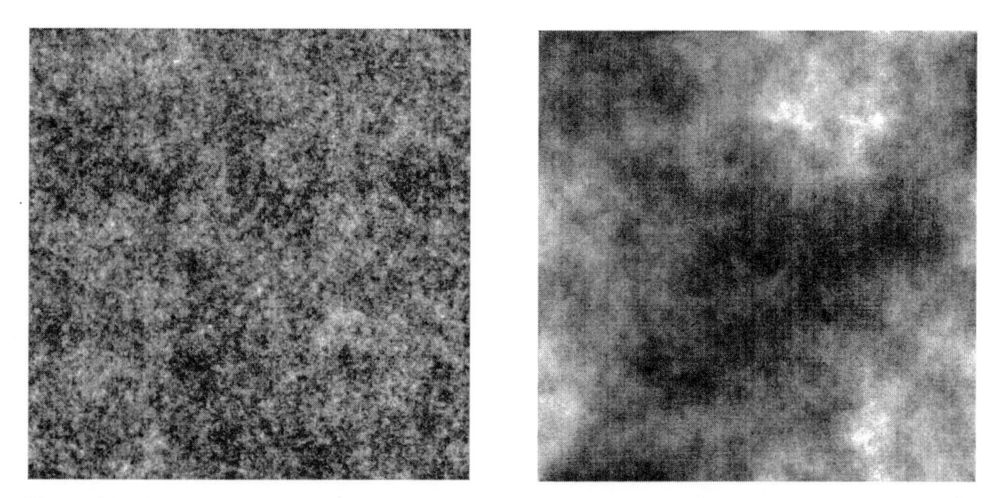

Figure 21. 1/*f* noise image ($D = 2.5$) and addition of 10% of the noise signal to the fractal surface shown in Figure 8.

The surface image containing 10% noise was measured using the log *P* vs. log *A* dimensional analysis method, the slit island method, and the Korcak method, as described above, and the FT method, as described in the next chapter.

For the dimensional analysis or log *P* vs. log *A* method, data from both bright and dark islands (features higher and lower than the level of the section plane) were obtained at eight different threshold settings and combined, and the slope fitted to data for features with sizes from 20 to 400 pixels as shown in Figure 22. Smaller features were bypassed because of the potential error in measuring the perimeter, and larger ones were bypassed to avoid bias due to the increased likelihood of intersecting an edge, as discussed above. The dimensional analysis method was used both with the net area of features and the total internal and exterior perimeter, and using the filled area (including any holes) and the external perimeter.

The slit island method uses the slope of a Kolmogorov (box count) plot for the boundary line from a threshold at the midpoint of the elevation range. The Korcak method uses the net area of islands with areas up to 400 pixels, obtained by thresholding at four uniformly spaced

Table 1. Effect of Adding Noise on Measured Fractal Dimension

	Original $\alpha = 0.7$	+10% noise	1/*f* noise
Surface measures			
log *P* vs. log *A*			
net area	2.387	2.494	2.681
filled area	2.303	2.411	2.614
Slit Island	2.261	2.362	2.455
Korcak	2.281	2.512	2.571
FFT	2.347	2.364	2.556
Profiles (vertical/horizontal)			
Minkowski	1.200/1.211	1.285/1.281	1.475/1.470
Korcak	1.290/1.320	1.379/1.381	1.532/1.559
FFT	1.324/1.337	1.335/1.339	1.529/1.540

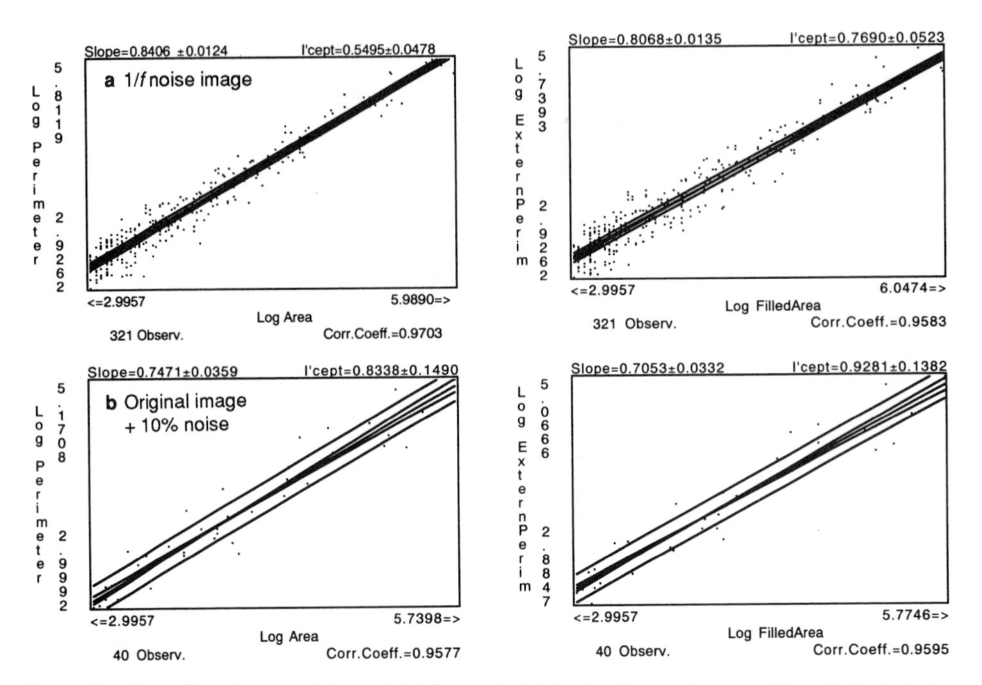

Figure 22. Plots of log P vs. log A for both bright and dark islands with areas between 20 and 400 pixels from the images in Figure 21. Plots are shown both for the net area and perimeter, and for the external perimeter and filled area, and can be compared to Figure 9.

elevation values and combining both the bright and dark (high and low) islands. Figure 23 shows the plots for all three images. There are fundamental differences between these methods, the dimension they actually measure, and the way they respond to the discrete array of data points. Consequently, the numeric values for the original isotropic surface differ slightly from the example value shown above, but these differences are within the expected accuracy for the techniques. Based on the standard deviation of the slope of the fit, the uncertainty (one sigma) in the reported values is typically about 0.06 for the log P vs. log A value, about 0.01 for the Korcak value, and about ±0.03 for the slit island and the FT results.

It is interesting to note that for the isotropic original surface and for the isotropic noise surface, each of which was generated with a single fractal dimension ($\alpha = 0.7$ and $\alpha = 0.5$ respectively), the reported surface dimension values are reasonably consistent, considering the differences between the various measuring techniques. The "expected" dimensions of these two surfaces are 2.3 and 2.5, respectively, and line profiles across the surface should have dimensions of 1.3 and 1.5. The combination with 10% added noise does not produce a straightforward interpolation of the two end cases, and the values reported by the different methods are quite different.

As for the case of the line profile, the FT method is largely insensitive to the addition of 10% noise while the other methods show a much greater shift. The change in reported value is in all cases (except the FT case) greater than the uncertainty in the method. This suggests that the presence of noise, even when smaller in amplitude by a full order of magnitude, will strongly affect the measured fractal dimension of surfaces if these techniques are used.

As a check on the results determined on the surface images, dimensions were also measured on line profiles across the same images. These elevation profiles were taken in both

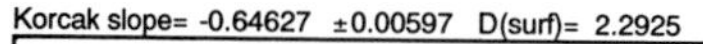

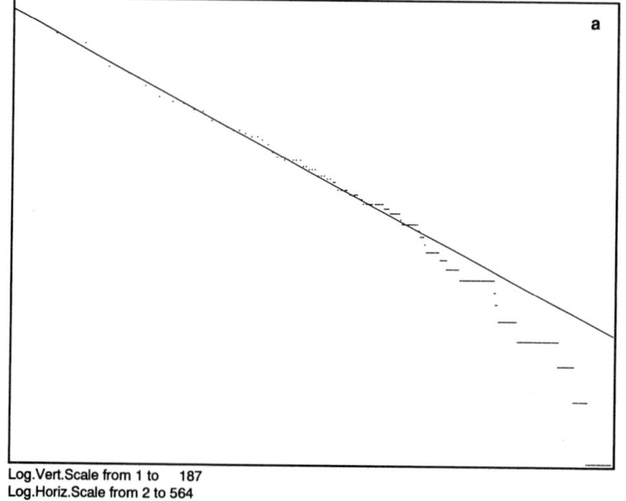

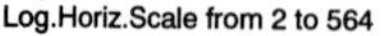

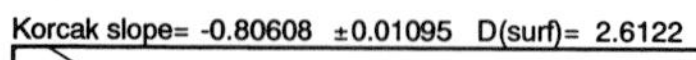

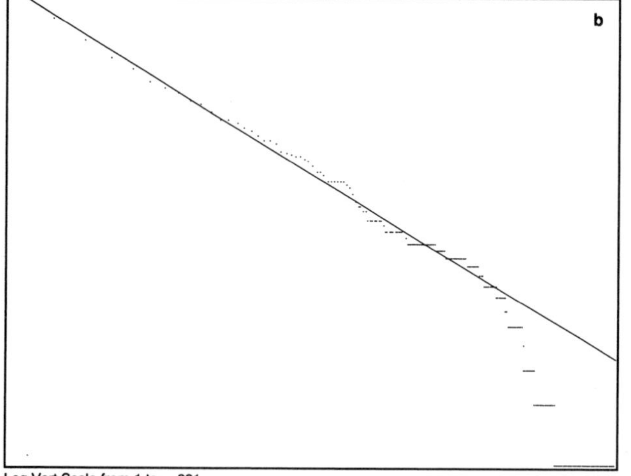

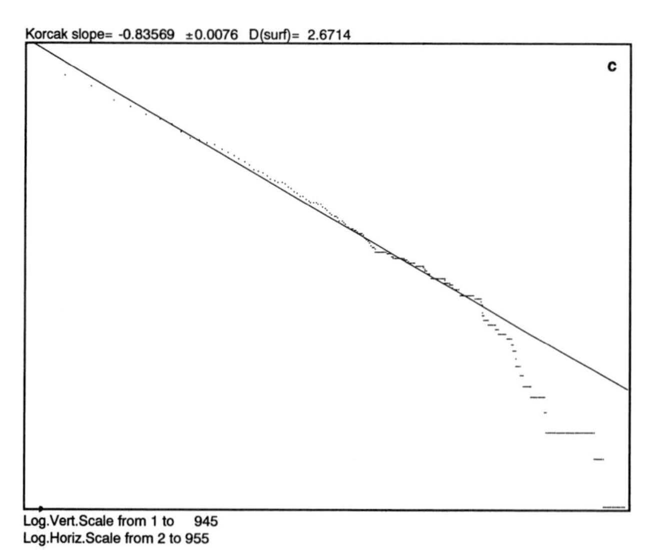

Figure 23. Plots of the log of the cumulative number of islands with areas greater than A as a function of log A for the original isotropic image (Figure 8), and the noise and sum images (Figure 21): **(a)** original image, **(b)** original image + 10% noise, and **(c)** $1/f$ noise image.

the horizontal and vertical directions across the center of each image. They were measured using the Minkowski or variational method, the Korcak method, and the FT method. The results are also shown in Table 1. The values are in all cases in close agreement between the horizontal and vertical directions, confirming the expected isotropy of the surface. They are also consistent with the surface values (of course, the line profile values have a dimension between 1 and 2 whereas the surface values are between 2 and 3).

The surface fractal dimensions obtained from this series of measurements confirm the results from the line profile case shown above. The FT results from the slope of the log ($Magnitude^2$) vs. log (Frequency) plot are least affected by the presence of the noise. Methods which either count or measure the perimeter of small features become biased by the presence of even small amounts of noise because the additional surface roughness produces many more small islands as a "froth" near the boundaries produced by thresholding the original image.

4

Hurst and Fourier Analysis

Time-Based Data

Although the primary focus of this text is on fractal surfaces, there are some tools which can be best introduced using other data types. Fourier analysis is most often applied to time-varying signals. Likewise, Hurst, or rescaled range analysis of fractals was initially performed on time-based historical data (Hurst, Black et al. 1965). Consider a record of some temporal phenomenon, such as the annual flood of the Nile river, or daily temperature readings, or the Dow-Jones industrial average. These examples and many others exhibit the same kind of self-similarity we have been discussing in earlier chapters.

As shown by the example in Figure 1, the irregular appearance of the plot of stock market data is independent of scale. Expanding the scale to look at weekly, daily or hourly prices instead of monthly values shows the same irregularity in a statistical sense. These profiles are of course self-affine rather than strictly self-similar, because the data are inherently single-valued and because the vertical and horizontal axes have completely different units and meaning, and scale differently.

Chapter 3 presented several measurement techniques which can be applied to self-affine profiles. For instance, the Minkowski dimension can be determined for a self-affine profile by using a structuring element consisting of a line parallel to the time axis, instead of a circle. A log–log plot of the volume which this sweeps out as it is moved along the line vs. the log of the line length will have a slope which gives the fractal dimension, as was shown in Chapter 3.

The Hurst Plot

Hurst analysis provides a simple tool for measuring plots of this type. As shown in Figure 2, this can be thought of as finding the minimum and maximum neighbor values within a range (along the time axis). It is equivalent to the horizontal dilation used in constructing the Minkowski sausage or ribbon for a self-affine profile. However, the Hurst plot is somewhat different from a Minkowski plot because it is not the total area swept out that is plotted, but the greatest difference between the maximum and minimum values. In other words, to construct a Hurst plot, you must scan through all of the data in the record to find the two points within a distance λ whose difference is greatest, and then plot that difference as a function of λ. In order to normalize the vertical scale of this plot, which would otherwise be in whatever units of measure (flood height, temperature, stock prices, etc.) the data set uses, the difference is rescaled by dividing it by the standard deviation of the data. This also makes the result independent of the size of the data set, as otherwise a very large data set might be expected to

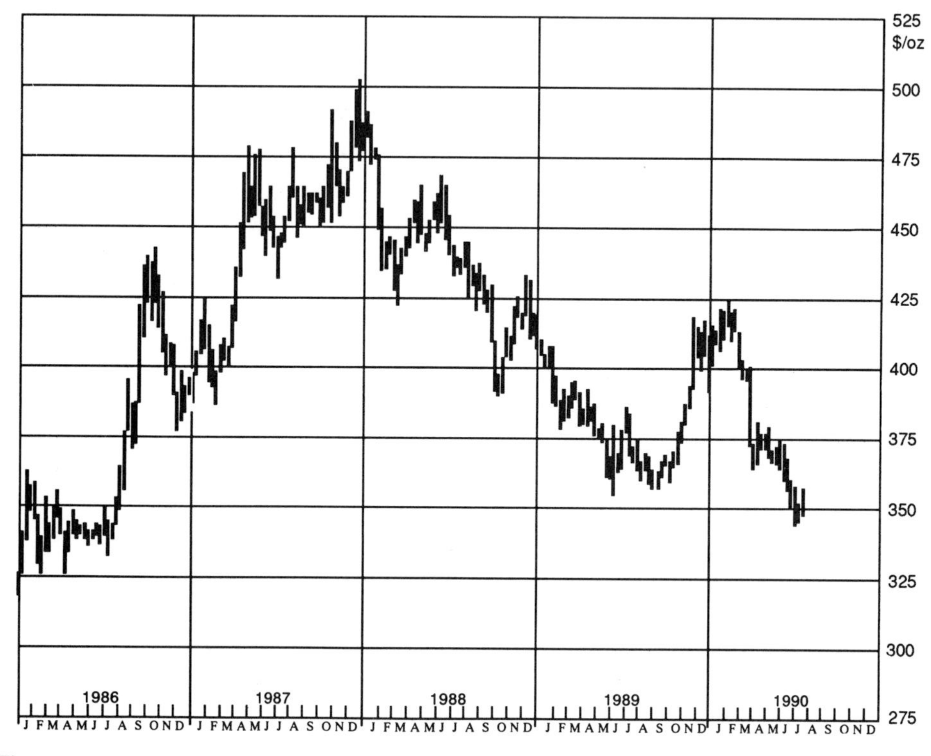

Figure 1. Weekly prices for gold. Examining the data on a daily or hourly basis reveals finer detail in this self-affine record.

contain somewhere within it a larger difference. It was pointed out in Chapter 2 that the standard deviation of the data set increases with its size.

Figure 3 shows the application of the Hurst method to an elevation profile. The points marked on the graph show those locations which have the greatest difference within horizontal distances from 1 to 4 neighbors. Each horizontal window finds the greatest difference wherever it lies within the profile. A plot of the log of the greatest difference in the vertical direction as

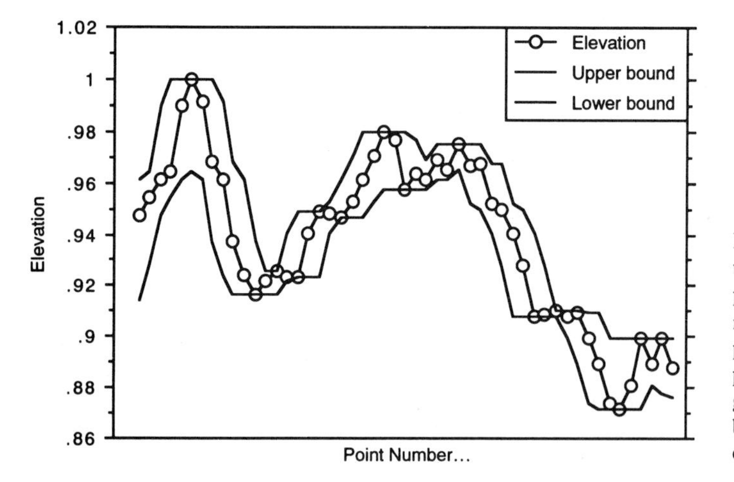

Figure 2. Portion of a fractal profile showing the upper and lower bounds for a neighbor distance of ±2 points along the horizontal position or time axis. The greatest vertical separation between the lines becomes one point on the Hurst plot.

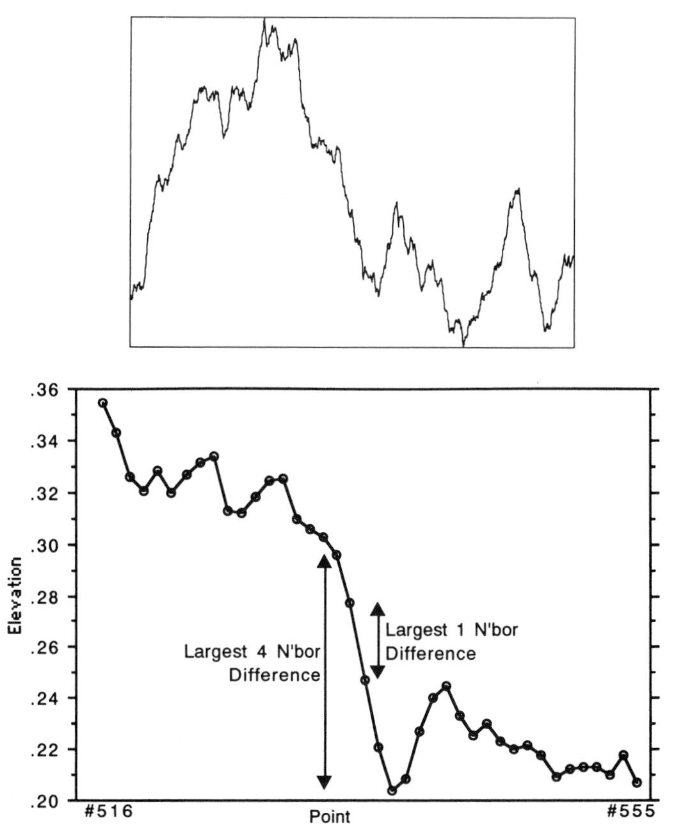

Figure 3. A fractal elevation profile (1024 points) showing an enlargement of the central region where the largest vertical difference for window widths from 1 to 4 neighbors happens to lie.

a function of the log of the width of the horizontal window is shown in Figure 4. The result is a straight line whose slope m gives the fractal dimension of the profile as $D = 2 - m$. Since the slope can vary between 0 and 1, the fractal dimension of the profile lies between 1 and 2, as expected.

For the profile in Figure 3, the fractal dimension determined by some of the other techniques appropriate for a self-affine elevation profile is different from the Hurst value. Figure 5 shows several of these methods. The Minkowski, rms vs. window width, and Korcak techniques were described in Chapter 2, and the Fourier technique is discussed later in this chapter. The Minkowski and Korcak values are rather similar to each other, as are the Hurst, Fourier, and rms vs. window width results, but the two groups disagree significantly as to the actual value. The best agreement between the techniques is for a profile with fractal dimension equal to 1.5.

The case in which the slope is exactly 1/2 merits special attention. This situation arises in the classic drunkard's walk. Consider plotting the path of motion of a point along a line where at each time increment the point can move one step to the right or left, with equal probability. This corresponds to Brownian motion in one dimension. The mean distance which the point moves away from the origin increases linearly with the square root of time. Each step is an independent sample from the probability distribution, with no knowledge of the history of past steps. Consequently the Hurst plot for this case has a slope of 0.5 and the fractal dimension of classic Brownian motion is 1.5.

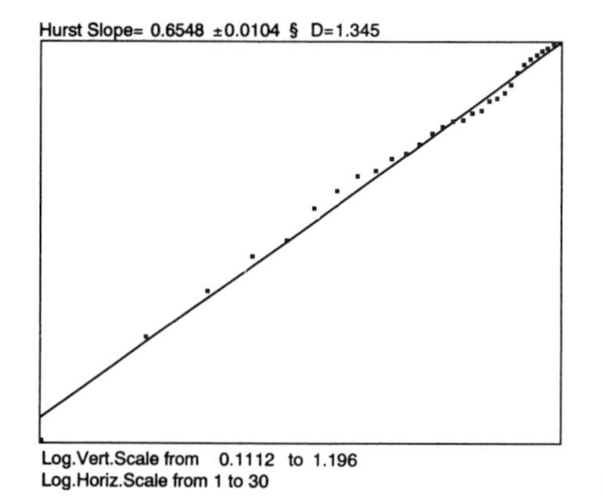

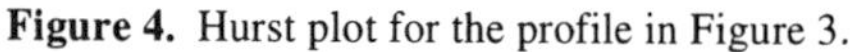

Figure 4. Hurst plot for the profile in Figure 3.

There are some other interesting properties of this plot, which were mentioned in earlier chapters. One is that the plot crosses the origin from time to time, and whenever it does there is a high probability that there will be many such crossings clustered together. The zero crossings of the line also constitute a fractal, which is in this case a fractal dust (or Cantor dust) whose dimension is 0.5. Just as intersecting a fractal surface with a horizontal plane to produce intersections whose boundary fractal dimension is one less than that of the surface dimension,

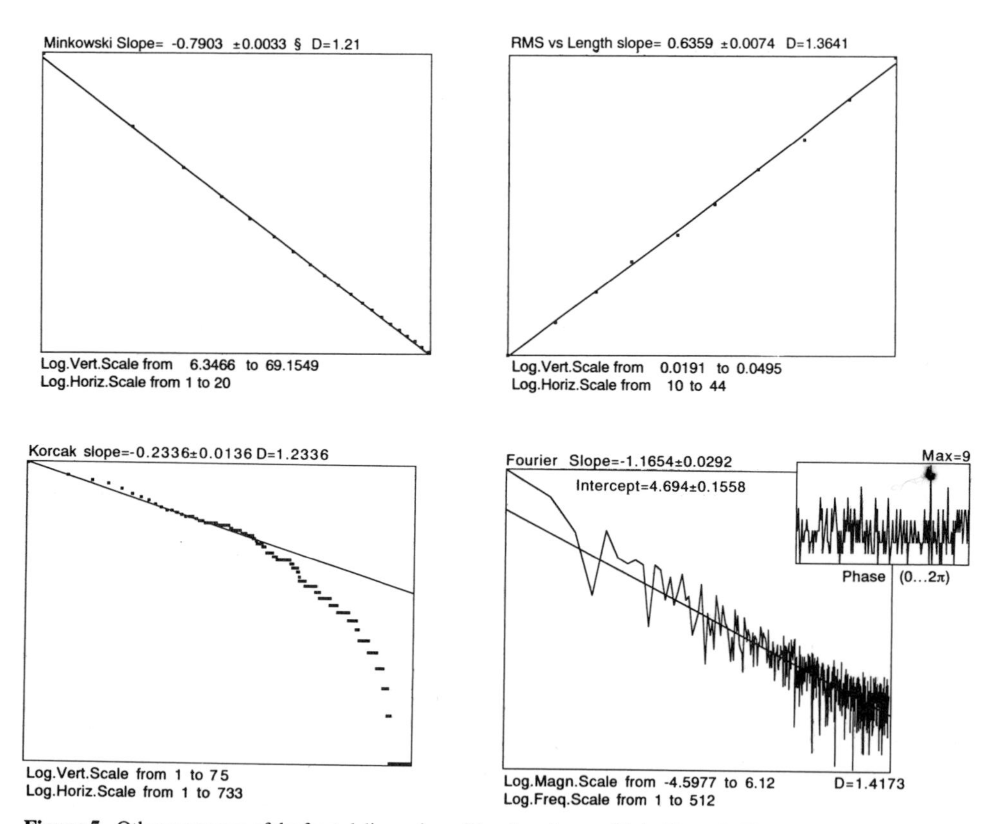

Figure 5. Other measures of the fractal dimension of the elevation profile in Figure 3. The Korcak and Minkowski values are quite different from the Hurst, rms vs. window, and Fourier results.

so this intersection of the Brownian profile with a straight line produces an intersection (the points) whose dimension is reduced by 1.

In addition, the distances between successive intersection points form a fractal. This is another example of the Korcak fractal, mentioned before in connection with the area of islands. Just as sectioning a fractal surface with a plane produces a straight line plot for the distribution of the total number of islands whose area is at least A, versus A (on log–log axes), so does a plot of the distribution of the number of segments as a function of length. The magnitude of the slope of this plot is the fractional part of the fractal dimension of the profile.

One of the consequences of this relationship is that there are a great many very small segments where intersections are clustered together. Inspection of such a profile reveals that there are a few quite long intercepts when the profile is far from zero, but whenever a crossing does occur, there are likely to be many crossings close together. By increasing the resolution and enlarging the region near a crossing, many more such crossings are likely to be revealed, producing many short segments. It is for this reason that one of the common measurement parameters for elevation profiles on real surfaces, the correlation length, is actually meaningless. This is supposed to be the average distance between zero-crossings on the profile. But for a fractal profile, there are infinitely many such points in clusters so that the average distance becomes an artifact of the measurement resolution. The same thing happens if instead of considering zero-crossings, we use the distance between local maxima (or minima) along the profile.

Fractal Brownian Motion

Physical processes which can produce classical Brownian motion, and hence a fractal dimension of 1.5, are well known. If the slope of the Hurst plot is greater or less than 0.5, it indicates some memory or auto-correlation in the data. For the case of the Nile river flood, a Hurst coefficient of 0.8 (fractal dimension of 1.2) indicates that there is temporal correlation from one year to the next. A large flood in one year is more likely to be followed by a large flood in the next year, and vice versa. This fact has been known and remarked upon since Biblical times ("seven fat and seven lean years"). Each year's flood is not simply an independent random sample from a master distribution. This may result physically from weather patterns in the region which the Nile drains which take several years to change, or from the fact that the Nile draws upon water reservoirs of many sizes that take various lengths of time ranging over many years to fill or to be depleted. The fact that there are a great many processes, each with a different "characteristic" time, that combine to produce the observed temporal pattern gives rise to the fractal character of the plot, and thus to the Hurst relationship.

Most of the other temporal patterns which have been subjected to Hurst analysis also show temporal correlation, and have slopes greater than 0.5. Batting averages are a good example. Every sports buff who follows baseball knows that when a player or team is in a slump, it is hard to break out of it, and vice versa, that hitting streaks are likely to continue. Hurst plots of many natural phenomena such as growth rings in trees, the thickness of sediments in lakes, the height of waves, etc., all show this tendency (Feder 1988).

There are a few examples in which the Hurst plot slope is less than 0.5 (fractal dimension greater than 1.5). A common one is the dripping faucet. Careful timing will reveal that this is not a perfectly regular phenomenon but a chaotic one whose "attractor" is a regular oscillation but which actually has continual minor variations that never exactly repeat. Vibration of the liquid/air surface and the dynamics of the drop release complicate the physics and produce the chaotic behavior. A record of the size of each of the drops shows an anticorrelation charac-

teristic or self-correcting response. A slightly larger drop depletes the reservoir so that the next drop is a bit smaller, and vice versa.

Mandelbrot (1982) calls these sequences whose Hurst plots have a slope different from 0.5 "fractal Brownian motion." The simulation of such processes is one of the methods that can be used to generate profiles with various fractal dimensions. Feder (1988) has described a particularly simple (manual) experiment in which a sequence of coin tosses or rolls of dice can be used to construct a profile plot of the number of heads or the total of the spots on the dice. By flipping all of the coins or rolling all of the dice for each time increment, a classic Brownian motion plot with slope of 0.5 is produced. If some of the coins or dice are held over from previous rolls at each turn, then the system has memory and the resulting plot will have a fractal dimension less than 1.5.

Feder also shows in detail that the Hurst relationship $R/S \propto \tau^H$ is an approximation to a more complex dependence of R/S expected for the superposition of many independent random processes. The consequence of this is that the slope of the Hurst plot obtained by a linear fit underestimates the slope H (and hence overestimates the fractal dimension $D = 2 - H$) when H is less than 0.72, and vice versa. This is part of the reason that measured Hurst dimensions do not necessarily agree numerically with other fractal measurement tools, although they remain an extremely efficient tool for comparing one data set to another.

One of the many interesting potential uses of Hurst analysis is in the quality control field. A record of the number of defective parts made each day (or each hour, or each shift, etc.) exhibits some variation. If the Hurst plot has a slope which is greater than 0.5, then the system which causes these defects has some memory—making one mistake makes it more likely to make additional ones soon thereafter. This indicates that an improvement in the process is possible, and it may be possible to find the cause of the mistakes and correct it. On the other hand, if the Hurst plot has a slope which is less than 0.5, then after one mistake has been made it is less likely to make another one (for instance, because the workers are more careful). This indicates that it is possible to make fewer mistakes, and so once again the process can be improved and defects reduced. When the Hurst slope is exactly 0.5, then each error or defect is an independent occurrence which does not depend on other events. In that case, there is no guarantee that improvements can be found.

Elevation Profiles

Application of the Hurst method to elevation profiles is straightforward. The plot shows the log of the greatest difference in elevation found anywhere along the profile as a function of the log of the width of the search window. The elevation value is normalized by dividing by the standard deviation of the elevation data. This scaling removes the dimensionality in the vertical direction. The method may be applied to fractal vertical elevation profiles, which are self-affine. Enlargement of the horizontal axis by a factor $1/r$ requires enlarging the vertical axis by $1/r^H$, where $H = 2 - D$ is the Hurst coefficient.

As noted above, there is a certain similarity between the Hurst measurement of a profile and the variational approach to the Minkowski measurement. The latter sums the differences between local minima and maxima over the entire profile, and plots the result versus the width of the local neighborhood used for finding the min and max. The Hurst method finds the greatest difference between minimum and maximum elevations anywhere in the profile, and plots the result as a function of the width of the neighborhood used for finding the min and max. The Hurst method involves less work, but is primarily sensitive to the greatest local deviation and will therefore be more sensitive to noise, as discussed in Chapter 3.

When applied to a series of elevation profiles, the Hurst plots give slopes that vary systematically with visual roughness and with the other fractal measures. The values are not identical, but they vary monotonically with profile irregularities. There is some scatter in the data points for each Hurst plot, but the fit to a straight line is relatively straightforward and the slope of the line is easily obtained. The mathematical straight line fit is generally poorer than the Minkowski method, for a profile of any given size. The uncertainty in the slope shown for these and all of the other linear fits is one standard deviation.

Since the slope m is related to the fractal dimension as $D = 2 - m$, and this is the correct dimension for the self-affine profile, it is possible to relate the surface dimension to this as $D_S = 3 - m$. Of course, this assumption that $D_{surface} = 1 + D_{profile}$ is subject to the usual concerns about surface isotropy. The extension of the Hurst method to deal directly with 2D arrays of elevation data is presented below.

Hurst Analysis of Range Images

For a two-dimensional array of elevation data, the Hurst method consists of a log–log plot of the difference in elevation between the highest and lowest points within a distance λ of each other regardless of direction. In other words, λ corresponds to the diameter of a circle moved across the image. Figure 6 shows an example of such a plot for an isotropic fractal surface. Figure 7 shows comparison plots of fractal dimension measurements for the same surface using some of the other tools (Minkowski comforter, Korcak islands, rms vs. area) that can also be applied to isotropic fractal surfaces. It is interesting that they agree with each other better than they do with the Hurst result.

Another comparison that can be made is to the fractal dimension of profiles across this image. In spite of the relatively small number of points and correspondingly poor measurement precision, the values in Table 1 indicate general agreement for the Hurst, Minkowski, and Korcak methods, but not for the rms vs. window width or Fourier techniques. In all cases, the results in the two perpendicular directions are the same within the precision of the data.

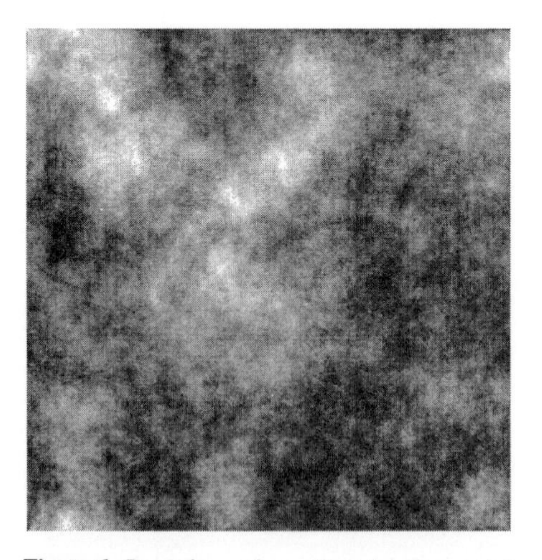

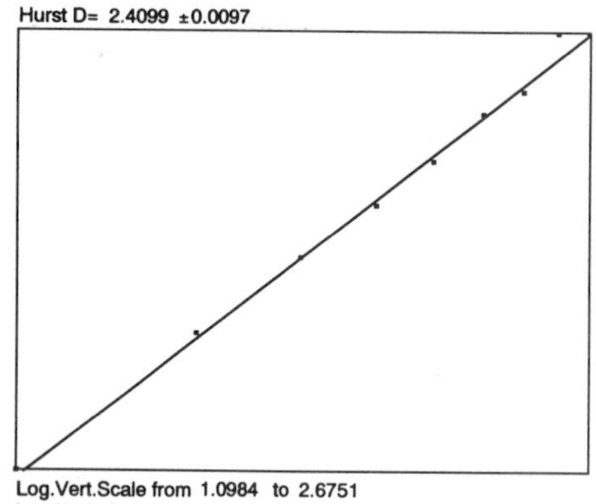

Figure 6. Range image from a fractal surface, and the result of applying a Hurst measurement. For each distance along the horizontal axis, the greatest elevation of difference of points within a circular disk of that diameter is plotted.

Table 1. Fractal Dimension of Profiles across the Center of the Image in Figure 6

Direction	Vertical	Horizontal
Minkowski	1.291	1.296
Korcak	1.251	1.243
Hurst	1.278	1.281
rms vs. width	1.596	1.613
Fourier	1.530	1.535

The Hurst plot is generally quite linear for a fractal profile over a considerable range of distances, so long as there is no single large step or cliff which would dominate the plot. One of the drawbacks of the Hurst method is that because only the single maximum elevation difference is saved for each horizontal distance, the presence of a single large step will overwhelm the rest of the data and produce a horizontal shelf in the plot that obscures the finer detail. For some surfaces, this makes it necessary to discard any large discontinuities, provided they can be assigned to some particular cause which is different from that which produced the remainder of the surface. For instance, in examining fractures, the presence of secondary cracks which leave crevasses in the surface like those in a glacier may be ignored in calculating the fractal dimension of the remainder of the surface.

On the other hand, the Hurst dimension can often be calculated with fair accuracy for relatively small data sets. This means that it is possible to evaluate it locally using only a small range of distances, making it very much quicker to calculate. That, in turn, has led to the use

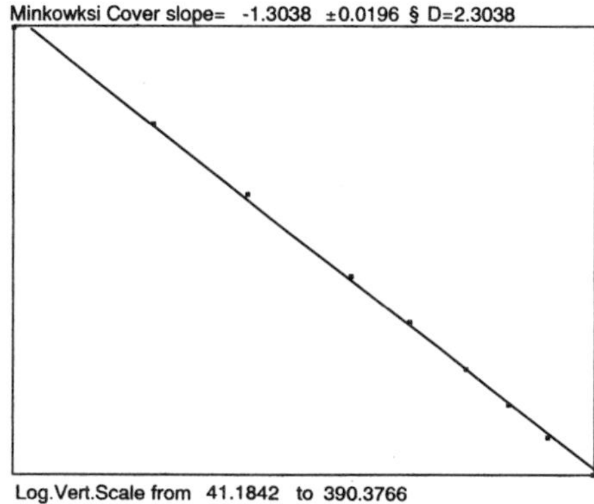

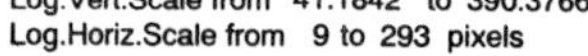

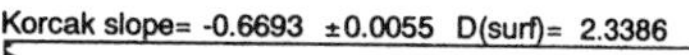

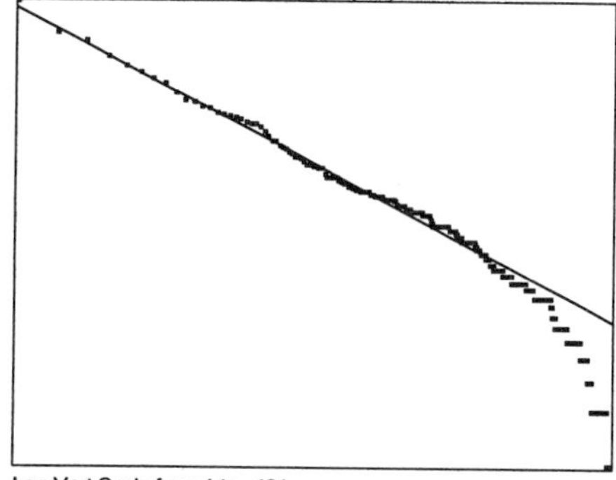

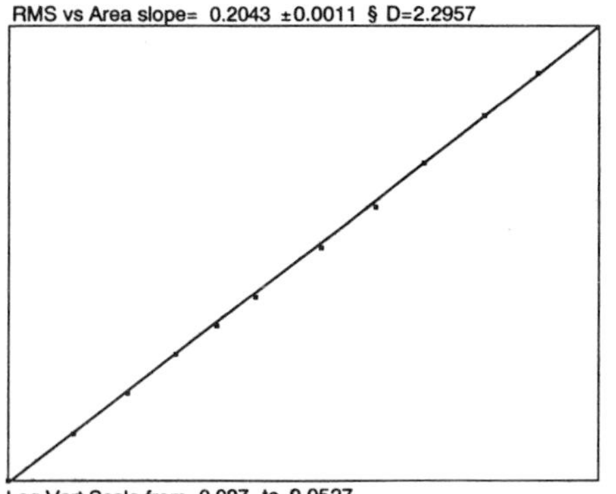

Figure 7. Fractal dimension measurement for the isotropic surface shown in Figure 6 using the Minkowski, Korcak and rms vs. window methods.

Table 2. Distance of Pixels Labeled in Figure 8 from the Center of the Neighborhood

Pixel class	Number	Distance from center
a	1	0
b	4	1
c	4	1.414 ($\sqrt{2}$)
d	4	2
e	8	2.236 ($\sqrt{5}$)
f	4	2.828 ($\sqrt{8}$)
g	4	3
h	8	3.162 ($\sqrt{10}$)

of the Hurst method on a local basis to extract the texture from a surface image (Russ 1990c). At each point in the image, a Hurst plot is constructed from the pixels within a small circular region. Figure 8 shows the pixels in the region, labeled according to the distance from the central pixel. Table 2 gives the distances for each of the classes of pixels.

Figure 9 is an electron microscope image of a thin section of liver tissue. The visually "smooth" and "rough" portions correspond to different intracellular structures, which can be segmented using the local Hurst coefficient. Figure 10 shows two tiny fragments of the image with the pixel brightness values for pixels in neighborhoods in each of the regions. Constructing a list of the brightest and darkest pixels as a function of pixel distance from the center (as shown in Table 3) allows constructing regression plots for the local Hurst value (Figure 11). The slope of this plot measures the local texture at the location of the central pixel. A new image is created in which the slope value is stored for that pixel address. This procedure must be carried out at each pixel in the image, which takes tens of seconds in the computer. This image is shown in Figure 9.

Figures 12–14 show another example. The specimen is a fracture surface in a steel. The upper part of the fracture is relatively smooth, and was formed by fatigue. The final tearing fracture in the lower part of the specimen is rougher. In the original image, it is not possible to distinguish the regions simply based on brightness. The histograms of pixel brightness for two 100-pixel-square regions shown in the figure overlap. However, the results of applying the local Hurst texture operator produces a new image (Figure 14) in which the two areas have distinctly different brightnesses, so that they can be separated by thresholding. The method has also been applied to aerial photographs of landforms, demonstrating the ability to distinguish rough, rocky terrain from smooth (but not necessarily horizontal) regions such as alluvial fans (Figure 15).

Modifications of this general approach are used that require less computation to provide texture segmentation of images, particularly for robotics vision. One method calculates a

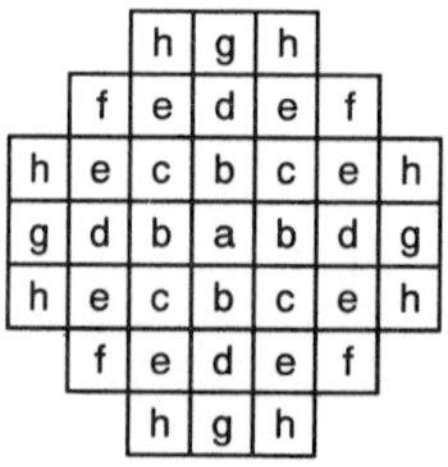

		h	g	h		
	f	e	d	e	f	
h	e	c	b	c	e	h
g	d	b	a	b	d	g
h	e	c	b	c	e	h
	f	e	d	e	f	
		h	g	h		

Figure 8. Octagonal 7-pixel wide neighborhood (37 pixels total) used for local Hurst coefficient calculation. Pixel labels identify groups with the same distance from the central pixel.

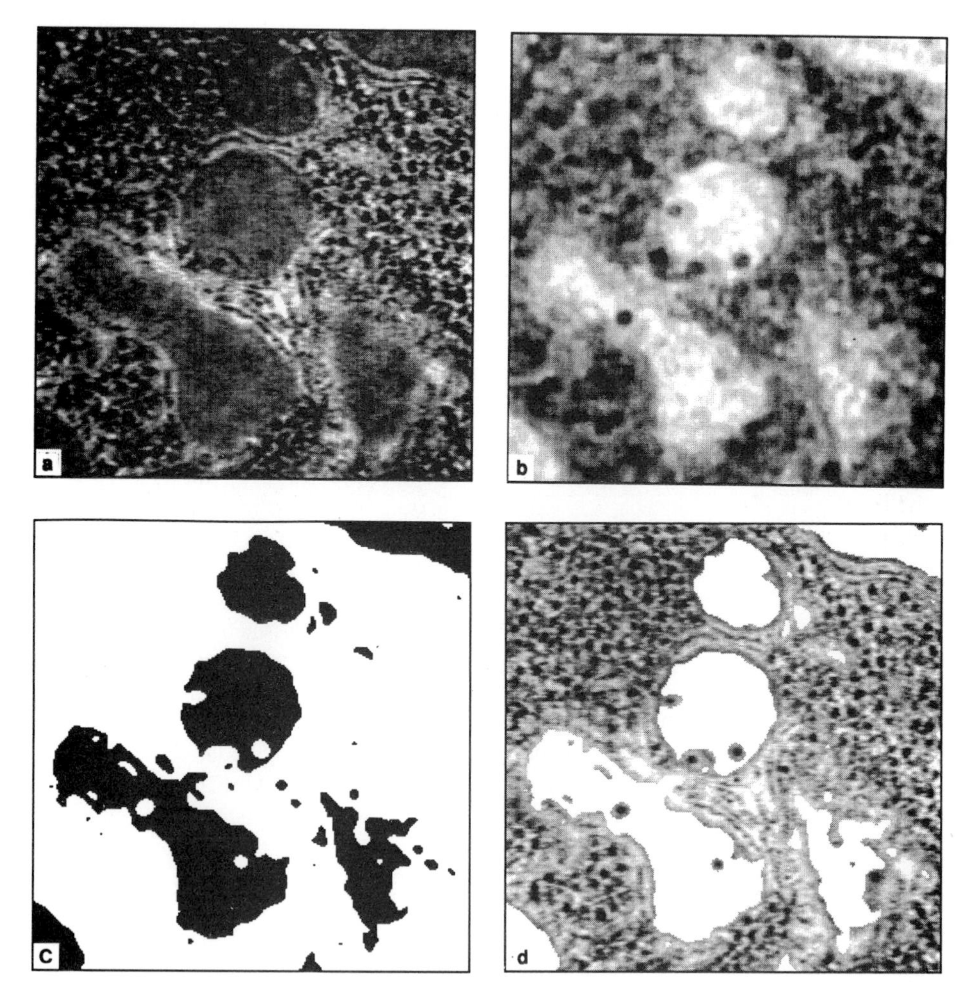

Figure 9. Segmentation based on texture using a Hurst operator: **(a)** liver (transmission electron micrograph of thin section); **(b)** Hurst transform image; **(c)** thresholded image from (b); and **(d)** high-texture regions in original using image (c) as a mask.

Table 3. Distance and Brightness Data for the Neighborhoods in Figure 10

Distance (pixels)	1	$\sqrt{2}$	2	$\sqrt{5}$	$\sqrt{8}$	3	$\sqrt{10}$
Image fragment a							
brightest	110	107	104	104	104	102	102
darkest	125	125	126	130	134	134	137
range	15	18	21	26	30	32	35
Image fragment b							
brightest	178	176	137	120	97	85	73
darkest	223	235	235	237	237	237	237
range	45	59	98	117	140	152	159

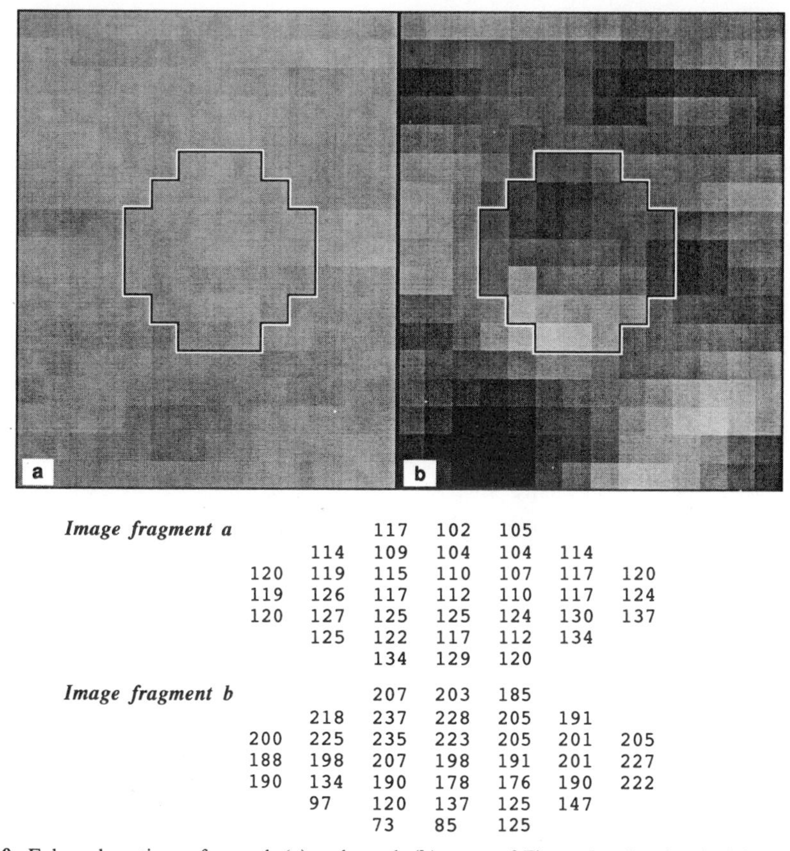

Image fragment a

		117	102	105		
	114	109	104	104	114	
120	119	115	110	107	117	120
119	126	117	112	110	117	124
120	127	125	125	124	130	137
	125	122	117	112	134	
		134	129	120		

Image fragment b

		207	203	185		
	218	237	228	205	191	
200	225	235	223	205	201	205
188	198	207	198	191	201	227
190	134	190	178	176	190	222
	97	120	137	125	147	
		73	85	125		

Figure 10. Enlarged portions of smooth (**a**) and rough (**b**) areas of Figure 9a, showing individual pixels in representative 7-pixel-wide octagonal neighborhoods with the numerical values of the pixels (0 = white, 255 = black).

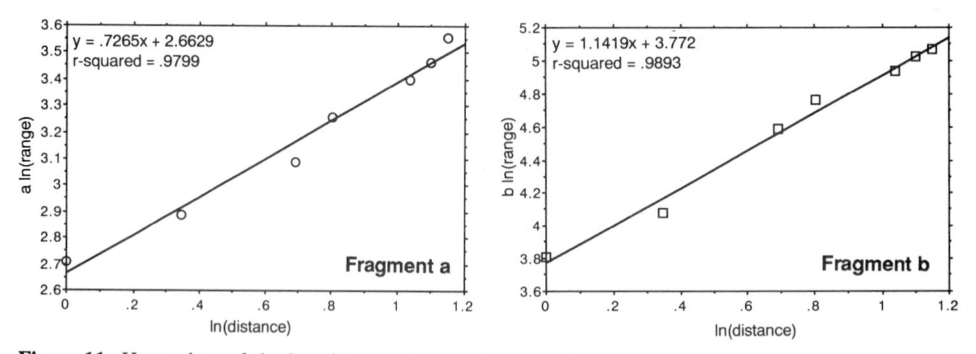

Figure 11. Hurst plots of the logarithm of the maximum brightness range vs. log of distance for the two neighborhoods shown in Figure 10.

Figure 12. Fractal surface of steel test specimen. The smooth portion of the fracture occurred by fatigue, and the rougher portion by tearing.

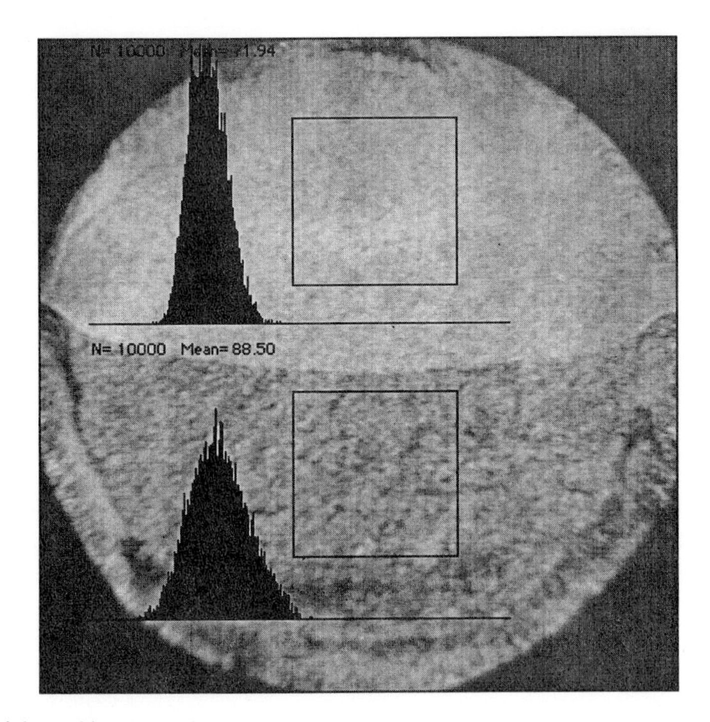

Figure 13. Brightness histograms of two 100×100 pixel regions (marked) in Figure 12 showing greater variation in the rough area, but extensive overlap of brightness values with the smooth region.

Figure 14. Application of the local Hurst operator to the image in Figure 12 showing the larger values (darker pixels) for the rough portion of the fracture surface.

normalized value for the autocorrelation function of the image (Luo and Loh 1991; Luo, Potlapali et al. 1992). Another technique constructs a matrix containing the number of pixel pairs whose absolute difference is less than a value V as a function of V and the distance between the pixels (Ait-Keddache and Rajala 1988; Xiao, Chu et al. 1992).

The Hurst Orientation Transform

The two-dimensional generalization of the Hurst measurement of fractal dimension is properly only applicable to an isotropic surface. Of course, this is also true for the Minkowski comforter, the Korcak plot, dimensional analysis (plotting log P vs. log A) and most other generalizations of measurement tools developed for elevation profiles to surface. If a surface is anisotropic in a way that produces different fractal dimensions in different directions, it is said to be strongly anisotropic. Examples of such surfaces will be examined in Chapter 7. The two-dimensional Hurst measurement of such a surface finds only the maximum and minimum elevation values within each neighborhood, and so responds to the greatest fractal dimension present. If a surface is "rougher" in one direction than another, then only the roughest direction will influence the Hurst measurement as described above. This is unfortunate, since for anisotropic surfaces it may be useful to characterize the direction and degree of anisotropy.

A modification of the Hurst approach which makes this possible is called the Hurst Orientation Transform (HOT). By comparing pixel values for all pairs of pixels in a neighborhood which is scanned systematically across the image, a table is constructed containing the greatest pixel difference as a function of the distance and orientation. From this table, the Hurst

coefficient can be calculated in many directions and plotted as a function of orientation to reveal and characterize the anisotropy (Russ 1990b).

The table constructed from this search can itself be displayed as an image using pixel brightness to represent the magnitude of the differences. In the example shown in Figure 16, the horizontal axis represents distance between the pixels and the vertical axis represents angle. The sparseness of the data is due to the small size of the neighborhood used (16 pixels square) and the finite number of angles and distances for pixel pairs within that region.

Finding the slope of the log (Difference) vs. log (Distance) data along each horizontal line through the transform image produces a family of Hurst plots whose slope represents the directional fractal dimension and can be plotted as a rose to show the anisotropy. For the case of an isotropic surface, the plot is a circle. This method is not equivalent to constructing a series of Hurst plots along elevation profiles across the surface at different angles, because it uses all of the pixels in the image as compared to only those along one profile. It is thus much better from a statistical point of view.

The application of the Hurst Orientation Transform to measure the directionality of anisotropic surfaces will be shown in later chapters. While comparatively slow to compute, it responds somewhat differently to anisotropic surfaces than the Fourier method, which is discussed next. Surfaces that are strongly anisotropic (have different fractal dimensions in different directions) with gradually varying directional properties can be characterized with either method, and will generally show an elliptical pattern in a rose plot. This is not the case for surfaces that have weak anisotropy, such as an extruded surface produced by moving a fractal profile across the surface, or ones that have one or a few strong directions of high or low dimension, such as those produced by machining operations. In those instances, the Fourier

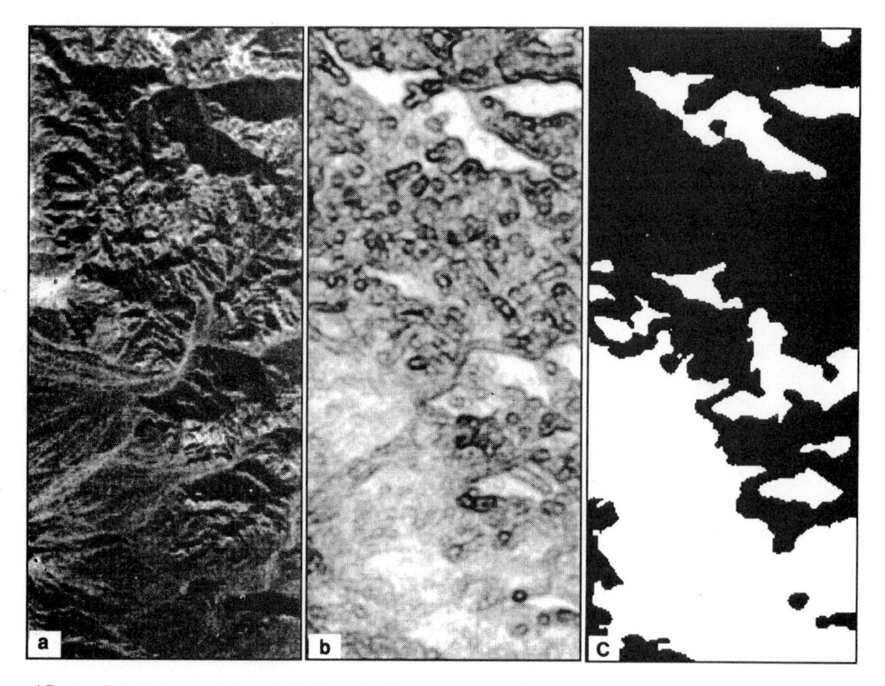

Figure 15. Aerial photograph (Death Valley, California): **(a)** original, showing differences in local texture; **(b)** application of the Hurst operator; and **(c)** binary mage formed by thresholding image (b) to select high values of brightness.

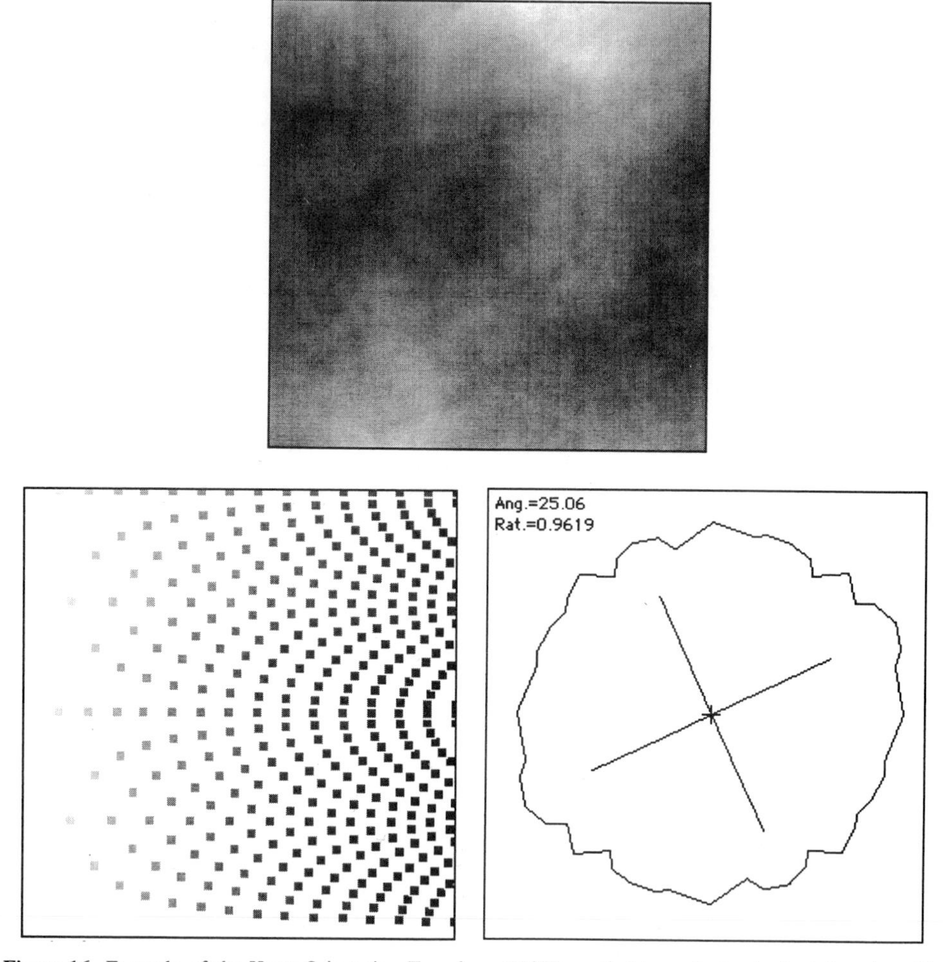

Figure 16. Example of the Hurst Orientation Transform (HOT) applied to an isotropic fractal surface. The greatest elevation differences between pairs of points are plotted as a function of distance and orientation, and the slope of the resulting one-dimensional Hurst plots drawn as a function of orientation.

method will respond to each strong direction separately, while the HOT plot will show a more gradual variation of properties with direction. For some purposes, this is a more descriptive characterization of the surface roughness.

Fourier Analysis

Analysis of complex time-based data is very commonly performed in the frequency domain, using a Fourier transform (FT). Joseph Fourier's theorem states that any complex motion can be broken down into a series of superimposed sinusoids as illustrated in Figure 17. The series is in principle infinite, but in practice the highest frequency of importance corresponds to the spacing of pixels in the image or along the elevation profile, or to the lateral resolution of the measurement technique. Likewise, the lowest frequency of importance is set

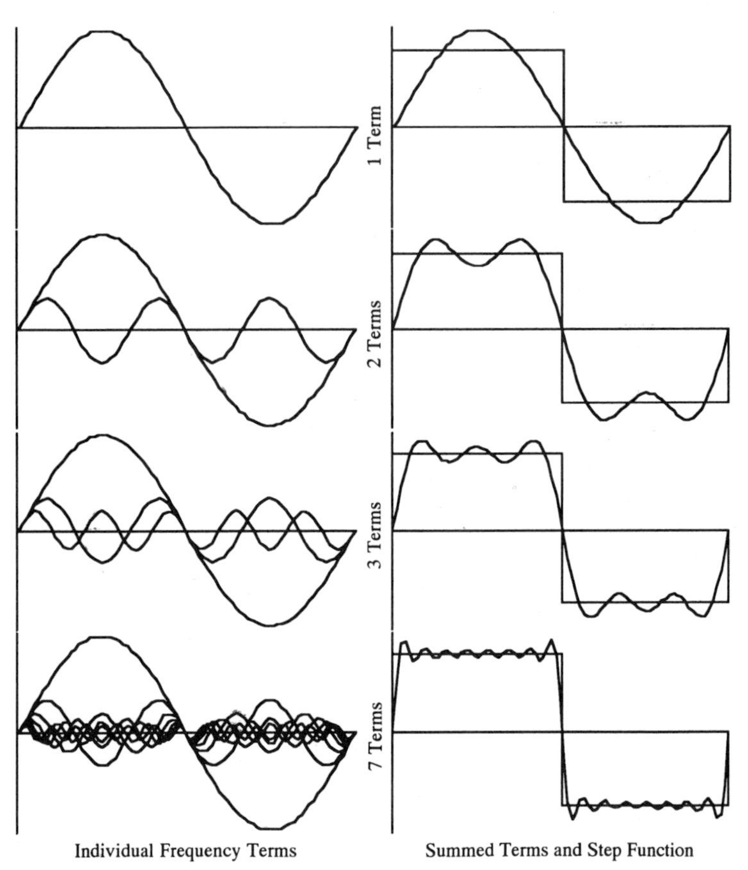

Figure 17. Summation of Fourier frequency terms to fit a simple step function.

by the width of the image or length of the profile. For a function of time, the Fourier equation becomes

$$f(t) = \sum_{k=lowest}^{highest} a_k \cdot \sin(2\pi kt - \vartheta_k)$$

Substituting x for t allows the same expression to be used for a line profile of elevation. The a_k coefficients in the expression give the magnitude of the kth frequency, and the ϑ values give the relative phase of each term. One of the useful characteristics of the Fourier method is that the coefficients for each term do not change as other terms are added to the series. Whatever set of frequencies are used, the resulting summation is the "best approximation." For most analysis purposes, the phase data are ignored. The magnitude values, or more commonly the square of the magnitudes, can be plotted against frequency. When the squared values are used, the result is the power spectrum of the original profile or signal. For time-based data, this is the amount of power in each frequency band of the signal, and has direct consequences interpretable to the electrical engineer.

Since a fractal profile by definition includes information at all frequencies, it might seem that this method would be difficult to apply to fractals. In fact, the relationship is surprisingly simple. The power or magnitude spectrum shows a linear variation between the logarithm of

the magnitude^2 and the logarithm of the frequency. The slope of the line β is related to the fractal dimension D as $D = (4 + \beta)/2$. This provides one of the most direct, easily understood and powerful techniques for the analysis of fractal profiles and surfaces, which can be applied to self-affine as well as self-similar data sets.

Figure 18 shows a fractal elevation profile with dimension $D = 1.5$, generated using the same midpoint displacement technique introduced in Chapter 1. Because the original profile has 1024 points, which is an exact power of 2, Fourier analysis of this profile can be performed with the Fast Fourier Transform (FFT) technique. The resulting series has 512 terms, and a power spectrum plot of the log (Magnitude^2) vs. log (Frequency) is shown in the figure. The individual data points fit a straight line rather well, and the slope of the linear least-squares fit line is close to -1.0, which agrees with the expected value for the profile. Note that a slope of -1 thus corresponds to the case of Brownian motion, and to $1/f$ noise.

The figure also shows a plot of the histogram of the phase values for the terms in the Fourier series. They are uniformly random, which they should be for a fractal. This is an important criterion that is not always tested in measurement procedures. The same series of amplitudes would correspond to a decidedly nonfractal profile if the phase values were not randomized. Figure 19 shows an example in which all of the phase values are set to zero. For a fractal, the phase values must be randomly independent of frequency, and, when the method is extended to surfaces, independent of orientation angle as well.

There is a rich literature on the mathematics and numerous applications of the Fourier transform, particularly in the area of signal and image processing, and it is not intended to duplicate that here. A few points may be useful in considering the use of the method for the analysis of fractals. In addition to checking that the phase of the terms is properly random, there may also be a problem in dealing with the presence (or absence) of a few frequencies in the spectrum. The process of fitting a linear least-squares line to hundreds of points in the Fourier spectrum tolerates having a few points widely scattered from the main trend. Figure 20 shows an example. The presence of the peak in the Fourier transform is visually evident but ignored by the linear fit. In terms of the physical profile, the peak corresponds to a single sinusoid superimposed on the otherwise fractal surface. Visually, the sinusoid dominates the elevation profile. However, it is only a minor component of the log (Magnitude^2) vs. log (Frequency) plot.

The slope of a linear fit line is affected more strongly by the displacement of points at one end of the data set. The discussion of the effects of instrument noise on the measurement of fractal dimension deals with this in some detail. High-frequency noise can cause an increase

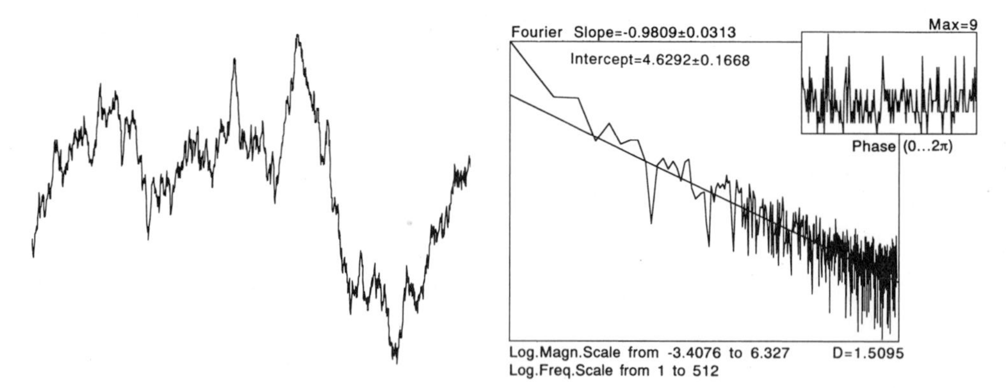

Figure 18. A fractal profile generated by iterated midpoint displacement, and the power spectrum of its Fourier transform, plotted as log (Magnitude^2) vs. log (Frequency).

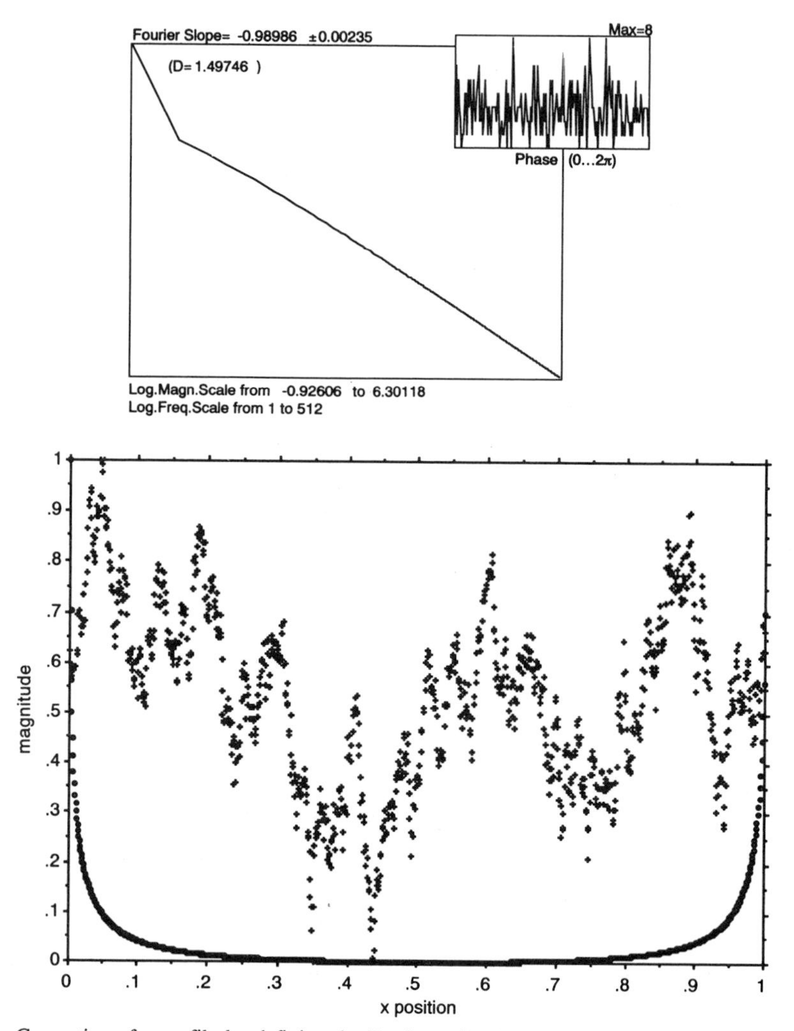

Figure 19. Generation of a profile by defining the Fourier series. The magnitude of successive terms drops according to $f^{-\alpha}$ but in one case the phase is randomized producing a fractal profile, and in the other case the phase is set to zero producing a non-fractal profile.

in the values at the right of the log (Magnitude^2) vs. log (Frequency) plot which would displace the points upward and reduce the apparent slope of the curve. Conversely, lack of resolution will limit the amount of information at high frequencies, reduce the amplitude of the power spectrum at the high frequency end, and increase the slope. Remember that a slope of 0 corresponds to white noise, while a slope of –2 corresponds to a deterministic or Euclidean shape.

A more subtle but perhaps even more common problem has to do with the points at the left end of the plot. The low frequency terms are fewer in number and more widely spaced on the logarithmic axes, and so they control the slope of the line even more. But the low frequency terms are likely to include information on the overall shape of the profile or surface based on design and fabrication. Many real surfaces are not intended to be ideally flat, but to follow some designed curvature or "figure." The "zeroth" term in the Fourier series is simply the

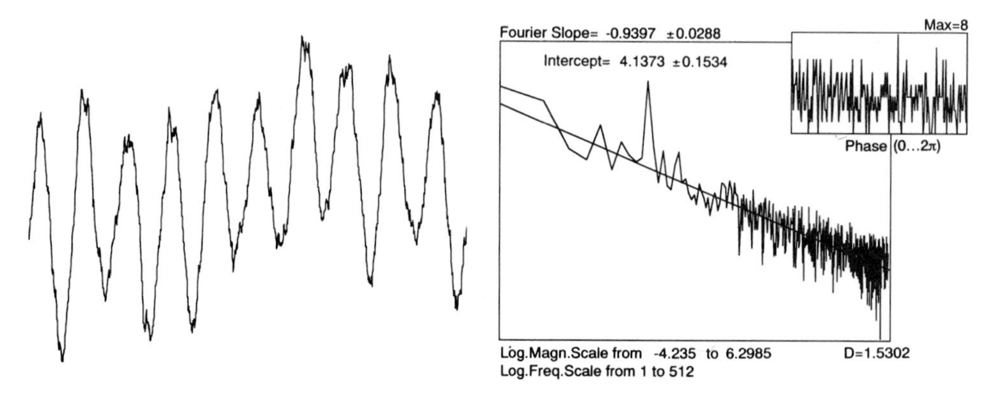

Figure 20. Superposition of a sine wave and fractal profile, and the Fourier transform results showing one spike in the power spectrum.

average height of the surface, which is arbitrary depending on the choice of reference point. The next few terms are quite sensitive to the slope and low-order curvature of the surface upon which the fractal details of interest are superimposed. In order to avoid contaminating the slope of the line with these points, it may be desirable to bypass them in performing the fit, and to arbitrarily begin the summations for the fitting equation with (e.g.) the tenth term in the Fourier series, or to choose a dimension representing the smallest "controlled" distance on the surface and begin the fit at the corresponding frequency.

Fourier Analysis of Boundary Lines

It is also possible to apply Fourier analysis to the boundary lines or coastlines around features where a surface has been sectioned by a horizontal plane. It has already been pointed out that these coastlines are fractal with a dimension that is exactly 1.0 lower than the dimension of the surface for the case of a well-behaved and isotropic, but not necessarily strictly self-similar, surface. In order to convert the boundary line of a feature to a profile suitable for Fourier analysis, consider the unrolling operation shown diagrammatically in Figure 21. A vector from the centroid of the feature to the boundary is swept through 360 degrees and the radius vs. angle is plotted. This waveform is well suited to Fourier analysis since it is periodic (repeats every 2π). Fourier or "harmonic" analysis of such unrolled shapes has been a major tool for classification of particle shapes in sedimentology and other fields for some time.

If the shape is more irregular than the one shown, the unrolling procedure may encounter difficulty. Wherever the radius (angle) function is multiple-valued, it cannot be evaluated. Even using a simplifying rule to take the outer radius in such cases introduces abrupt steps in the function which are difficult to model because they require many high-frequency terms with fixed phases. Consequently, another approach is used. By plotting the slope of the boundary line in some arbitrary x, y coordinate system, as a function of the distance along the boundary line, another periodic function can be generated for analysis.

In image processing terms, this can be done by using the so-called "chain code" for the boundary. Chain code is simply a series of one-pixel-long links in a zig-zag line along the feature periphery. Since each pixel which is part of the boundary must be connected to two others, and they must occupy the eight possible locations for touching pixels, a chain of digits from 1 to 8 can describe the sequence of connections. Figure 22 shows one way this can be done. Stringing together the chain code digits creates a numeric representation of the feature

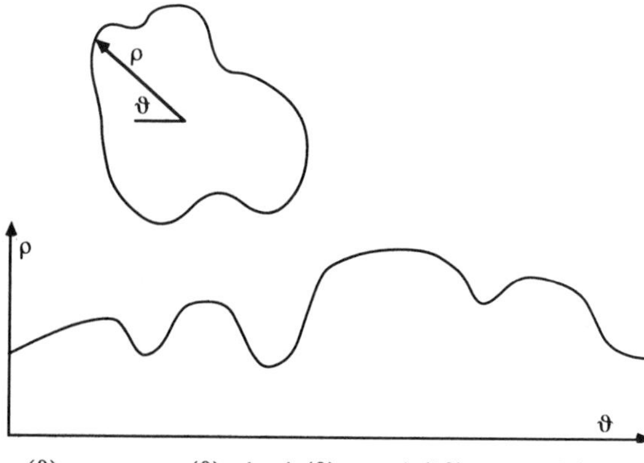

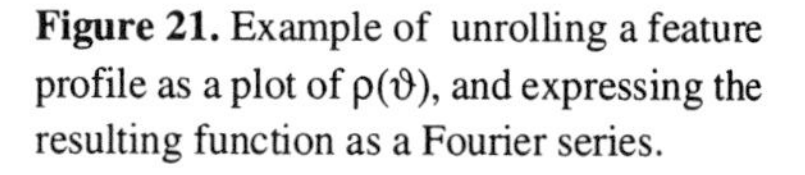
Figure 21. Example of unrolling a feature profile as a plot of $\rho(\vartheta)$, and expressing the resulting function as a Fourier series.

shape. Some image analysis operations are routinely performed using this code. For instance, the perimeter length of a feature may be taken as simply the number of digits in the code. A somewhat more accurate value can be obtained by adding the number of odd digits plus $\sqrt{2}$ times the number of even digits, to take into account the greater distance between a pixel and its corner-touching neighbors as compared to the edge-touching ones.

Plotting the values of the digits along the chain code would seem to provide a periodic function to which Fourier analysis could be applied. However, there is one difficulty: the points are not uniformly spaced along the boundary. This is due to the same variation in link length mentioned above. It is necessary to have evenly spaced points to perform the Fourier transform. One way to accomplish this is to replace each even digit (a link in the "short" or 90-degree directions) with five shorter links of the same value, and each odd digit (a link in the "long" or diagonal directions) with seven shorter links of the same value (Russ 1989). The ratio 7/5 = 1.40 is satisfactorily close to the $\sqrt{2}$, so that the points become uniformly spaced. Figure 23 shows an example.

White and 1/*f* Noise

The Fourier spectrum shown above in Figure 18 had a slope of $\beta = -1.0$, corresponding to a fractal dimension of 1.5 for the profile. In the earlier discussion of Brownian motion, it was pointed out that an ideal random walk also has this dimension. This phenomenon is so common in many physical processes that it is called 1/*f* noise, meaning that the Fourier power spectrum has an amplitude that varies as frequency to the (−1) power, in agreement with the log plot shown. 1/*f* noise is thus the signpost of random processes with no temporal correlation. It is sometimes called pink noise, to distinguish it from white noise (equal power at all

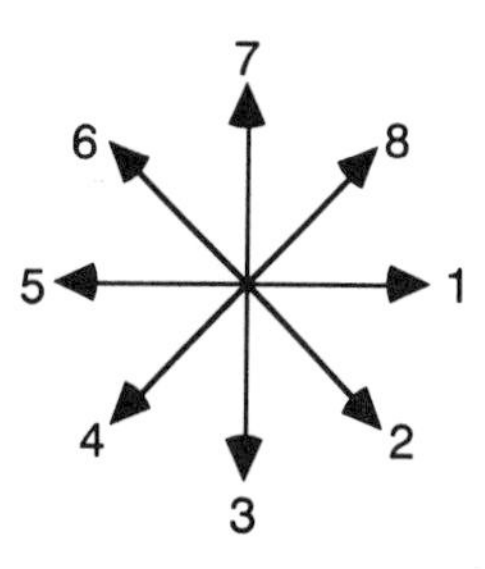

Figure 22. Diagram of a convention for assigning values to directions to construct a chain code.

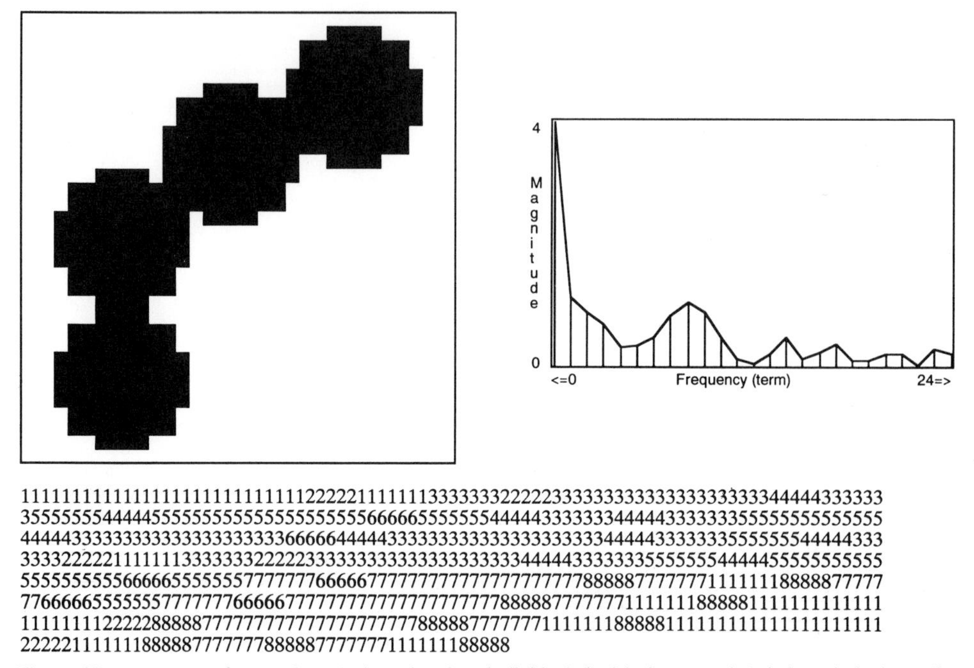

Figure 23. A re-entrant feature shape (enlarged to show individual pixels), the expanded chain code for its outline (starting at the topmost line of pixels), and a plot of the Fourier transform of the chain code.

frequencies, no slope to the power spectrum plot). However, somewhat confusingly, the term "brown noise" is sometimes used to describe a power spectrum whose slope with frequency is $1/f^2$, corresponding to a purely deterministic (or in mathematical terms, Euclidean) case. Slopes that vary in value from 1.0 indicate that there is some correlation along the profile, which may be either temporal or spatial depending on the quantity being measured. Figure 24 shows examples of profiles with quite different fractal dimensions. For both, the log (Magnitude^2) vs. log (Frequency) plots are quite linear, but the slopes are very different.

Aguilar, Anguiano et al. (1992) and Pancorbo, Aguilar et al. (1991) measured one component of STM instrument noise as $1/f$ (exponent equal to -1.0 ± 0.1) independent of scan rate and other instrument parameters, by recording an image without any sample present. This is only true along the scan lines and ignores line-to-line variation, tip effects, etc. Knowing this noise contribution, they were then able to introduce an asymmetric Wiener filter to remove it from images. However, it is not clear that this is all of the noise present, nor how to separate this $1/f$ component from the surface roughness (which may also be close to $1/f$ for surfaces with very small total relief).

The relationship for a profile is $D = (4 + \beta)/2$, where β is the slope of the power spectrum. Hence, a slope of β with value greater than 1 (between 1 and 2) corresponds to a value of D less than 1.5, meaning that the system has memory. This is the same as the cases discussed above in which the Hurst coefficient is greater than 0.5. Conversely, a slope β with value less than 1 (between 0 and 1) corresponds to a value of D greater than 1.5. In the limit when β approaches 0 the Fourier spectrum shows a uniform amplitude at all frequencies (white noise), which is not fractal.

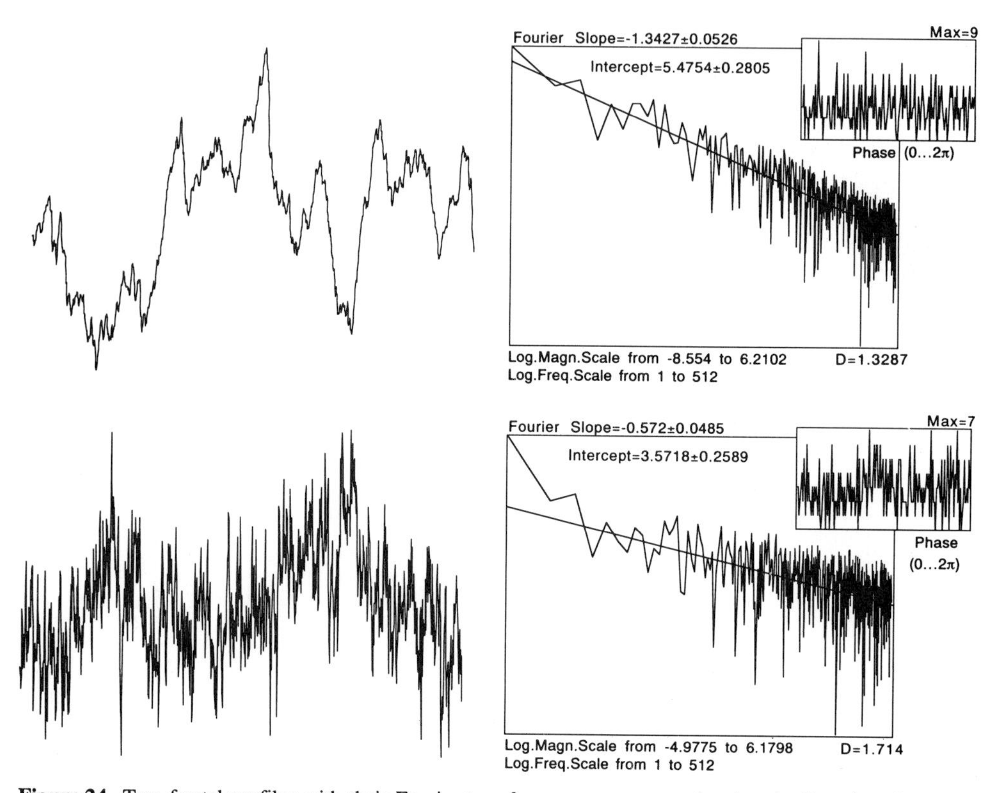

Figure 24. Two fractal profiles with their Fourier transform power spectra, showing the linearity of the log (Magnitude^2) vs. log (Frequency) plot and the variation of slope with fractal dimension.

Fourier Analysis in Two Dimensions

Fourier analysis can also be performed on two-dimensional data sets. In this case, the transform consists of magnitude and phase data covering a two-dimensional array where each point represents a frequency and an orientation in the original image. Figure 25 shows an example with several pure sinusoidal waves with different orientations and wavelengths. The frequency space representation of each consists of a single point corresponding to the orientation and wavelength. The frequency space representation of magnitude actually shows two points, but the representation is redundant and duplicates the upper half plane on the lower half, because the orientation of the sinusoidal pattern can be considered as being in either of two directions. Again, as for the case of a line profile, a convenient and fast implementation of the transform method (an FFT) can be carried out if the image is square with a dimension in pixels that is a power of 2. Most of the examples shown are thus either 256×256 or 512×512 pixels.

Superimposing several frequencies in the spatial domain is equivalent to adding the corresponding components in the frequency domain. It is this property of Fourier analysis of images which is exploited in filtering out, or removing, noise that consists of one or a few discrete components, as may arise due to image acquisition or transmission problems. This is illustrated in Figures 26 and 27. An image printed in a magazine shows a half-tone screen (the slightly angled array of dots) due to the printing technology employed. In the 2DFT image, these high-frequency regularly spaced dots produce a few spikes or peaks in magnitude. Removing these spikes by setting the magnitude to zero allows the image to be reconstructed

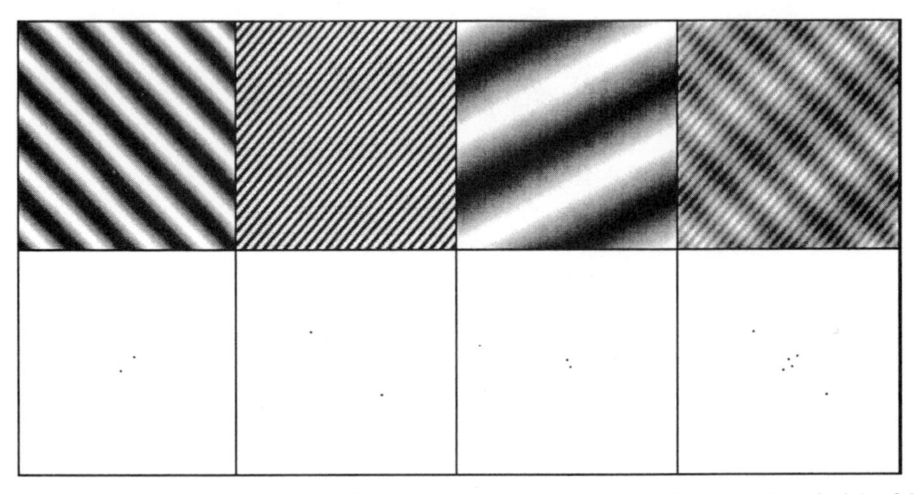

Figure 25. Three sinusoidal patterns, their frequency transforms, and their sum, illustrating the principle of the Fourier transform in two dimensions.

by an inverse FFT to recreate the original image with the screen removed, but all of the other image information (which consisted of different frequencies and orientations) intact.

When a range image of a fractal surface is Fourier transformed, the expected result is that the log (Magnitude^2) will fall off linearly with log (Frequency). Since the usual representation of the image in frequency space uses a linear scale for frequency, radial from the center of the image, with a logarithmic scale for the magnitude or power, the result is a transform image as shown in Figure 28 in which the pixel brightness drops off smoothly but nonlinearly with radius. For measurement purposes, it is not difficult to average the values and plot log (Magnitude^2) vs. log (Frequency) to determine a fractal dimension. For a surface, the dimension D is related to the slope of the plot β as $D = (6 + \beta)/2$. As for the line profile case, it is also necessary to check the phase values for the transform to verify that they are random if the original surface is to be considered a fractal. Figure 28 also shows the phase image for the

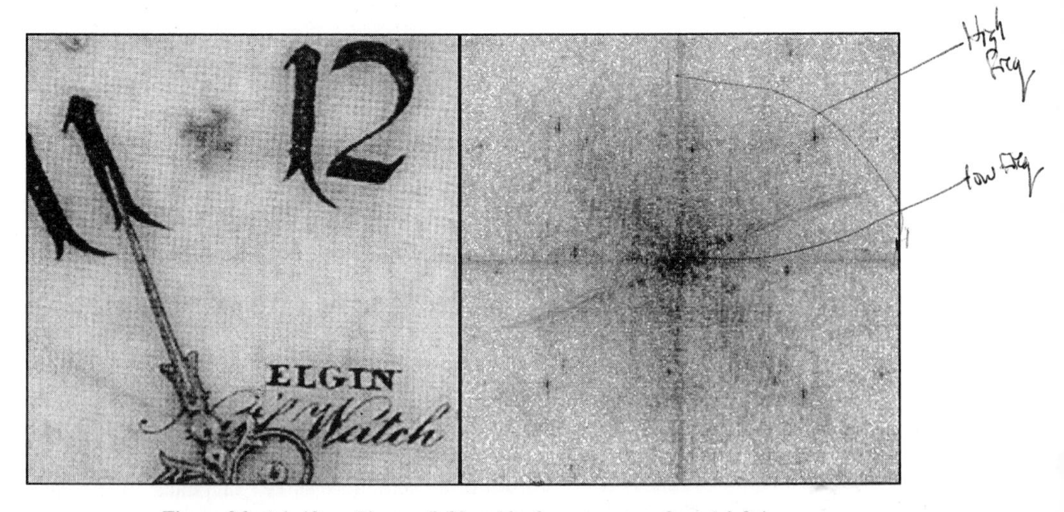

Figure 26. A halftoned image **(left)** and its frequency transform **(right)**.

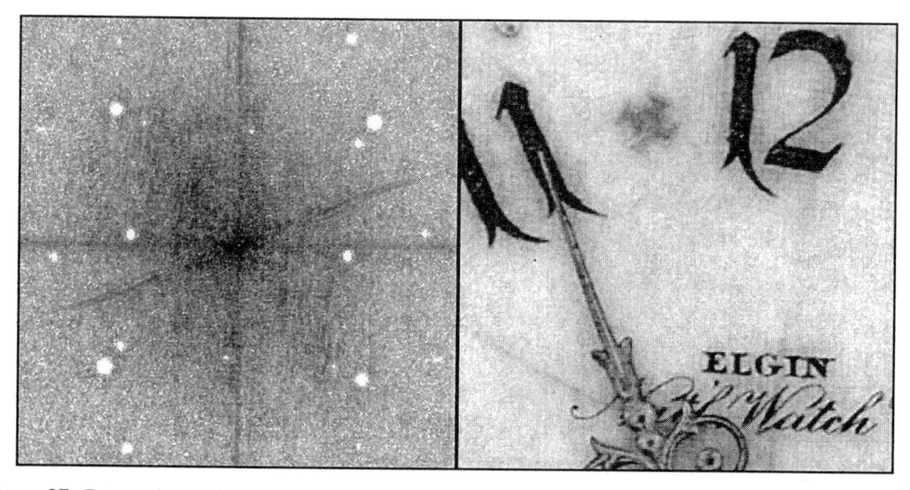

Figure 27. Removal of periodic information from Figure 26 by reducing the magnitude of the high frequency spots corresponding to the halftone **(left)** and retransforming **(right)**.

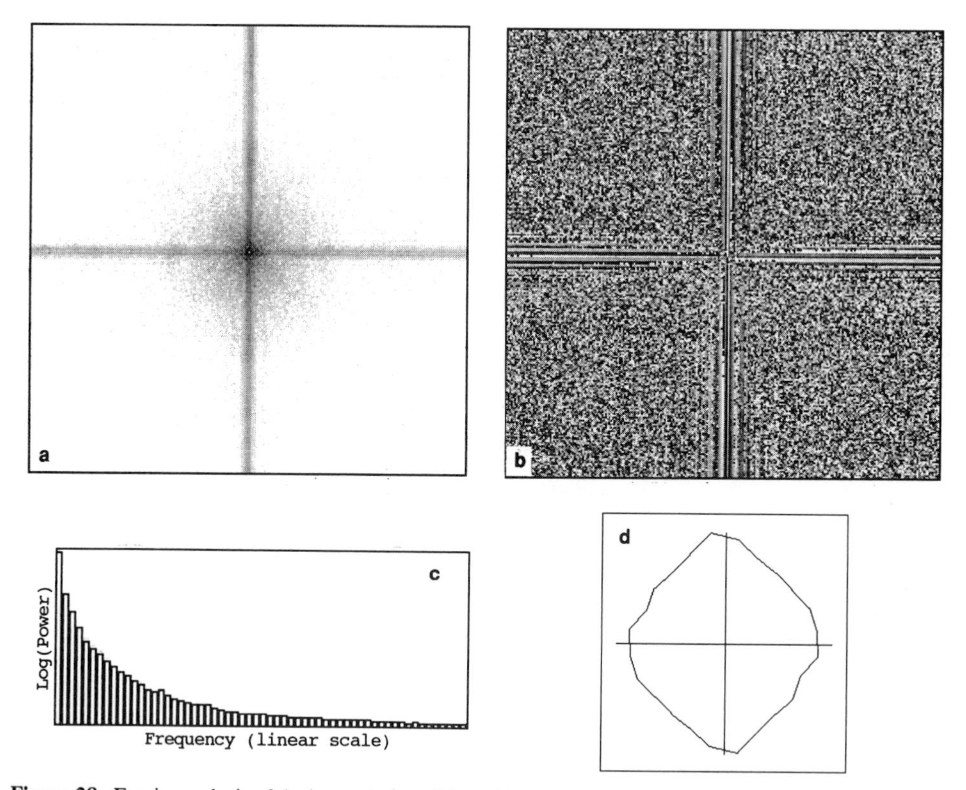

Figure 28. Fourier analysis of the isotropic fractal from Figure 16. The cross evident in the power image results from mismatch at the edges **(a)**. The phase image is random **(b)**. The value of the power spectrum averaged over all directions declines with frequency **(c)**. A rose plot of the slope of the log (Magnitude^2) vs. log (Frequency) relationship in various directions shows an approximately circular shape, indicating isotropy **(d)**.

surface from Figure 16, confirming that this is the case, and the slope of the log (Magnitude^2) vs. log (Frequency) plot as a function of direction.

For both the magnitude and phase images, the cross pattern of horizontal and vertical lines corresponds to the mismatch along the edges of the images. The Fourier transform method makes the assumption that the signal being processed is periodic. For a line profile, this means that the signal repeats itself along the x or t axis. For an image, it means that the image is one tile in an endlessly repeating array. This in turn means that the upper and lower edges of the image, and the left and right edges of the image, should match. If they do not, it requires many high frequency terms to handle the abrupt step. These show up as a cross in the image, as shown in Figure 28. It is necessary either to ignore these terms, or to pad the original image out to a larger dimension before performing the transform. The cross is an artifact of the finite image, and does not contain information on the fractal dimension. These values are bypassed in all of the analyses shown for real surfaces.

The dimension D determined from Fourier analysis of an image or from a profile is not the same as the other dimension values discussed in previous chapters. Compared to the Hausdorf dimension (e.g., as determined by a Richardson plot), the frequency analysis method produces a dimension value that is theoretically less than or equal to the Hausdorf value. As usual, the "equal" limit would only be expected for a "well-behaved" fractal. But, as for the other cases, the ideal relationship does not apply to the common situation of discrete data arrays. Gomez-Rodriguez, Baro et al. (1992) compared FT analysis of profiles to dimensional analysis of surfaces (log P vs. log A for lakes) and state that the FT method overestimates D for $D < 2.5$ and underestimates it for $D > 2.5$. Specific situations and anisotropic surfaces may produce other differences.

Anisotropy

When the original surface image is anisotropic, the two-dimensional Fourier transform (2DFT) reveals this. The slope of the log (Magnitude^2) vs. log (Frequency) plot can be evaluated as a function of direction by using all of the points within a pie-shaped wedge in the transform image. Performing a least-squares fit to these data points provides a direct measure of the fractal dimension as a function of direction (Russ 1990b). Plotting these values as a rose plot vs. orientation angle shows the degree and direction of the anisotropy.

Figure 28 shows that for the case of an isotropic surface, the rose plot is a circle. Anisotropic surfaces may have a variety of noncircular patterns. Figures 29 and 30 show two examples (more will be presented in subsequent chapters). The surface images were generated using algorithms discussed in detail in Chapter 6. Figure 29 is a cross-extrusion in which a fractal profile with dimension D_1 is translated along a second profile with a different dimension D_2. The rose plot clearly shows the two directions, but does not reflect the roughness present in intermediate directions on the surface. The cross evident in the Fourier transform in Figure 29 arises from the directional anisotropy of the surface, not from the edge mismatch discussed above.

The second example (Figure 30) shows a surface generated with anisotropic Fourier functions. The rose plot exhibits a gradual variation of slope (and hence of fractal dimension) with direction. In both cases, the plot of log (Magnitude^2) vs. log (Frequency) averaged over all directions shows a linear trend, but the slope does not by itself characterize the anisotropic surface. Chapters 6 and 7 show additional examples and suggest some methods for interpreting surfaces with mixed fractal behavior. Chapter 8 presents Fourier analyses of several measured surface range images from a variety of applications.

For comparison, Figures 31 and 32 show the Hurst Orientation Transform results for the same images from Figures 29 and 30. While these also show the surface anisotropy, there are some clear differences between the two approaches. Because the FT rose patterns average

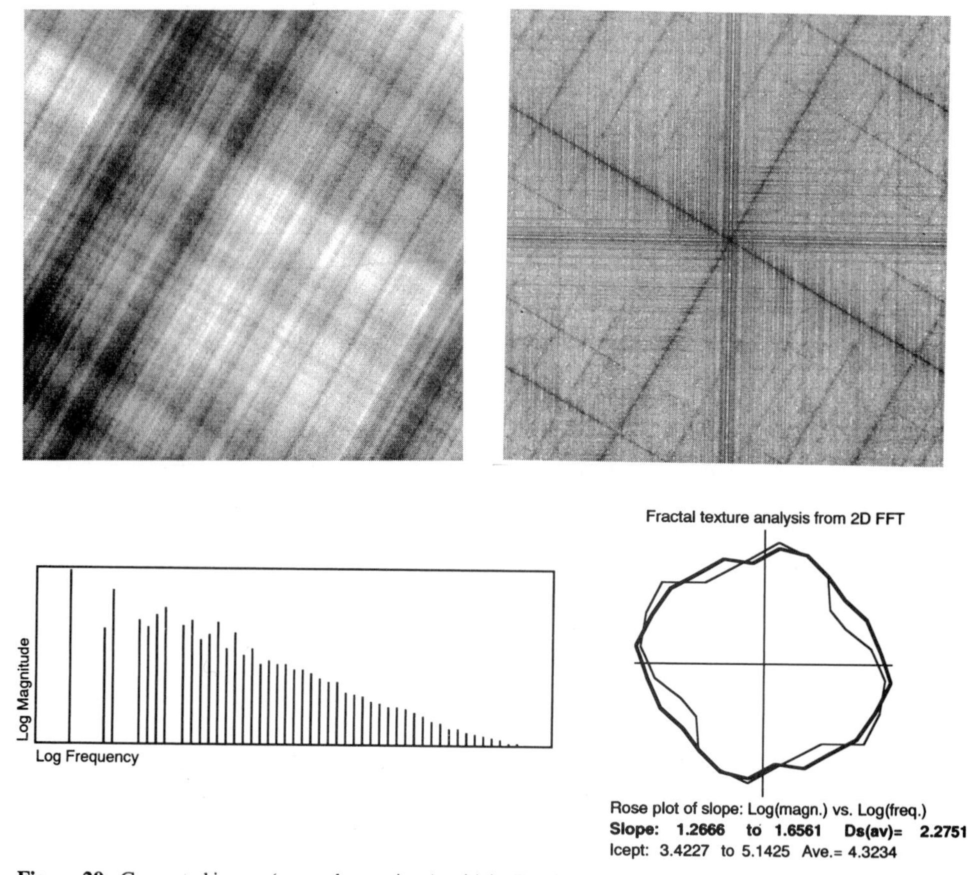

Figure 29. Generated image (crossed extrusions), with its Fourier transform image, a directionally averaged plot of log (Magnitude^2) vs. log (Frequency), and a rose plot of the slope of the log (Magnitude^2) vs. log (Frequency) curve as a function of direction.

values over a range of directions to construct the plot, they are generally smoother than the HOT plot. The HOT rose pattern shows the two principal axes of these generated anisotropic surfaces more clearly than the FT, but may not estimate the extent of the anisotropy as precisely.

The Fourier method has several advantages for studying fractal surfaces. It reveals and can characterize anisotropy, as indicated above. As shown in Chapter 3, the Fourier method is relatively insensitive to the presence of the noise in images due to the operation of the measuring instrument. The calculation method is well understood, easily programmed, and not too slow to compute. Provided that the image array is square with a dimension that is an exact power of 2, there is a well-known "fast" algorithm, the FFT, and a particularly efficient implementation, the "butterfly." The entire basis of the Fourier method is that of orthogonal and independent functions. For a two-dimensional array of pixels, it is possible to compute the transform of each row of pixels, and then of each column, to produce the complete transform (although this is not the fastest way to accomplish the overall transform). The same method can be extended to higher dimensions, as in fact is sometimes done to perform three-dimensional reconstruction in computed tomography (CT) scans.

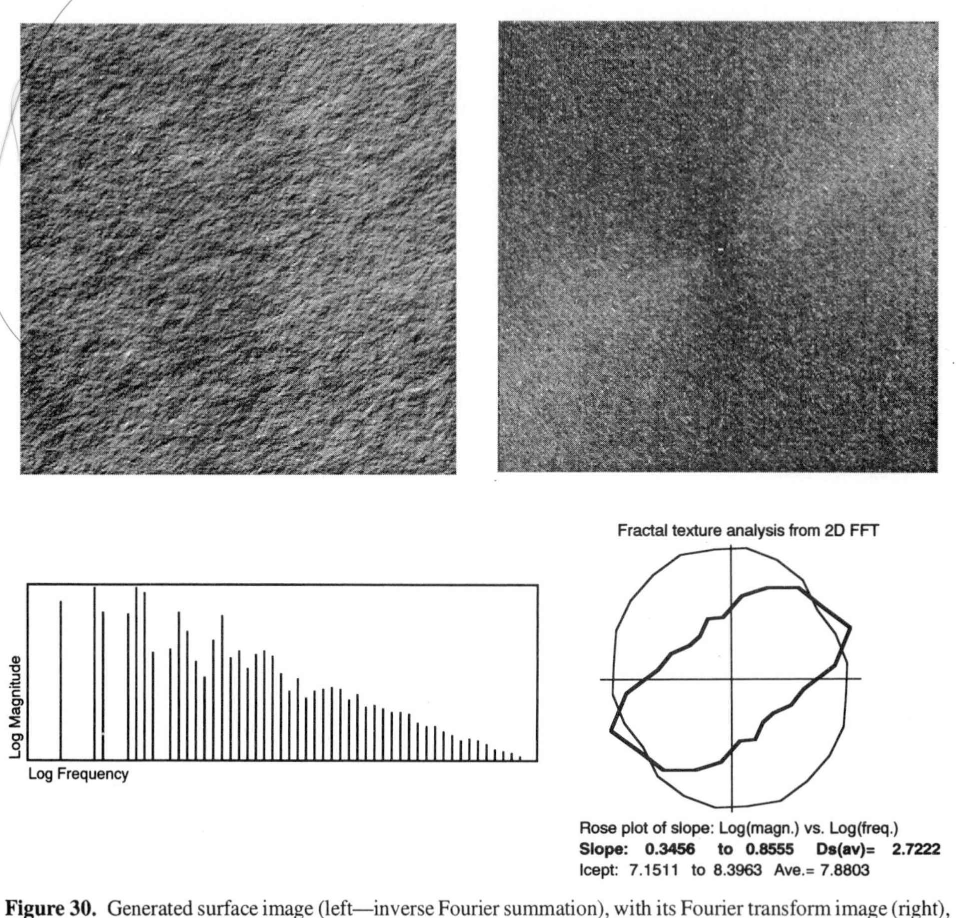

Figure 30. Generated surface image (left—inverse Fourier summation), with its Fourier transform image (right), a directionally averaged plot of log (Magnitude^2) vs. log (Frequency), and a rose plot of the slope of the log (Magnitude^2) vs. log (Frequency) curve as a function of direction.

On the other hand, the Fourier method has a few shortcomings which limit its usefulness and provide niches for the other measurement methods which have been discussed. The relatively large scatter of the points in the log (Magnitude^2) vs. log (Frequency) plot limits the precision of the slope of the fitted line. By comparison, most of the plots from methods such as Minkowski, Richardson, and Korcak techniques show very little scatter and give extremely good straight-line fits with modest numbers of points. Secondly, it is quite necessary also to check that the phase values are uniformly random, and this is not often done.

Third, since few real surface images will have upper and lower edges that match, or left and right edges, the presence of a "cross" of large magnitude terms at all frequencies must be avoided in performing analysis and interpretation. Fourth, if the surface measuring technique does not produce an image which is square and has a dimension which is an integral power of 2 (e.g., 256 × 256, 512 × 512, 1024 × 1024), then the data must be padded out to the next larger size.

Fifth, if the surface has average elevation, slope, or curvature values that are not zero, this can bias the low-frequency terms in the transform, creating the possibility of bias in the interpretation. Likewise (sixth), at high frequencies the presence of noise or lack of real resolution can shift the values and create bias.

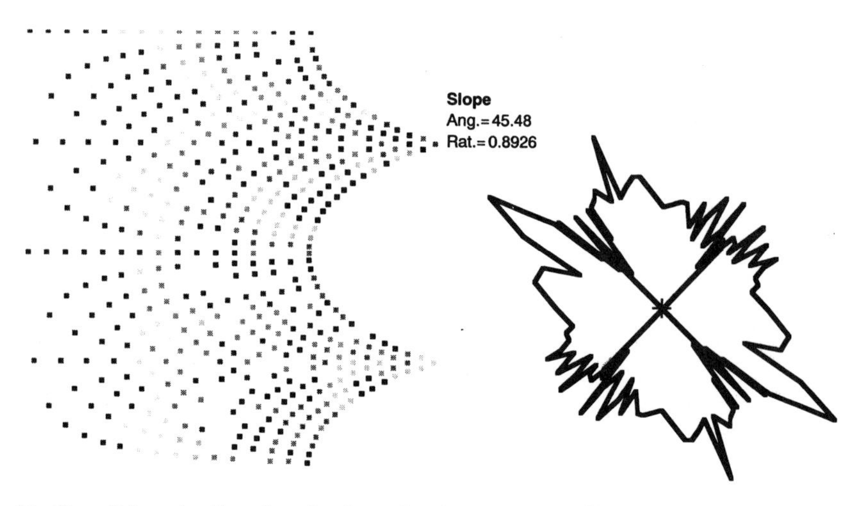

Figure 31. Hurst Orientation Transform for the surface image in Figure 29, and a rose plot of the slope as a function of direction.

Characterizing the Magnitude of the Roughness

All of the measurement techniques discussed this far have used some kind of plot on log–log axes to determine the fractal dimension of a line profile or of a surface from the slope of a straight line fitted to the data. But two parameters are required to describe a straight line, the slope and the intercept. Consider two profiles with the same fractal dimension, whose Hurst and Fourier plots show the same slope but are displaced vertically as shown in Figure 33. What is the meaning of this offset?

Clearly it has to do with the magnitude of the roughness of the surface. The second of these two profiles was generated from the first using a sine function, which nonlinearly

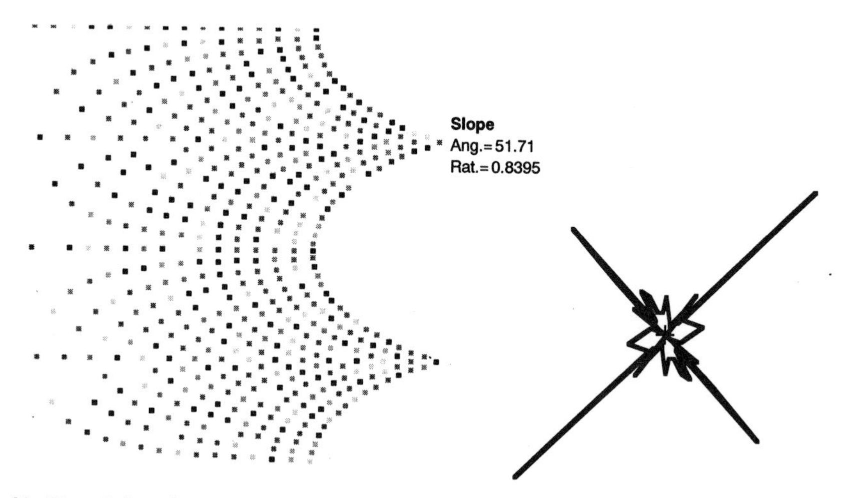

Figure 32. Hurst Orientation Transform for the surface image in Figure 30, and a rose plot of the slope as a function of direction.

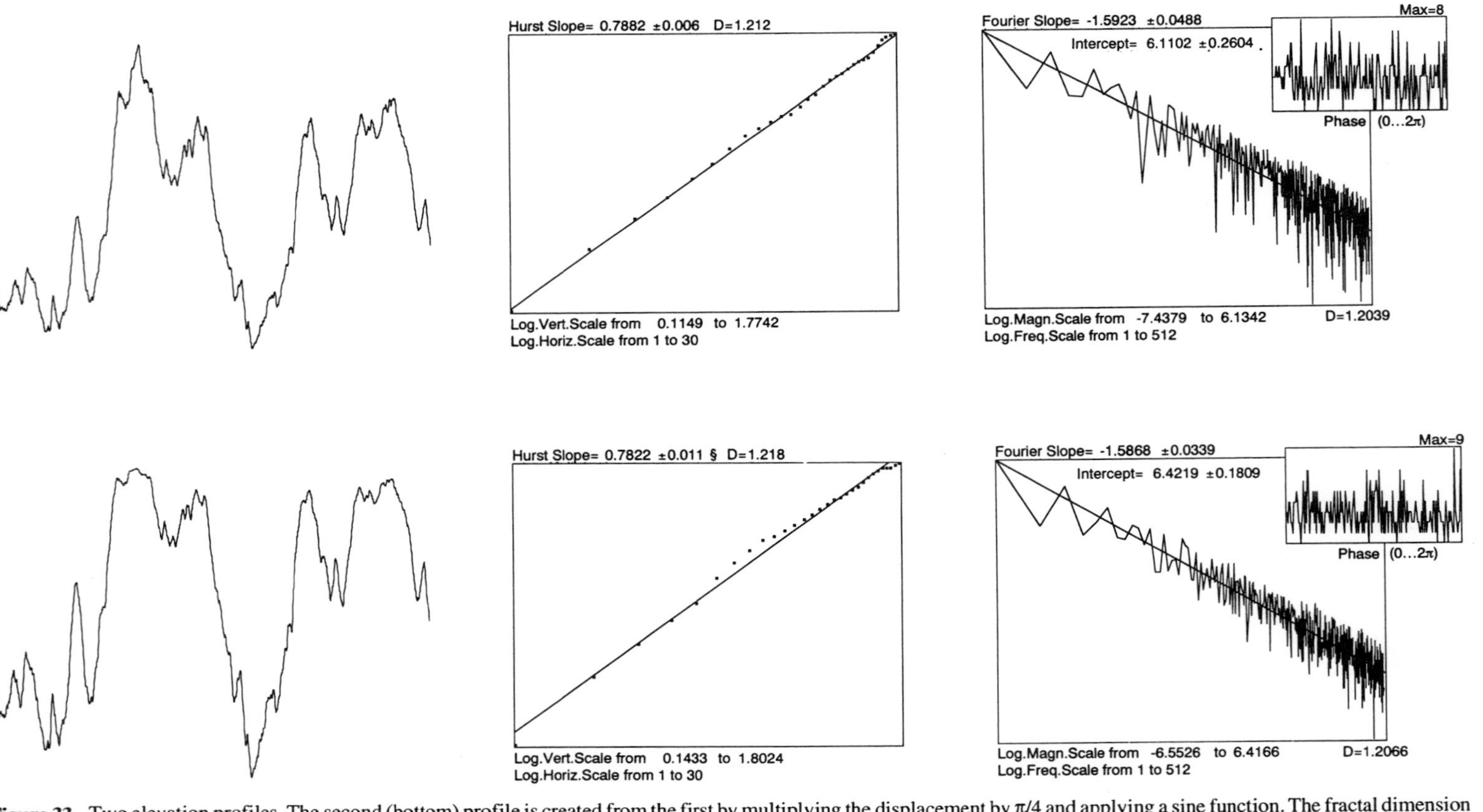

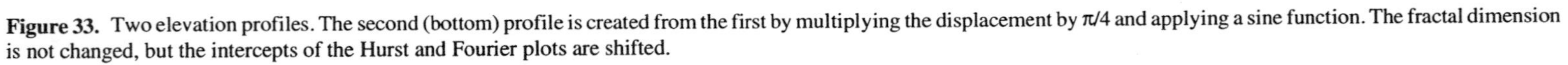

Figure 33. Two elevation profiles. The second (bottom) profile is created from the first by multiplying the displacement by $\pi/4$ and applying a sine function. The fractal dimension is not changed, but the intercepts of the Hurst and Fourier plots are shifted.

increases some local differences and decreases others. In the case of the Hurst plot, the intercept of the line is at log (1) = 0, or in other words the distance corresponding to the adjacent pixel, and has a value equal to the maximum nearest-neighbor difference value, normalized by the standard deviation of the data. If the surface were simply expanded vertically, like blowing up a balloon, that value would not change. However, changing the shape of the distribution of the random numbers used to generate the profile can alter the intercept without affecting the fractal dimension in any way.

Similarly, the Fourier representation of a surface or profile has the form

$$\text{Magnitude}^2 = \text{Constant} \cdot \text{Frequency}^{-\beta}$$

where the exponent β is related to the fractal dimension. The constant of proportionality in this expression is the vertical position of the line, and also describes the overall magnitude of the roughness. Clearly this is an important parameter, since two surfaces which differ in this parameter will have different properties, and may be produced in different ways.

A conventional measurement of surface roughness is the "rms" or root-mean-square roughness. For an elevation profile or surface image, this is determined by standard descriptive statistics. The mean value of the elevation values is the average surface elevation. The variance is the sum of squares of the deviations of individual values from this mean divided by the number of points, and the standard deviation or "rms" value is the square root of the variance.

It may seem at first glance that this is a robust measure of surface roughness, and certainly it has many years of use behind it. But unfortunately the value that is determined in this way depends on the length of the profile and the instrument resolution, rather than on the surface itself. While it may be acceptable as an ad hoc tool for comparing similar surfaces using the same measurement device, it does not provide a real characterization of the surface.

Without going through an elaborate proof of the statement above, a simple thought experiment will suffice to indicate the dependence of the rms measurement on the measurement procedure. Consider a profile with a fractal dimension of 1.5, corresponding to the case of Brownian motion discussed several times before. It has already been pointed out that the path of Brownian motion will occasionally stray far from the origin but will always return, and that when a zero crossing does occur, there will be many more close to it. The result is that the plot of the frequency of various deviations from the origin will have a normal or Gaussian shape.

Determining the "average" displacement from the origin is meaningful only over a finite range of dimensions. In other words, the rms value which is determined by any particular measurement tool will depend on the range of dimensions which it can measure. It may be useful to compare such averages determined on different surfaces using measurement data from the same tool, but the value has no independent meaning. In fact, the variation of the "rms" roughness value with measured area provides another tool for determining the fractal dimension, as has been shown previously both for elevation profiles and for surfaces. This is a little-used method because it is time consuming and not too precise, but it certainly indicates that the standard deviation is a function of the area of the surface used for measurement.

The Topothesy

One suggestion for overcoming this limitation has been to introduce a new quantity called the topothesy. This is formally defined as the distance along the profile for which the expected angle between two points would be one radian. For most real fractal surfaces this corresponds to a distance that is very small, generally far below the resolution of the measurement tools and often smaller than the atoms which comprise the material. This dimension has no physical

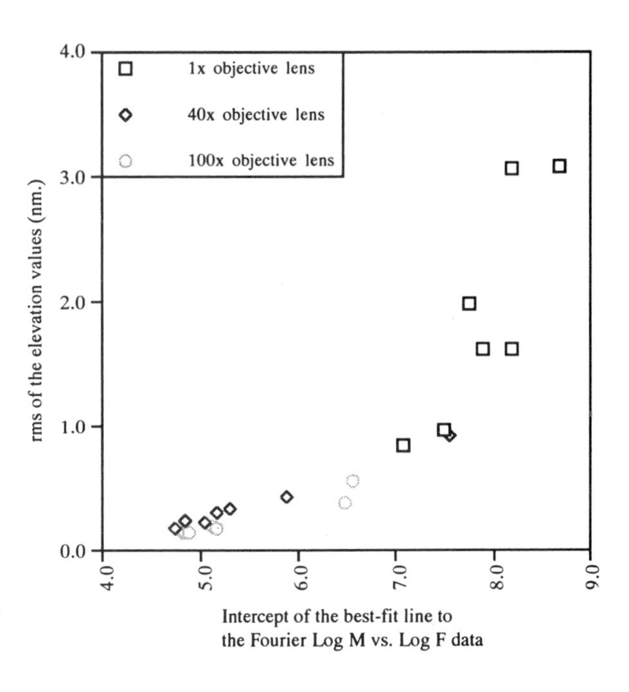

Figure 34. Correlation between classical rms roughness value and intercept of the straight line fit to the log (Magnitude^2) vs. log (Frequency) plot from the FT, for a series of interferometer images of low-relief polished surfaces, obtained with different objective lenses and thus covering different areas of the samples.

meaning, but does serve as a possible way to represent the absolute scale of the surface structure.

The topothesy Λ is formally defined in terms of the power spectrum

$$\text{Magnitude}^2(\text{Frequency}) = \text{Constant} \cdot \text{Frequency}^{-\beta}$$

as

$$\Lambda^{3-\beta} = \frac{(2\pi)^\beta \cdot \text{Constant}}{2 \cdot \Gamma(\beta) \cdot \cos(\beta\pi/2)}$$

This value contains both the constant and the exponent from the power spectrum expression, and so it depends on the fractal dimension as well as the magnitude.

The advantage of the topothesy is that it can be determined from a line profile and its structure function. The structure function is defined and discussed in Chapter 5. Briefly, it is a plot of the average value of the square of the deviations of points along the profile as a function of the lateral distance between the points. For a fractal profile, the structure function $S(\tau)$ is proportional to $\tau^{\beta-1}$, and $S(\Lambda)/\Lambda^2 = 1$ according to the definition of topothesy.

Sayles and Thomas (1978) used the topothesy to normalize the roughness data from a variety of surfaces (machined surfaces, roadways, etc.) so that a plot of the normalized power spectral density $G(1/\lambda) = k\lambda$ fell on the same power law relationship over eight decades on a log–log plot, where k is the topothesy. The paper discusses the effect of many independent variables with different length scales and any distribution producing a random fractal due to the central limit theorem, and the dependence of sigma on the sampled length of a profile.

Because of the simplicity of applying Fourier analysis to both profiles and 2D range images, it is usually simpler to use the constant that defines the vertical position of the line. This is independent of the slope (and hence the fractal dimension), represents the magnitude of the roughness directly, and is simpler to calculate. Regardless of which method is used, either parameter will represent the magnitude of surface roughness in a way that is independent

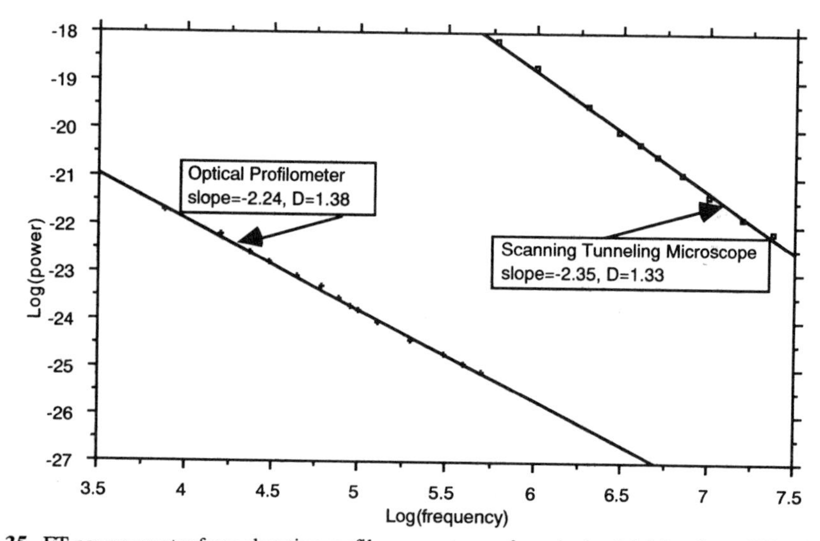

Figure 35. FT power spectra from elevation profiles on a wear surface obtained (Majumdar and Bhushan 1991) using an optical profilometer (viewed area 240 μm wide) and scanning tunneling microscope (viewed area 5 μm wide). The slopes of the curves and hence the fractal dimensions are similar, but the vertical positions and hence the topothesy values are quite different.

of the measurement technique or the range of dimensions over which the measurement is performed.

A series of surface measurements using an interferometer were subjected to 2D Fourier analysis to determine the intercept of the plot of the log (Magnitude^2) vs. log (Frequency curve). Figure 34 shows these values plotted against the traditional rms roughness. The correlation for each magnification indicates that the intercept value offers a way to characterize the "traditional" surface roughness that is independent of the image magnification and instrumentation. The slope of the Fourier plot and the resulting fractal dimension might then be expected to provide an additional surface characterization tool that describes the spatial organization of the roughness in a way that simple descriptive statistics (e.g., the rms value) does not.

However, this use of the intercept and slope to cleanly separate the magnitude and organization of the surface roughness is not always supported by observations. One series of measurements on the same surface using interferometry and STM (Majumdar and Bhushan 1991) is reproduced in Figure 35; it shows that the slopes from the two data sets are nearly the same, within the precision expected from the FFT method, but the two lines are vertically displaced by several orders of magnitude. This is unexpected and, thus far, unexplained.

5

Light Reflection and Scattering

Visual Appearance

There has been interest in characterizing the quality of surface finish for machined or otherwise man-made objects, and the intentional roughness of surfaces such as catalysts, since long before fractal geometry came along. The same measurement tools which are used for these purposes can also provide the raw data for a fractal interpretation of the surfaces. These include the scattering of light or photons of other energies (from X rays to radio waves), as well as various forms of microscopy. Basically, a surface that is rough will show a pattern of scattered intensity, or of measured brightness in an image, that has variations. Analysis of these brightness variations can be used to describe the roughness.

Even the visual appearance of rough surfaces contains information about the fractal dimension. Humans are very good at predicting how rough surfaces will feel from how they look, and if given a series of pictures of different surfaces, will rank them consistently in order of fractal dimension. This suggests that the light scattered from the surface contains the necessary information to quantify the texture. "Texture" is a word with diverse meanings used in different situations. We will not attempt to define it rigorously here. Nevertheless, most observers would describe a brightness pattern as either "smooth" or "textured" depending on whether the local variation in brightness was small or great. This has been used to classify various types of terrain and crops, for instance, from aerial and satellite imagery. Haralick, Shanmugam et al. (1973) defined more than a dozen different texture parameters which can be calculated from the brightness differences between neighboring pixels in regions of an image, and other researchers have constructed additional combinations.

Peleg, Naor et al. (1984) went one step further and constructed a table of the number of pixel pairs as a function of their brightness differences and the distance separating them. He and others (Pentland 1984; Larking and Burt 1983a; Larking and Burt 1983b; Werman and Peleg 1984) realized that from a mathematical point of view, the two-dimensional array of brightness values of light diffusely scattered from many natural surfaces was a fractal. This means that it is not simply the magnitude of the brightness variation that describes the roughness, but its spatial organization.

For methods that measure the surface elevation, knowing the standard deviation of elevations is not enough to describe the surface roughness. The fractal dimension includes the needed information about the correlation of values as a function of their separation. From the data in such a table, it is possible to compute a dimension as discussed in previous chapters. However, the units of brightness bear no absolute relationship to anything about the surface. For perfectly diffuse lighting, the scattered intensity follows a Lambertian law and varies as a

cosine function of angle from the surface normal. Consequently, the brightness variations represent a change in surface slope, not elevation.

Analysis of the relationship between surface elevations and slopes is best begun with a single elevation profile. Working only in the two-dimensional space of the plane containing the profile, the slope of the line is the first derivative of the elevation. If the elevation profile is a fractal, what can be said about the slopes? This turns out to be a pretty difficult question. In fact, one of the classic mathematical "monster" curves discussed in the introductory chapter was the Koch curve, which is continuous but has no defined derivative.

Fractal Brownian Profiles

In one particular case we can assign some meaning to the derivative, if only by reasoning backwards. Consider the case of Brownian movement. If the coordinate of a moving particle in one dimension is plotted as a function of time, the result is a profile which is self-affine and has a fractal dimension exactly equal to 1.5. One of the classic ways to construct such a profile is to generate a series of random numbers with mean value of zero and a Gaussian distribution, and add them together sequentially to generate the coordinates. This describes the probability that each "bit" of motion has an equal chance of being in the positive or negative direction, and its magnitude.

In other words, the trail of the Brownian motion is the integral of a Gaussian noise with a forest of positive and negative spikes, and more and more spikes are present at any level of time resolution. This means that since the positive and negative values must occur with independent values at infinitely fine spacing along the time axis, they effectively form a cloud that can only be described as a probability function. That must be the derivative of the fractal profile. Figure 1 illustrates this case. It turns out that it is not essential that the distribution of the magnitude of the positive and negative values have a probability distribution that is Gaussian. Any kind of distribution will do. The idea that different fractal profiles with the same dimension may differ in another property (sometimes called the "lacunarity" of the profile) will be used later.

For fractal Brownian motion that has a dimension less than 1.5 (smoother) corresponding to temporal correlation, the profile can be generated by a series of spikes which also come from a Gaussian random number generator with a mean of zero, but each spike is not an independent value. Instead, it is a running sum of several values to produce the temporal correlation. Still, the plot of the values is essentially a forest of noise describable only as a probability function. So too for the case of a profile with a dimension greater than 1.5. It is not usually possible to examine the derivative curve for a temporal function directly, but we will look at the power spectrum of a profile and of its derivative shortly.

For the case of light being diffusely reflected from a fractal surface, the derivative model hinted at by these curves suggests that individual facets on the surface should have orientations with widely distributed angles. Later in this chapter, the relationship of the power spectrum of an elevation profile to the standard deviation of elevation, of slope, and of curvature is shown. Further, the cosine function relating slope to brightness for the scattering of diffuse light must not be important, just as changing the shape of the distribution of spikes from Gaussian to something else does not alter the dimension of the resulting integral. All that is really needed is a monotonic or at least single-valued function between slope and brightness.

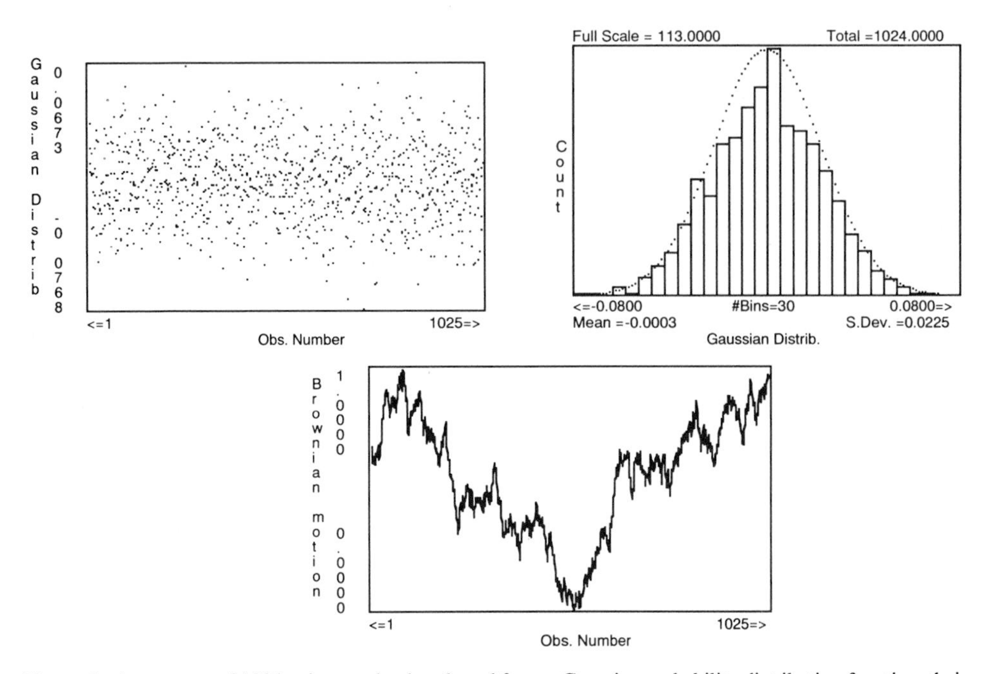

Figure 1. A sequence of 1024 points randomly selected from a Gaussian probability distribution function, their actual frequency distribution, and the cumulative running sum of the points, which is a one-dimensional trail of Brownian motion with a fractal dimension of 1.5.

Electrons, Radar, etc.

This means that other imaging modalities, such as the backscattered or secondary electrons used in the scanning electron microscope, or the scattering of synthetic aperture radar from rough terrain, will also have this character. Practically all of the imaging systems which rely on scattering or re-emission of light or other signals from surfaces generate signals which are related in some unique way to the local surface slope. Consequently, the brightness pattern which will be observed from a fractal surface will be mathematically a fractal.

This effect has been demonstrated for scanning electron microscope (SEM) images using either secondary electron or backscattered electron signals (Russ and Russ 1987). The physics of interaction between the incident electrons and the surface to produce these signals is entirely different from light scattering, and the functional dependence upon slope is somewhat different as well. Yet again, the fractal nature of the surface is revealed in the brightness pattern of the image. In the paper cited, a texture parameter which is essentially a Minkowski dimension of the brightness pattern was calculated for a series of surfaces with different physical roughnesses and shown to be proportional to the surface fractal dimension determined using a Richardson technique applied to slit islands.

The same technique was applied to electrodeposited coatings (Russ 1990a). Figure 2 shows the definition of surface "texture" used in those studies; it is the slope of the graph of mean pixel brightness difference vs. pixel separation distance. Except for the use of mean difference instead of maximum difference, this is the same as the local Hurst dimension shown in Chapter 4 for identifying image texture to allow segmentation. Figures 3 and 4 show low-power light microscope and scanning electron microscope images of several deposited surfaces whose surface roughness was characterized using this technique. Cross-sectioning

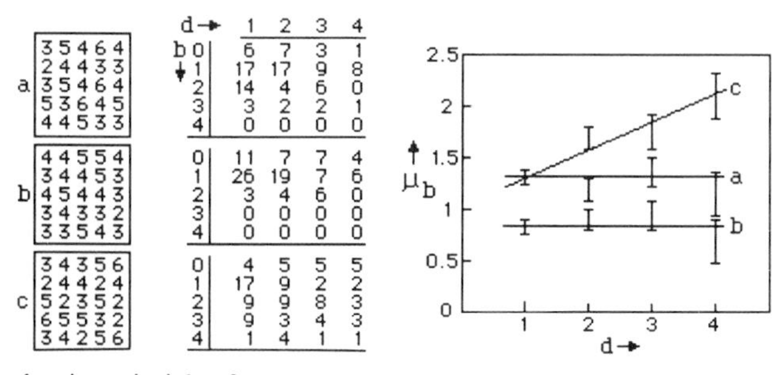

Figure 2. Diagram showing principle of texture measurement from pixel brightness values. The three image fragments (with single-digit pixel values) are used to construct tables of brightness difference (b) vs. pixel distance (d), from which the mean values are plotted. Images **a** and **b** have different mean values but no trend with distance. Image **c** has a slope which is interpreted as the image texture.

was also used (Figure 5) to obtain a physical fractal dimension for the surface (determined using a Minkowski technique).

The various texture values do not agree numerically with the physical dimension, because they depend on the details of the illumination, surface albedo (absolute reflectivity), camera response, etc. However, they vary in proportion to surface fractal dimension as shown in Figure 6. For comparison purposes, the image texture determined at either low or high magnification shows the change in physical surface roughness.

The same method was also extended to additional surfaces and shown, for instance, to correlate with fractal dimensions obtained from vertical sections, and with the strength of adhesives bonded to the surface (Russ 1990a). Figure 7 shows the correlation between the surface texture (determined as above from SEM images of the etched metal surfaces), the physical fractal dimension of the surfaces (determined by sectioning after all of the adhesive

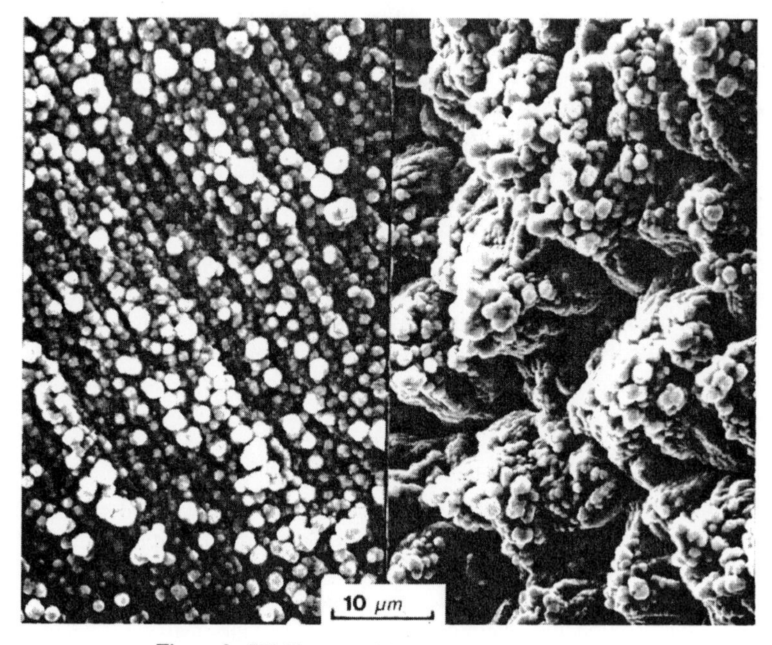

Figure 3. SEM images of deposited copper surfaces.

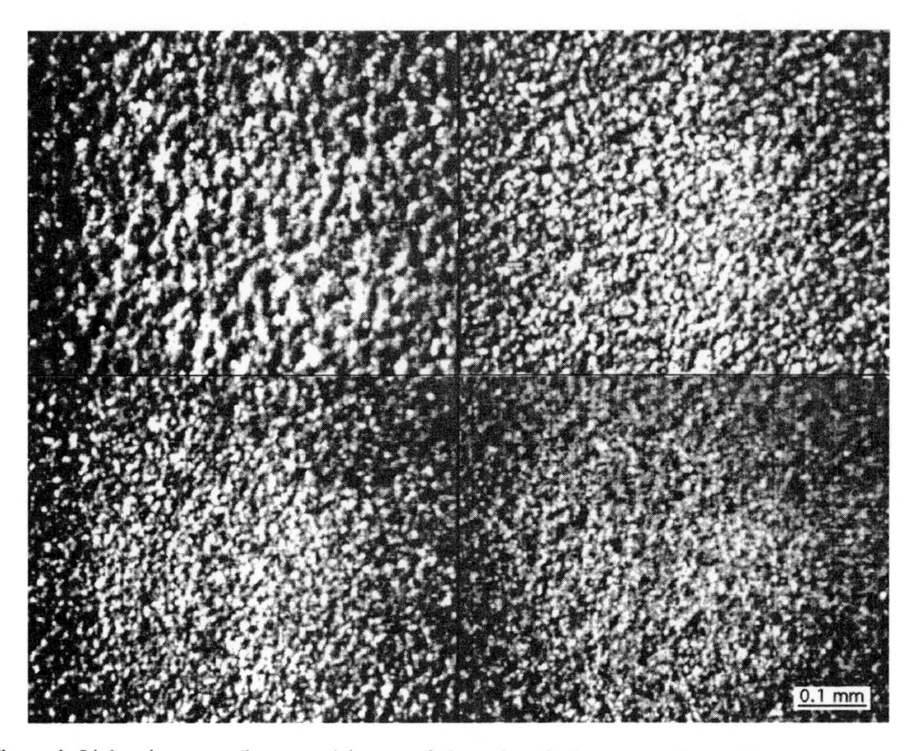

Figure 4. Light microscope (low power) images of electrodeposited copper surfaces with 5, 2, 1, and 0.5 oz coating weights.

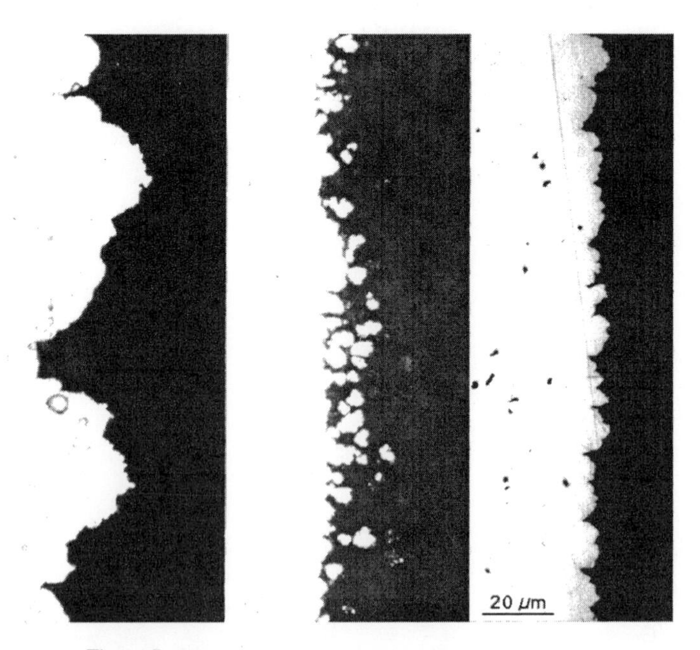

Figure 5. Cross-section images of the copper coatings.

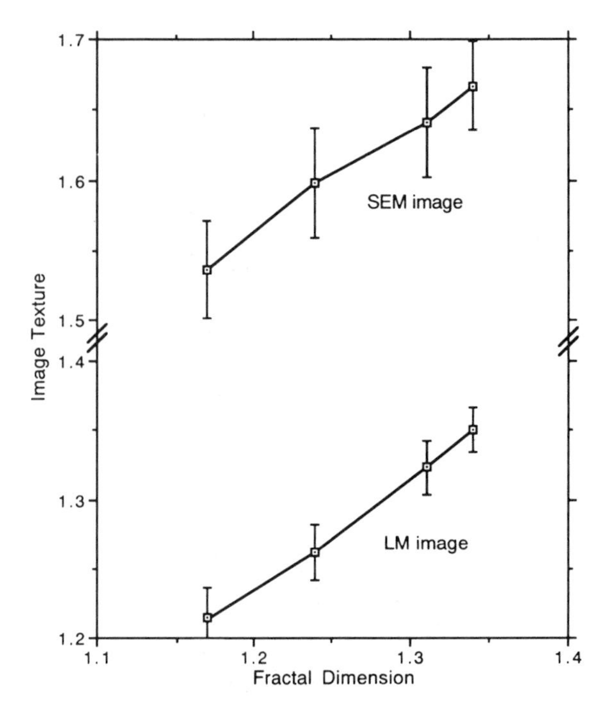

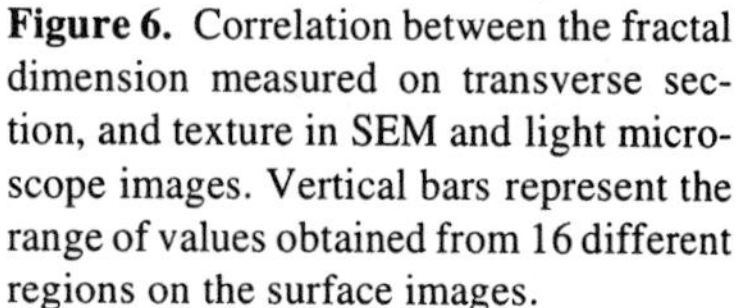
Figure 6. Correlation between the fractal dimension measured on transverse section, and texture in SEM and light microscope images. Vertical bars represent the range of values obtained from 16 different regions on the surface images.

tests were completed), and the strength of the bonded surfaces. It is not surprising that the strongest bond results from the "roughest" surface. However, before this study it was not understood why the electrochemical or chemical etching of the surface did not simply increase the surface roughness and bond strength progressively. Instead, the surface roughness (fractal dimension) first increased and then decreased with the length of etch time.

Since many other factors determine the absolute brightness relationship, and since the brightness fractal is self-affine rather than self-similar, the numerical correspondence between the two may not be accessible. Nevertheless, as a quick, noncontacting and nondestructive

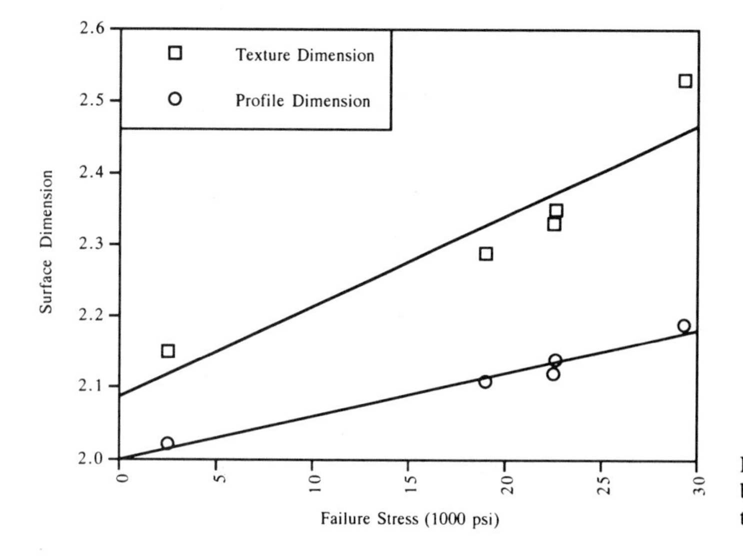

Figure 7. Plots of adhesive bond strength vs. *D* and texture.

method for observing relative changes in surface "texture," this approach has much to recommend it. Light scattering measurements or other forms of surface imaging are a very rapid way to obtain qualitative roughness data, by digitizing the image and measuring any of the one or two-dimensional fractal dimensions from the array of brightness values.

It is possible to compare values from surfaces of the same materials under similar lighting and viewing conditions, and this can be done with imaging tools such as the scanning electron microscope as well as with visible light. However, it is more difficult to obtain quantitative data since the proportionality coefficients arising from the illumination, the surface albedo (for visible light) or electron emissivity and atomic number (for the SEM), the physical processes involved in producing the image, and the detection process are complex. Only direct calibration seems to offer a practical solution, and even this will be susceptible to variations in the surface cleanliness, variations in material composition or color, etc.

Local Texture Measurement

The ability to distinguish different kinds of roughness lies behind the use of texture operators for image segmentation (Rigaut 1990). At all scales from the microscopic to satellite imagery, it is often necessary to classify each pixel according to some surface characteristic, collect together all of the touching pixels into regions, and perform various measurements on the resulting features. A local texture value which measures a fractal dimension from the brightness differences between each pixel and its neighbors, as a function of their distance, can produce a derived image in which the local pixel brightness is proportional to local roughness (Russ 1990c). This has been applied to aerial photos of mountain ranges, macroscopic photos of metal fractures (Chermant and Coster 1978; Coster 1978), and electron microscope images of structure within human cells, as discussed in Chapter 4.

There are two, not quite equivalent ways to produce these maps. Both use pixel values in an approximately circular neighborhood, and consider the distance of each pixel from the central one. One approach plots the average brightness difference between the central pixel and its neighbors as a function of their distance (as usual, on log–log axes). The slope of this line is then appropriately scaled and assigned as the brightness value for that pixel location in the derived image. The process is repeated for every pixel. Figure 2 illustrated the process.

The second method finds the brightest and darkest pixels within a disk as a function of radius. The description in Chapter 4 used this approach. The central pixel may not be one of the extreme values. The log–log plot of the range between darkest and brightest as a function of disk radius produces a slope value which is scaled and assigned to the pixel location. This is in effect a Hurst plot generalized from one dimension to two. Applied to an isotropic two-dimensional image in which pixel values represent elevation (a range image), scattered light intensity, etc., the result has the same meaning as the Hurst coefficient. The case in which the surface is not isotropic is more complicated (see Chapter 7). Repeating the process for every pixel produces a new image whose brightness variations show the local differences in surface roughness.

Both approaches give a measurement of a local dimension. If the pixel brightness values are imagined to represent an elevation, then the disks are the structuring element. Finding the difference between the minimum and maximum is functionally the same as calculating the thickness of the Minkowski comforter. Peleg, Naor et al. (1984) measured the fractal dimension of the brightness pattern in two-dimensional image arrays using this method, described in Chapter 3 as the Minkowski covering method generalized to two dimensions. For a surface, this comforter volume is equivalent to the thickness of the sausage along a boundary line. Because the process can be carried out quite efficiently by simply finding the minimum and

maximum values as a function of distance, the procedure is also the equivalent for surfaces of the Hurst measurement of profiles, which was discussed in Chapter 4.

An important limitation of these images is that for two reasons, they are self-affine rather than self similar. First, the array of brightnesses is single-valued. There are no re-entrant surfaces or multiply-connected paths possible with this kind of data. Secondly, there is no meaningful consistent relationship between the brightness values for the scattered light or other signal, and the horizontal dimensions which are typically in units of length, so the scaling of values horizontally and vertically is not related. Because the surfaces are self-affine and the values are not normalized in any way, the slope of the log–log plot may not be constrained to lie between 2.0 and 2.99+, which is the possible range of fractal dimensions for a physical surface. This frustrates any attempt at a direct interpretation of the numeric values of these dimensions.

Brightness Patterns from Rough Surfaces

To investigate the effects of surface (physical) fractal dimension on the brightness pattern of scattered light, an empirical approach was used which involves the modeling of fractal surfaces and of the scattering of light from them (Russ 1993a). A series of fractal surfaces was generated using an inverse frequency transform as discussed in Chapter 6. This produced range images which were analyzed using the Fourier, Minkowski, rms vs. area, and Korcak techniques discussed in previous chapters. The generated surfaces were then rendered using Lambertian light scattering rules with Phong shading (Foley and Van Dam 1984). Two surface conditions, one with predominantly diffuse scattering (a dull surface) and one with a strong specular component (a shiny surface), were modeled. For each, the light source was placed normally incident to the nominal surface orientation, and also at 30 degrees to the normal. Each image was rendered from a point of view perpendicular to the nominal surface. The resulting brightness patterns were then measured using the same fractal analysis tools listed above, and the results compared.

Figure 8 shows a series of isotropic fractal surfaces generated using an inverse Fourier technique. The mean magnitude of coefficients in the harmonic expansion was scaled as $\langle \text{Magnitude}^2 \rangle = (\text{Frequency})^{-2\alpha}$, where α varied from 0.5 to 0.95. The nominal fractal dimension of such a surface is $(3 - \alpha)$. Individual coefficients varied about this mean with a Gaussian distribution, and all phase values were uniformly randomized. The images in Figure 8 show these surfaces as range images, in which grey scale brightness represents elevation. Rendering these surfaces using the dull and shiny surface models mentioned above, with normal and oblique incident light, produces the images shown in Figures 9 through 12. The highly localized bright points of specularly reflected light from the shiny renderings of the rougher surfaces are especially apparent. The natural appearance of these surfaces, especially the less rough ones, shows why such fractal modeling is often used to create synthetic surfaces for movies, geometric modeling, computer-aided design, etc.

Comparison of the rms vs. area, Korcak, and Minkowski dimensions for the physical surface with the dimension determined from the FT slope is shown in Figure 13. None of the methods corresponds exactly to the dashed identity line, although the trends are generally similar. This is not unexpected for two reasons, discussed in earlier chapters: 1) the different measures evaluate somewhat different characteristics of the surface, yielding technically different "dimensions"; and 2) the discrete nature of the pixel array used for measurement may differently bias the different measures. However, all of the methods used here are correctly applied to range or elevation images, which are self-affine rather than self-similar and therefore inappropriate for some other kinds of measurement methods, such as box counting.

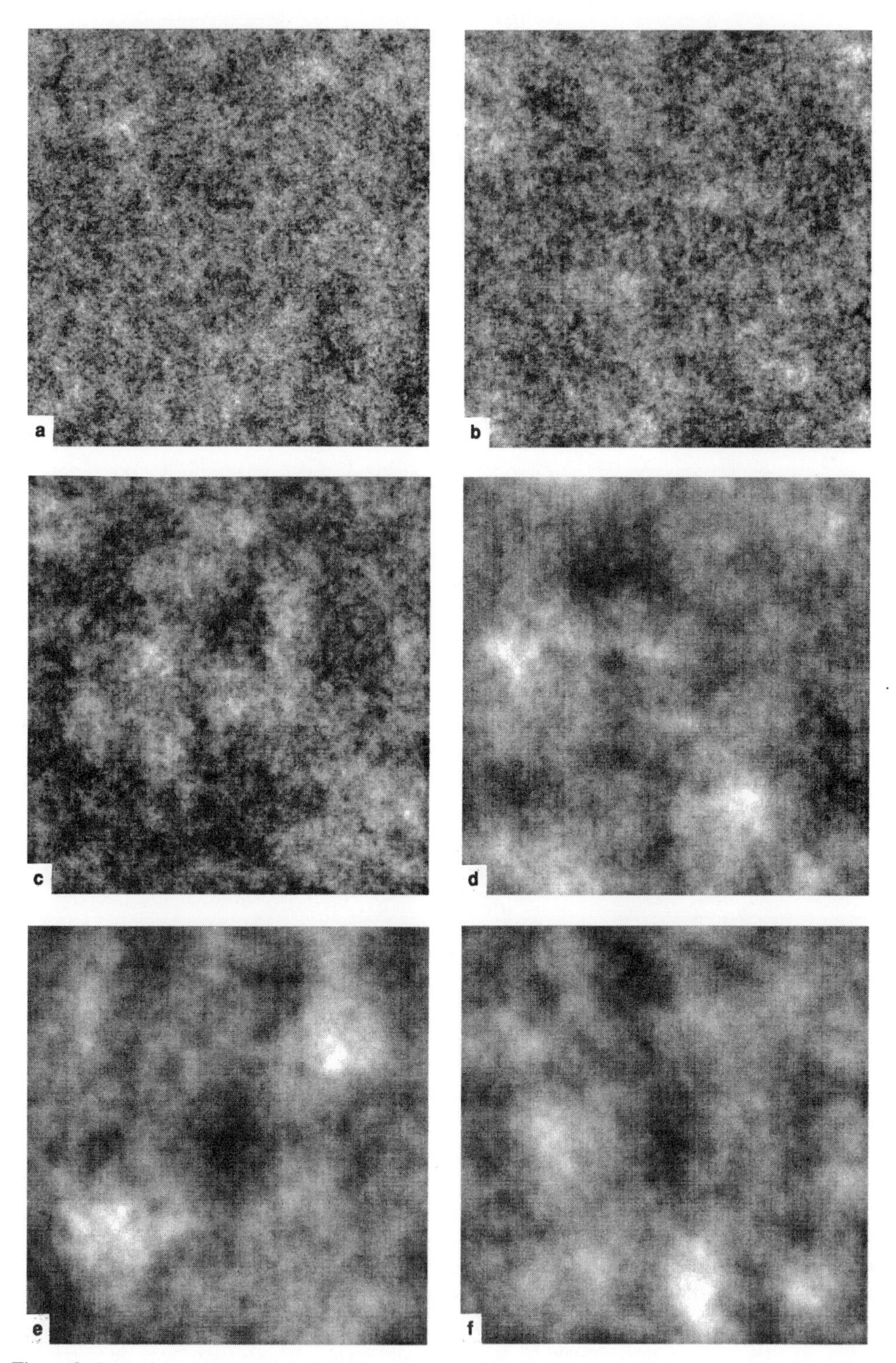

Figure 8. Range images for generated fractal surfaces. α = **(a)** 0.5, **(b)** 0.6, **(c)** 0.7, **(d)** 0.8, **(e)** 0.9, and **(f)** 0.95.

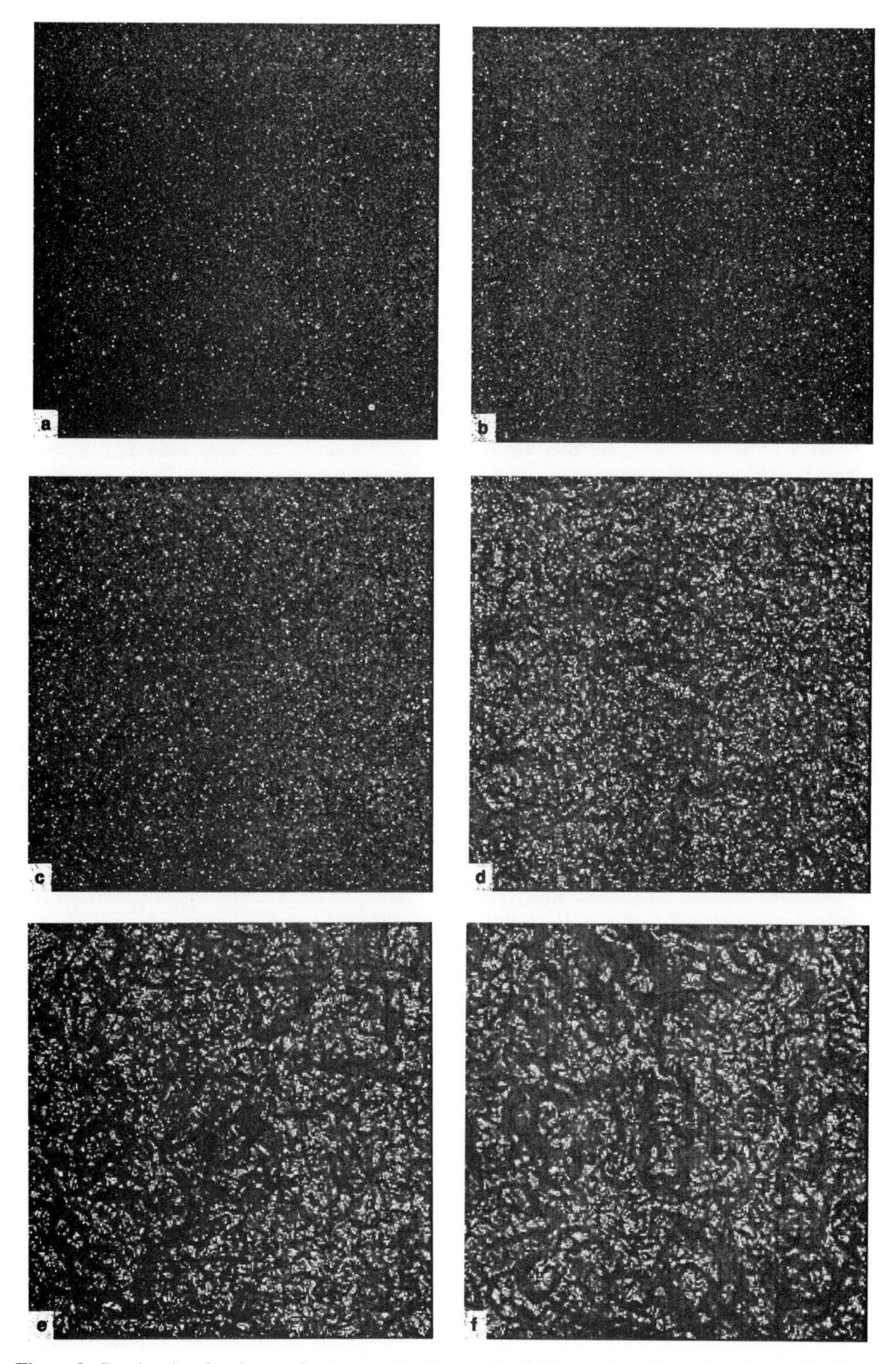

Figure 9. Rendered surface images for the case of a shiny surface (high specular reflection) and normal incidence. α = **(a)** 0.5, **(b)** 0.6, **(c)** 0.7, **(d)** 0.8, **(e)** 0.9, and **(f)** 0.95.

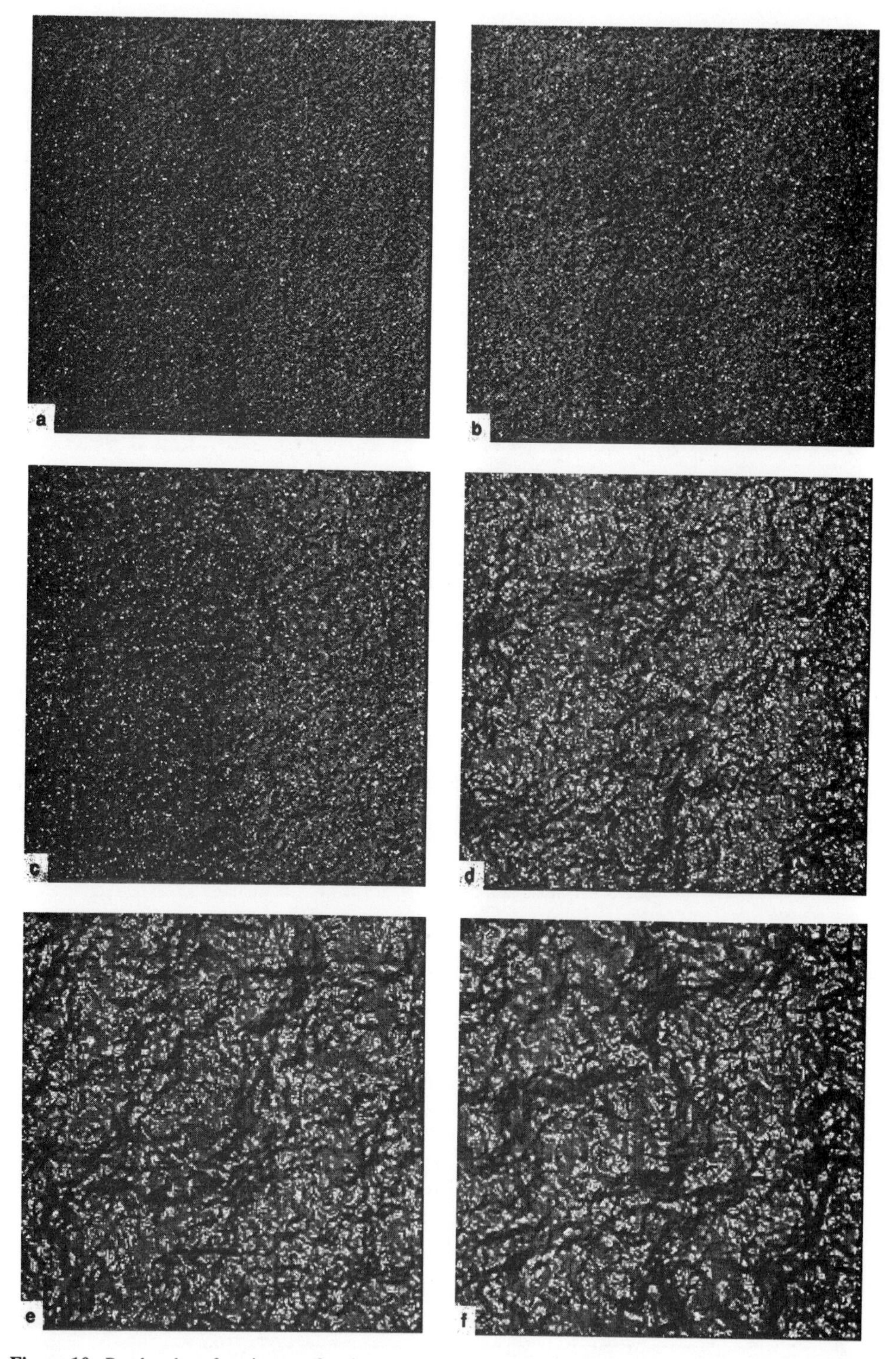

Figure 10. Rendered surface images for the case of a shiny surface (high specular reflection) and oblique incidence. α = **(a)** 0.5, **(b)** 0.6, **(c)** 0.7, **(d)** 0.8, **(e)** 0.9, and **(f)** 0.95.

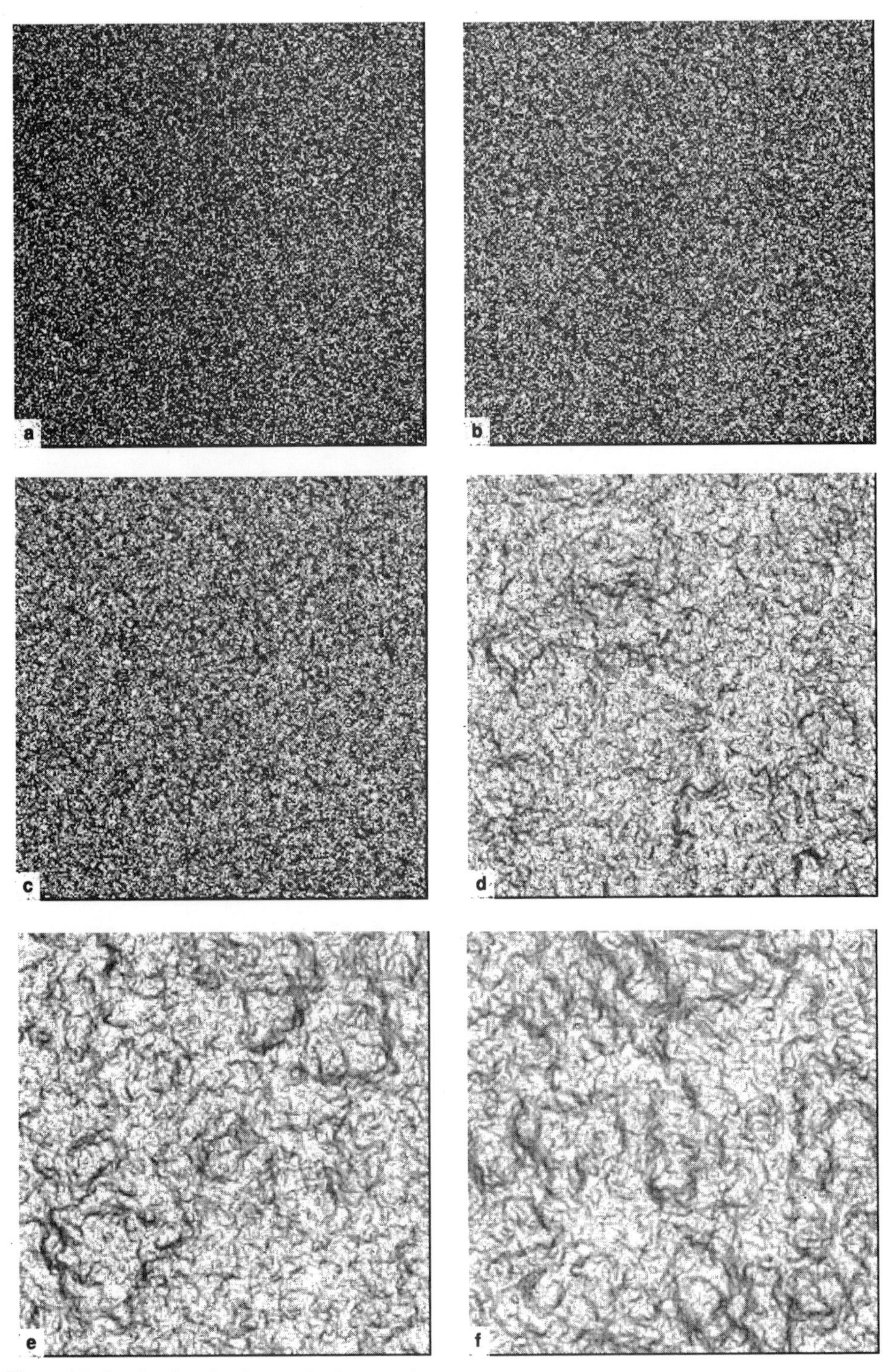

Figure 11. Rendered surface images for the case of a dull surface (low specular reflection) and normal incidence. α = **(a)** 0.5, **(b)** 0.6, **(c)** 0.7, **(d)** 0.8, **(e)** 0.9, and **(f)** 0.95.

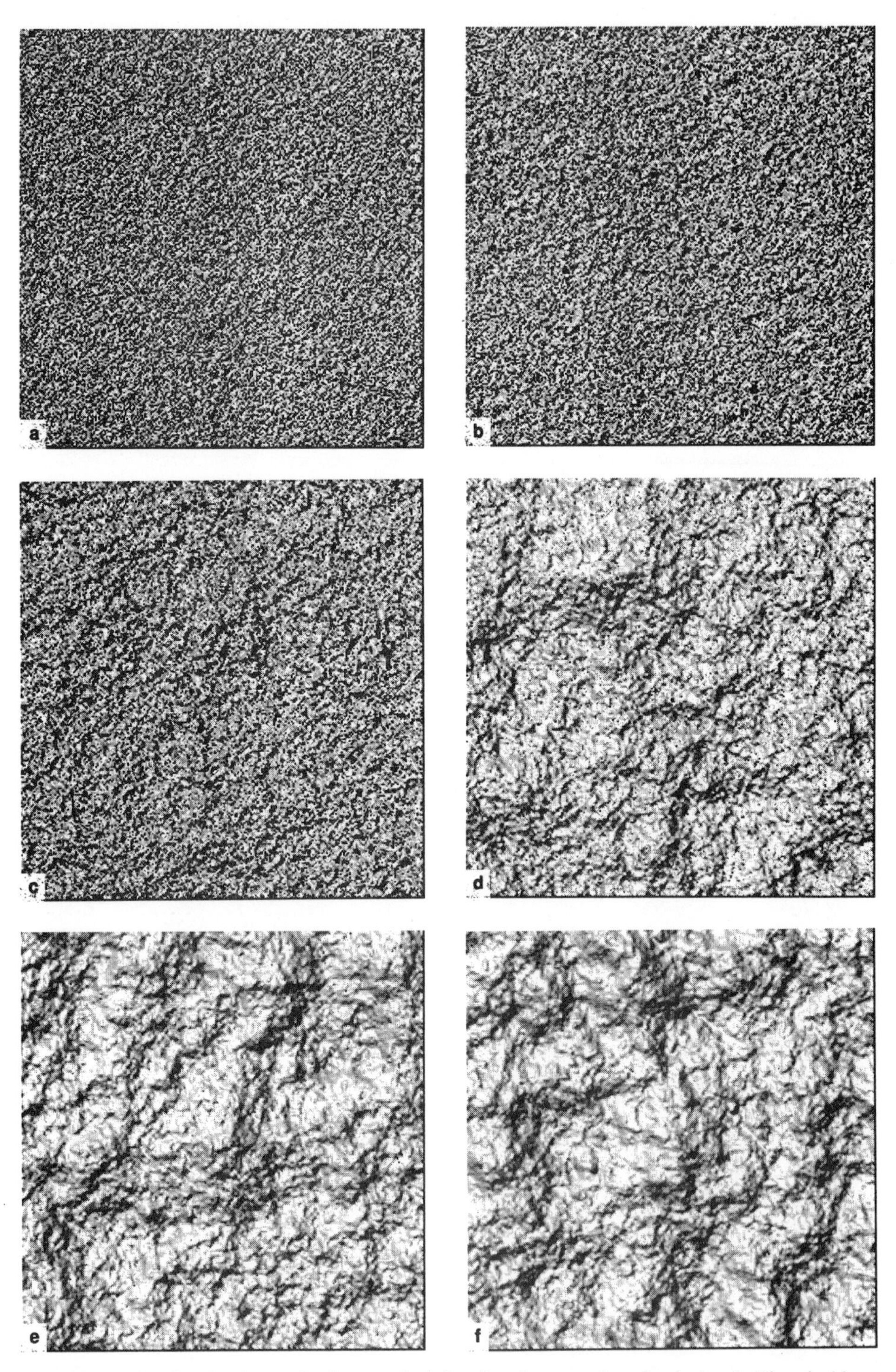

Figure 12. Rendered surface images for the case of a dull surface (low specular reflection) and oblique incidence. α = **(a)** 0.5, **(b)** 0.6, **(c)** 0.7, **(d)** 0.8, **(e)** 0.9, and **(f)** 0.95.

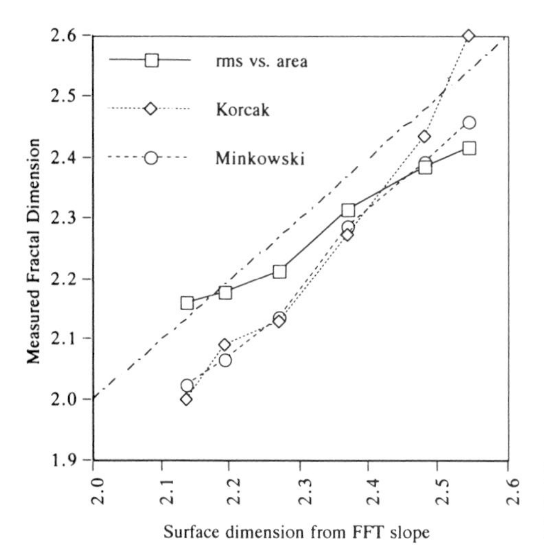

Figure 13. Comparison of fractal dimension values measured on elevation data from images in Figure 8 by various methods.

The rms vs. area, Korcak, and Minkowski methods do not detect any directional variation in the surface dimension that may be present. It is not clear whether they provide a meaningful "average" value for anisotropic surfaces. One effect of non-normal incident light is to produce an anisotropic brightness pattern as shadows are elongated in one direction, and this raises possible concerns about confusing the effects of an anisotropic surface with the effects on non-normal illumination. As a check on this, an anisotropic surface was also generated, as shown in Figure 14. In this case the slope of the log (Magnitude^2) vs. log (Frequency) curve produces an ellipse when plotted against direction. The rendered surface is shown in Figure 15.

The same measurement tools applied to the elevation data were also used for the rendered brightness images. Figures 16 through 18 show the rms vs. area, Korcak, and Minkowski plots, respectively, and Figure 19 shows rose plots of the slope of the log (Magnitude^2) vs. log (Frequency) plots from 2DFTs, for a representative surface. These can be compared to similar plots for the same physical surface, shown in Figures 20 through 23. Several differences can

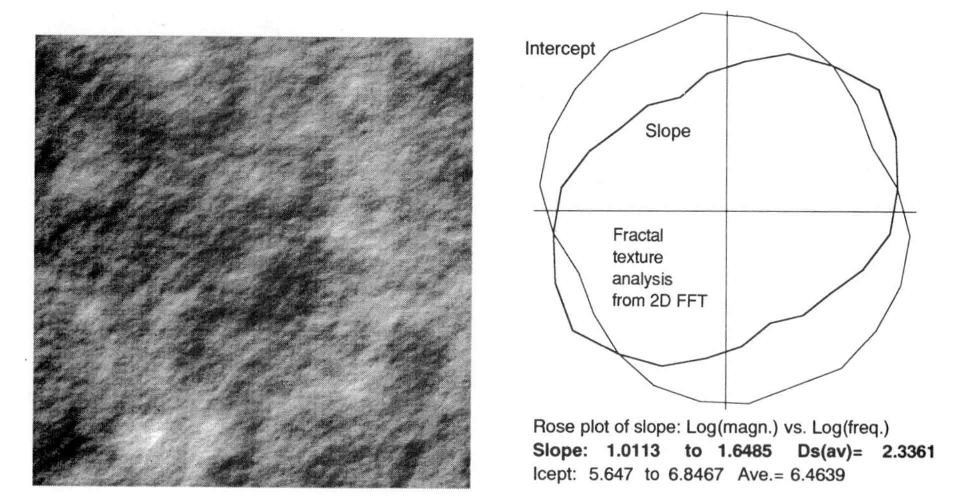

Figure 14. Elevation of generated anisotropic surface ($\alpha = 0.6 \times 0.95$) with rose plot of slope from 2DFT.

be noted, such as the greater spread of data points around the regression line in the rms vs. area plot, and the anisotropy of the 2DFT data for the rendered brightness images, as compared to the uniform circle shown for the isotropic elevation data in the original range image.

Figure 24 shows the range (maximum and minimum values) of the slope of the 2DFT data determined for each of the rendered brightness images, as a function of the mean fractal dimension. In spite of this anisotropy, there is still an essentially fractal character to the image, which confirms the earlier statements that scattered light from a fractal surface produces a fractal brightness pattern.

Relating the Surface and Brightness Dimensions

The question remains as to what (if any) quantitative relationship may exist between the dimension of the surface and that of the brightness patterns. Figure 25 shows plots of these

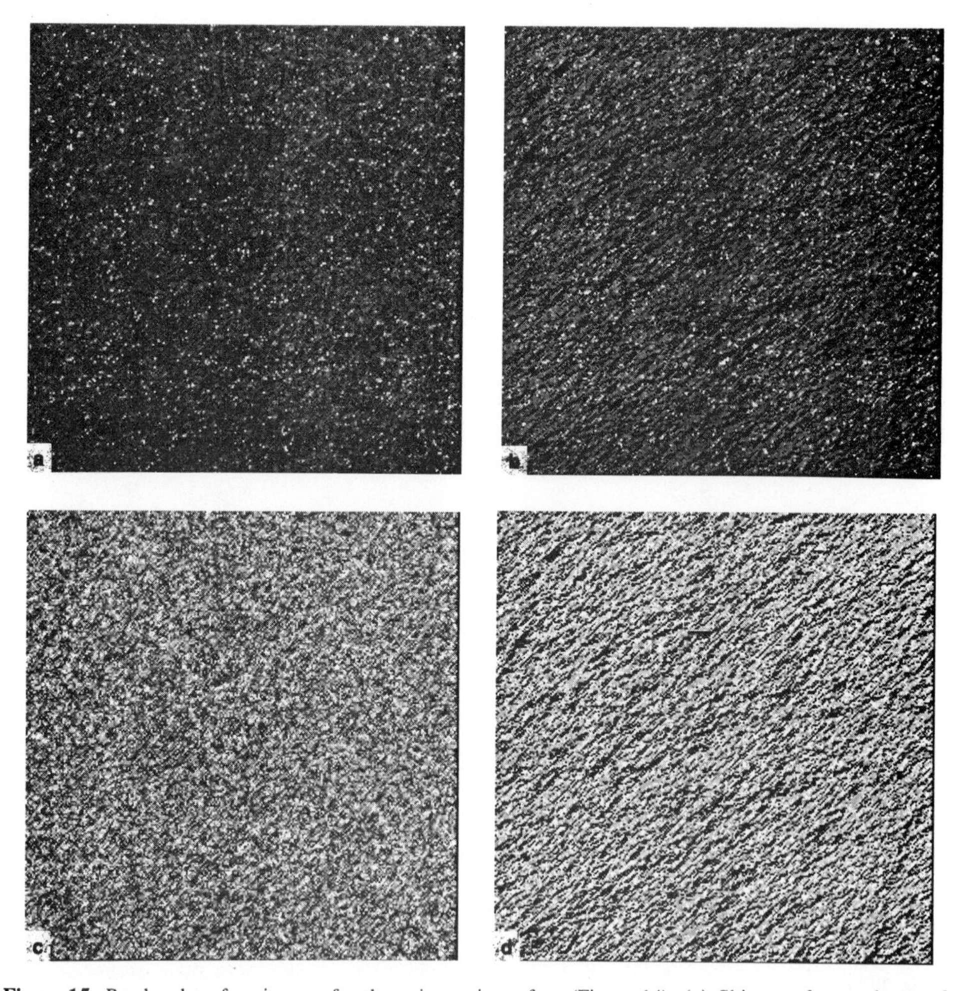

Figure 15. Rendered surface images for the anisotropic surface (Figure 14): **(a)** Shiny surface and normal illumination, **(b)** shiny surface and oblique illumination, **(c)** dull surface and normal illumination, and **(d)** dull surface and oblique illumination.

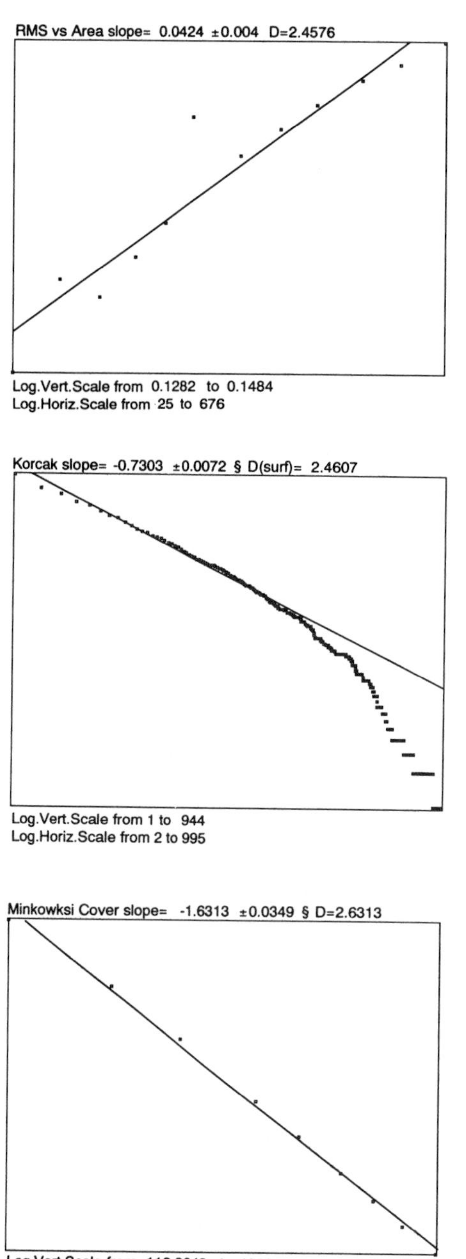

Figure 16. Plot of rms vs. area for brightness values for rendered image (dull surface, normal incidence, $\alpha = 0.9$).

Figure 17. Korcak plot (number of islands vs. area) for brightness values in same image as analyzed in Figure 16.

Figure 18. Minkowski plot of comforter thickness vs. dilation distance for brightness values in same image as analyzed in Figures 16 and 17.

results, comparing for each of the measurement techniques (2DFT, rms vs. area, Korcak, and Minkowski) the dimension from the rendered brightness images with that from the same method applied to the original elevation data. There are considerable differences evident amongst the various techniques.

For the Fourier plot based on the slope of log (Magnitude^2) vs. log (Frequency), the value used for the anisotropic brightness patterns was calculated by averaging the slopes for all

directions. This is not the same as using the slope based on the direction-averaged value of magnitude. The results shown in the figure correspond roughly to an offset of +0.5 in the measured dimension. That is, the dimension of the brightness pattern is approximately 0.5 greater than the dimension of the physical surface. This corresponds to an increase of +1 in the slope of the Fourier plot, which is in agreement with the expected relationship of slope (first derivative of elevation) to elevation.

However, this relationship is not rigorously observed. The shiny surfaces tend to lie above this offset and the dull ones below it, and the surfaces rendered with oblique illumination lie below the ones with normal illumination. In addition, the slopes of the lines are different, all meeting at a limiting dimension approaching 3.0 for the quite rough (high fractal dimension) surfaces.

These deviations appear to be systematic. However, it is not clear how a quantitative calibration of the brightness dimension or texture fractal against the physical surface dimension could be established except by measuring a few known surfaces of the same material using the

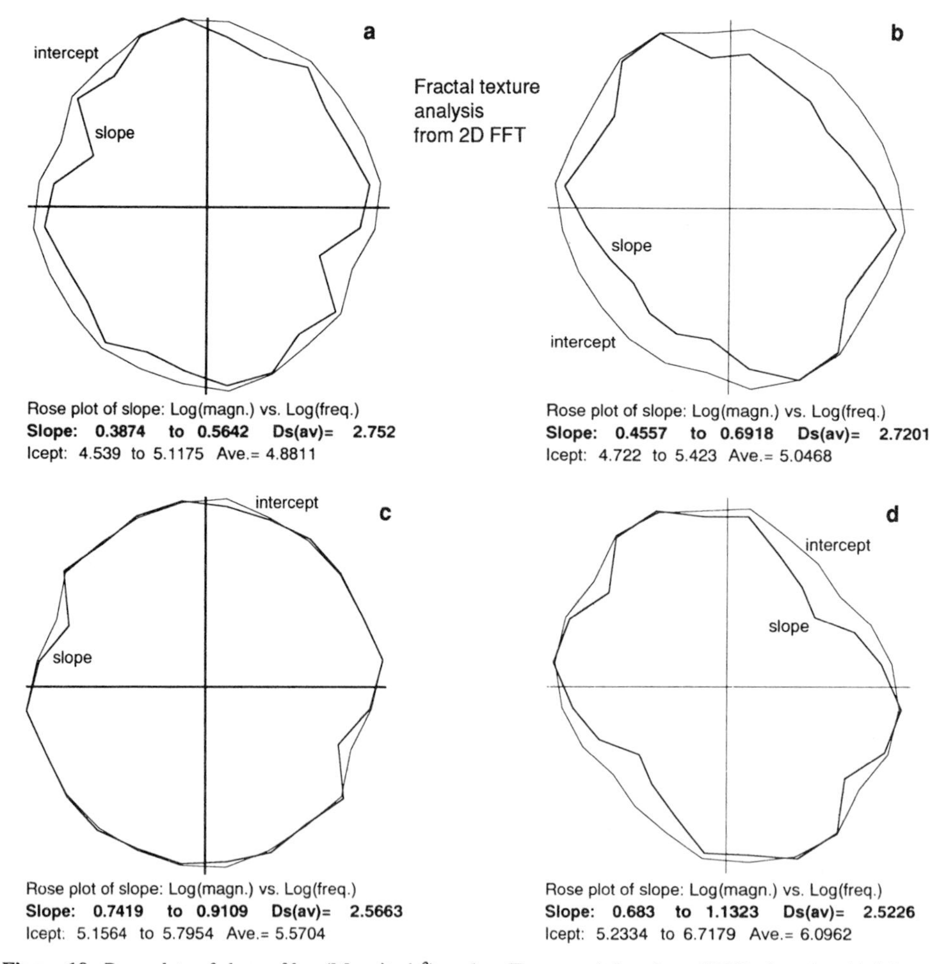

Figure 19. Rose plots of slope of log (Magnitude^2) vs. log (Frequency) data from 2DFT of rendered brightness images computed from the same generated surface ($\alpha = 0.9$) used in Figures 16 through 18; **(a)** shiny surface, normal illumination, **(b)** shiny surface, oblique illumination, **(c)** dull surface, normal illumination, and **(d)** dull surface, oblique illumination.

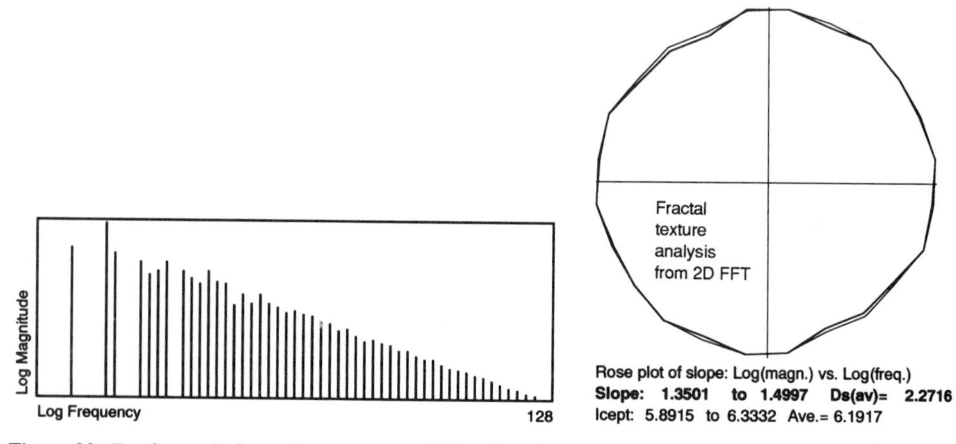

Figure 20. Fourier analysis results on generated fractal surface ($\alpha = 0.9$): plot of log (Magnitude^2) vs. log (Frequency) and plot of slope and intercept of log (Magnitude^2) vs. log (Frequency) as a function of direction.

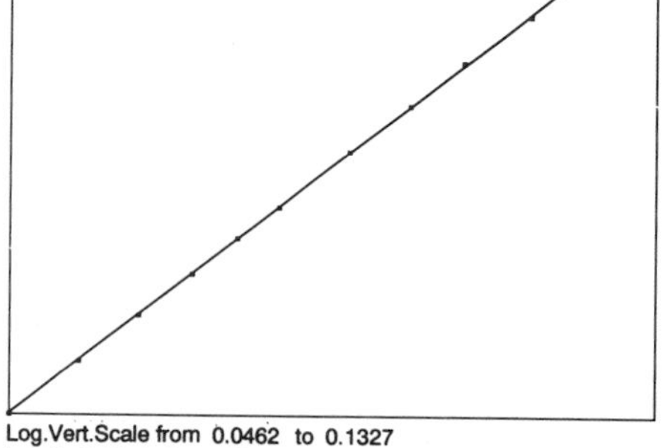

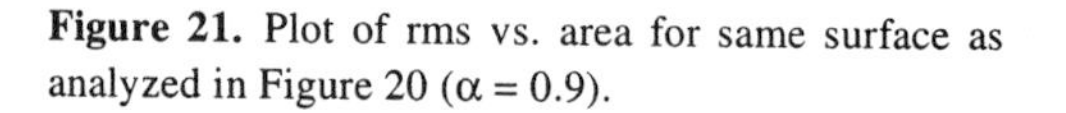

Figure 21. Plot of rms vs. area for same surface as analyzed in Figure 20 ($\alpha = 0.9$).

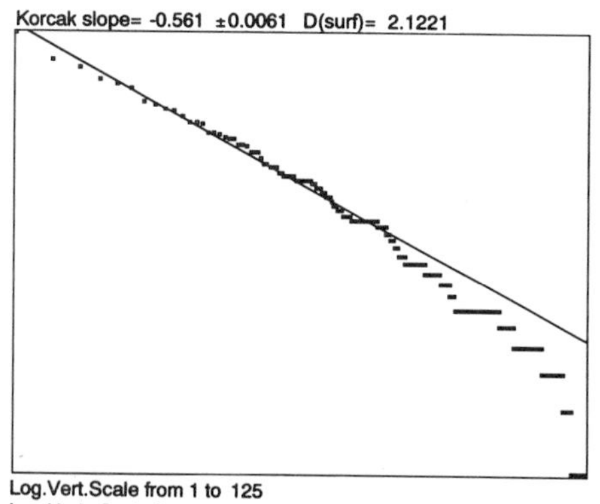

Figure 22. Korcak plot (number of islands vs. area) for same surface as analyzed in Figure 20 ($\alpha = 0.9$).

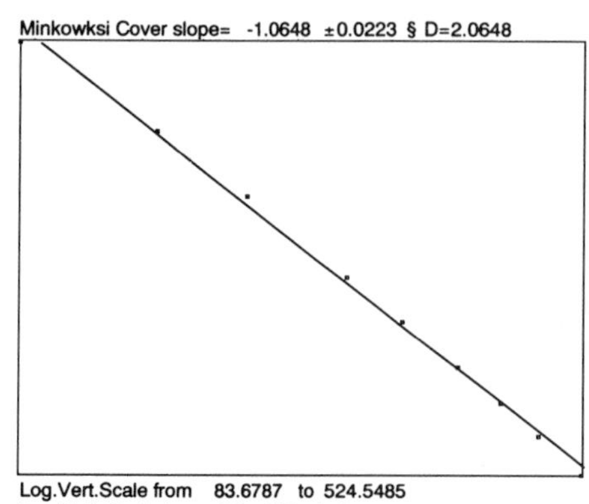

Figure 23. Minkowski plot of comforter thickness vs. dilation distance for same surface as analyzed in Figure 20 ($\alpha = 0.9$).

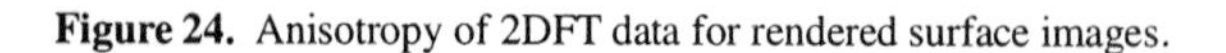

Figure 24. Anisotropy of 2DFT data for rendered surface images.

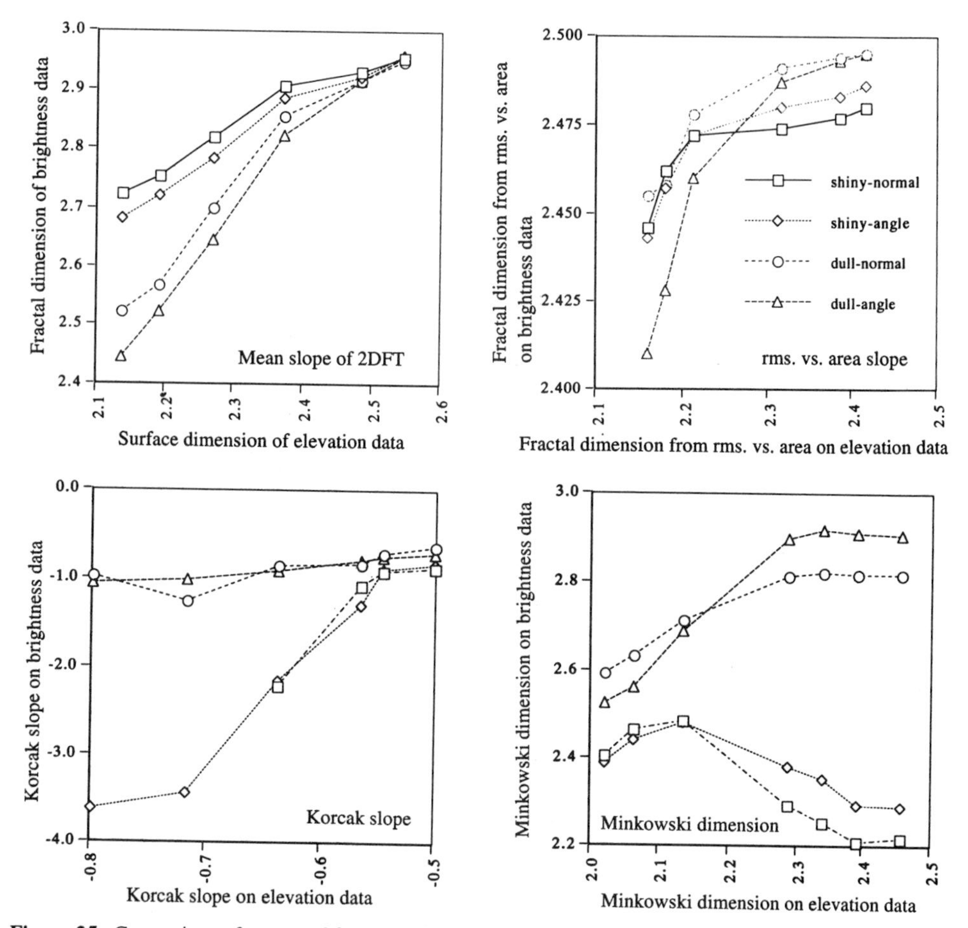

Figure 25. Comparison of measured fractal dimension values for the rendered brightness patterns with those of the original surface elevation data.

same illumination. With these restrictions, it still appears that good qualitative comparison of the roughness (fractal dimension) of similar surfaces can be made by comparing the brightness dimension (texture fractal) of light scattered from the surfaces.

The other measuring techniques exhibit additional problems. The rms vs. area plots give a maximum dimension of 2.5, which corresponds to a variance of pixel values that is constant with neighborhood size (slope = 0). The curves for the various surfaces approach this limit asymptotically and offer little discrimination between the surfaces with different physical (elevation) dimensions except for the relatively low dimension cases. The Korcak data do not even produce a meaningful fractal dimension because the slope of the lines for the number of features vs. area plot exceeds the range corresponding to a physical surface (maximum slope = 1). This results from the large number of very small (one or a few pixel) spots that arise from specular reflections from the surface. The data are therefore plotted as slope rather than fractal dimension. They show that the shiny surfaces do offer some discrimination, because the number of small spots is highly dependent on roughness, but the dull surfaces do not.

The Minkowski data are even stranger in appearance. The brightness dimension increases with physical dimension for the dull surfaces but decreases with physical dimension for the

shiny ones. Again, this is related to the specular reflection spots. In terms of brightness, these values "stick up" from their surroundings and control the thickness of the Minkowski comforter. The few specular reflections control the measurement. These specular reflections disrupt the otherwise well-behaved monotonic variation of reflectivity with surface orientation, and thus alter the relationship between the physical elevation dimension and the brightness dimension.

For the anisotropic surface, the data are tabulated in Table 1. The measurement data for the physical (elevation) data show good agreement between the various measurement techniques, and the expected variation in maximum and minimum slope for the 2DFT plot with direction. For the brightness patterns from the rendered images, the mean 2DFT values show the same +0.5 offset as noted above for the isotropic surfaces. The ratio of maximum to minimum slope is not interpretable, nor is the direction of orientation of the ellipse axes. There does not appear to be any reasonable way to separate the two effects of surface anisotropy and oblique lighting.

Measures of the brightness fractal dimension for the anisotropic surface show the same defects as noted for the isotropic examples. The rms vs. area plots give dimensions close to the limiting value of 2.5, while the Korcak slope covers a wide range with values greater in magnitude than 1.0, which cannot correspond to a fractal dimension. The Minkowski results vary widely and are controlled by the specular reflections, which mask the more interesting details of surface roughness.

Table 1. Results for Measurement of Anisotropic Surface

Elevation data		
Mean fractal dimension from 2DFT	2.336	
Maximum slope	1.649	
Minimum slope	1.011	
Dimension from rms vs. area	2.336	
Korcak dimension	2.329	
Minkowski dimension	2.339	
Shiny surface		
	Normal	*Angled*
Mean dimension from 2DFT	2.895	2.880
Maximum slope	0.380	0.434
Minimum slope	0.136	0.073
Dimension from rms vs. area	2.469	2.479
Korcak slope	−2.223	−2.085
Minkowski dimension	2.249	2.349
Dull surface		
	Normal	*Angled*
Mean dimension from 2DFT	2.867	2.848
Maximum slope	0.382	0.453
Minimum slope	0.151	0.195
Dimension from rms vs. area	2.492	2.490
Korcak slope	−1.219	−0.961
Minkowski dimension	2.819	2.915

To interpret the implications of these results on simulated surface images to real ones, the four images shown in Figure 26 can be compared to the images shown before. These are reflected light images of stone tools prepared in an archaeological study of tool wear (Bueller 1992), in which the wear surface on tools whose function and use were known was compared to wear on tools from archaeological sites. In this case, the four surfaces are those of an unused flint tool prepared by flaking, and similar tools after use for scraping bone, hide, and oak wood. In all cases, the appearance of the real surface images is strikingly similar to the synthesized images.

Table 2 summarizes the results of measurement of fractal dimensions of the brightness patterns from these images. Although all of the various measurement techniques produce numeric values for the dimension, the results above suggest that only the 2DFT data may reasonably be interpreted in terms of the physical fractal dimension of the surface. The rose plots of slope vs. direction, shown in Figure 27, indicate some anisotropy resulting either from the illumination, directionality of surface wear, or both.

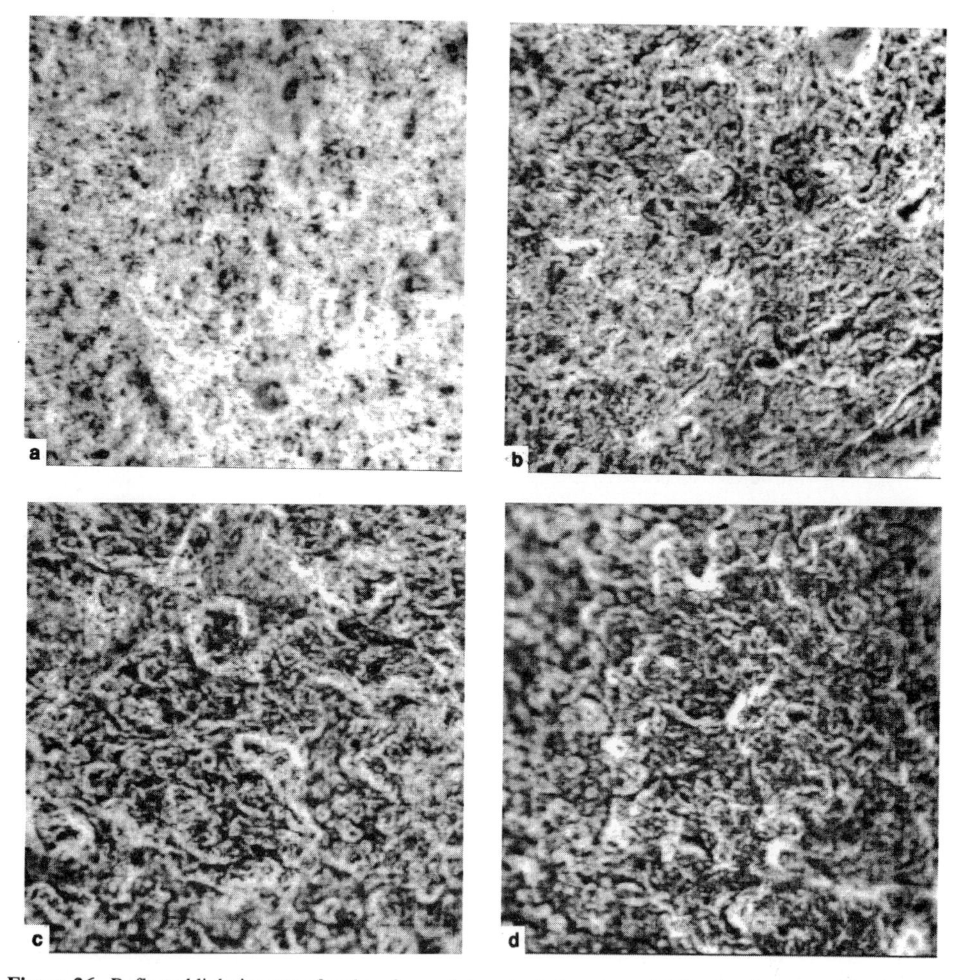

Figure 26. Reflected light images of real surfaces of stone tools (courtesy of J. Bueller, Institute of Archeology, The Hebrew University of Jerusalem); **(a)** unworn flint; **(b)** scraping bone; **(c)** scraping hide; **(d)** scraping oak wood.

Table 2. Measured Fractal Dimension Data for Reflected-Light Images of Flint Tools

Designation	2DFT (mean)	Minimum FT slope	Maximum FT Slope	rms vs. area	Korcak	Minkowski
Saw bone	2.382	1.100	1.308	2.412	2.331	2.489
Scrape hide	2.348	1.106	1.446	2.403	2.226	2.464
Scrape oak wood	2.295	1.227	1.582	2.385	2.247	2.419
Unworn flint	2.442	1.079	1.449	2.386	2.197	2.363

The numerical magnitudes of the slopes are too small to correspond to the +0.5 offset observed for the simulated surfaces, but it was shown above that somewhat larger or smaller offset values were possible depending on the surface albedo, specularity, and illumination orientation. Since these parameters are the same for all of the tool images, it is reasonable to interpret the results as indicating that the unworn tool has the "roughest" (highest fractal

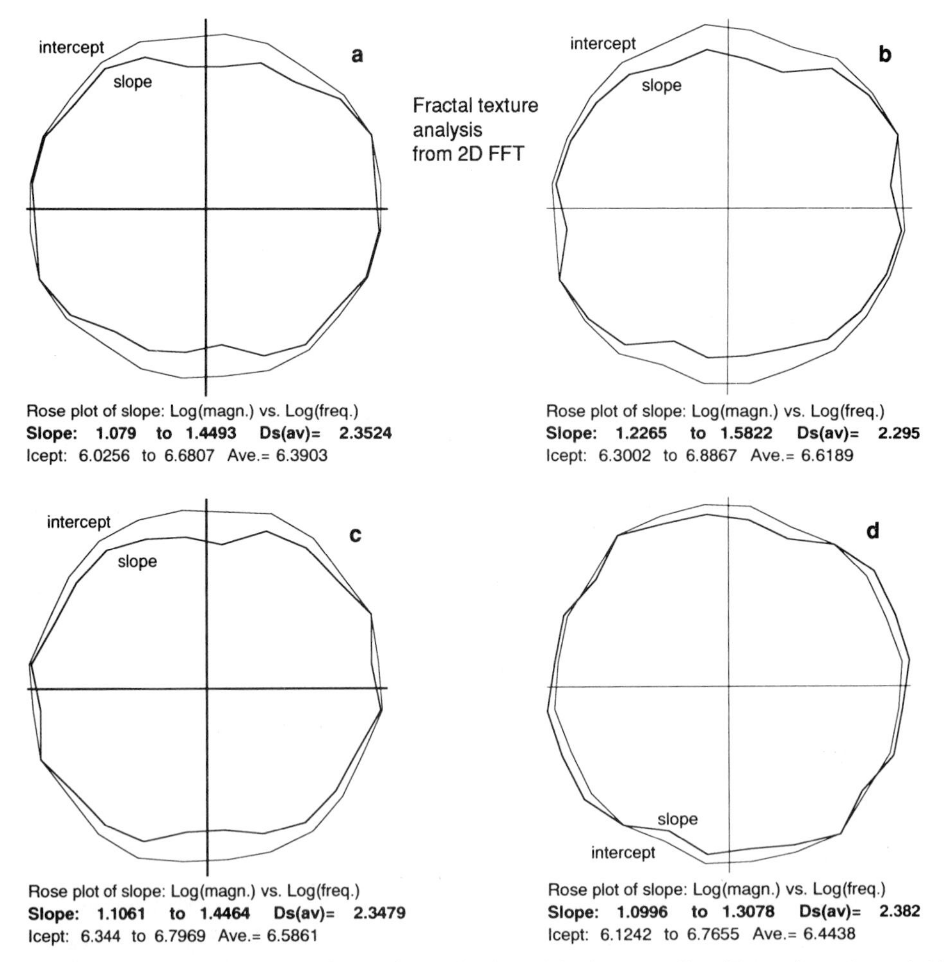

Figure 27. Rose plots of 2DFT slope for tool images in Figure 26: **(a)** unworn flint, **(b)** scraping oak wood, **(c)** scraping hide, and **(d)** scraping bone.

dimension) surface, the tool used for scraping oak wood has the "smoothest" (lowest fractal dimension) surface, and the other two tools are intermediate.

A word of caution is needed here: The fractal dimension of the surface describes the self-similar or self-affine reduction in the magnitude of detail with dimension, but is only one aspect of the surface characteristic that is loosely associated with the term "roughness." It may also be necessary to consider the other aspect, which is the magnitude of the vertical displacement of points on the surface. Two surfaces with identical fractal dimension but much different magnitude or topothesy values would not be distinguished by these measurement techniques, and it is not clear how the characterization of surfaces due to their history, properties, or appearance depends on each of these two independent parameters.

It is not proposed here that the fractal dimension offers a way to identify the use to which various tools may have been put, or that there is any unique relationship between surface roughness and tool use. However, it does seem reasonable to expect the amount of tool wear to be revealed by a progressive change in fractal dimension and surface roughness. The comparative use of brightness fractal dimension for this purpose is rapid and precise, and can be extended to other types of surfaces as well.

Light Scattering

In the preceding section, an image of light reflected from a surface was interpreted in terms of a fractal dimension. It is also possible to integrate the information from an entire surface or specimen by examining the light scattered from it. This technique is most fully developed and commonly applied to agglomerates, but can also be used with surfaces.

The wavelength spectrum of light scattered from fractal particle agglomerates has been shown to exhibit a power-law behavior which is formally fractal (Wilcoxon, Martin et al. 1986). In fact, the fractal character of surfaces is also revealed quantitatively and can be measured by photon scattering. Chapter 4 showed the relationship between fractal profiles in physical space and their frequency space representation.

This applies to surfaces as well. A 2D Fourier transform (2DFT) of an image in which the pixel values represent elevation of points on a fractal surface, or one in which scattered light from a rough surface has a fractal pattern, will exhibit a linear relationship of log (Magnitude^2) vs. log (Frequency) as discussed in Chapter 4. In Chapter 7 the directional variation of these 2DFT patterns will be used to describe anisotropic fractals, but that additional complexity will be ignored for the present. Assuming that the surface (or whatever the pixel values represent) is isotropic, the simple straight-line relationship produced by the Fourier transform can be used to evaluate the fractal dimension.

The same use of scattered signals to determine surface texture can be used at very small or large scales, provided that correspondingly shorter or longer wavelengths of radiation are employed. Side-scan sonar of the sea bed (Linnett, Clarke et al. 1991) shows textures that can be modeled as fractals, and synthetic aperture radar shows similar textures from both sea and land surfaces. At the other extreme, X-ray scattering and Raman scattering (Beckmann and Spizzichino 1963; Church 1980; Schmidt 1982; Bale and Schmidt 1984; Hogrefe and Kunz 1987; Schmidt 1989; Schmidt, Avnir et al. 1989; Sun and Jaggard 1989; Duval, Boukenter et al. 1990) have been applied to measure the fractal dimension of porous materials, agglomerates, or surfaces. Observation of a $1/f$ power spectrum (linear on a log–log scale) is interpreted to imply a fractal structure, and to measure its dimension.

Many surfaces of interest are quite smooth, and are intended to function as mirrors. Then it becomes possible to directly measure the power spectrum without having to calculate the FT. Physically, this applies if the magnitude of the surface irregularities is much less than the

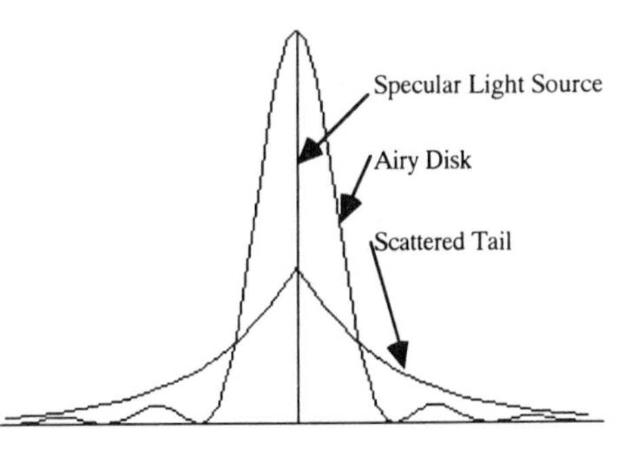

Figure 28. Airy disk + scattered tail produced by reflection of a specular light source.

wavelength of the light or other electromagnetic radiation being reflected. Then the intensity profile of the reflected spot from a specular source (usually a laser) contains the power spectrum information. As indicated in Figure 28, the original point source is spread out by three principal effects. First, the finite solid angle or aperture of the mirror causes a broadening of the main peak into the Airy disk, which also has some interference rings around it. Then any imperfections in the overall shape of the mirror, called its "figure," will further reduce the intensity. Finally, any surface texture will cause a tail of intensity to extend beyond the central peak (Jullien and Botet 1989).

This tail is simply the power spectrum, and in principle would provide the necessary information to determine the fractal dimension of the surface. Unfortunately, it is not very easy to measure this with any precision. The exact shape of the central peak is unknown because the defects in the overall shape or figure of the mirror are uncertain. This peak is much larger than the scattered tail due to surface texture, and tends to cover it up. Very little of the tail extends beyond the central peak, the intensity is low, and it is difficult to measure it accurately.

Measurement of the fraction of the reflected intensity which falls within a certain angular range is often used as a descriptor of the quality of a mirror and its finish, but this typically depends both on the figure and the fine-scale roughness. For instance, reports on the mirror for the Hubble space telescope state that only 14% of the light falls within 0.15 arc seconds, instead of the intended 85%. However, this is primarily due to imperfection in the shape or figure of the mirror. Even if the total amount of reflected light in the tail could be determined, that would not be enough information to determine the surface fractal dimension. The shape of the tail corresponds to the power spectrum and hence to the fractal dimension of the surface, and measuring the shape of the tail accurately is not easy to accomplish.

Scattering of Diffuse Light from Rough Surfaces

In addition to brightness patterns that result from the diffuse scattering of light from rough surfaces, there are other occurrences of brightness patterns that can be characterized by fractal dimensions. These dimensions in turn describe in a compact way the structures which give rise to the brightness pattern in the first place.

Measurement of a fractal dimension using what is effectively a Minkowski or covering dimension applied to arbitrary scenes and images has been used as discussed above to characterize the textures present, for classification and identification. These approaches ignore any physical meaning to the brightness patterns and do not attempt to answer the question of why the brightness values might be fractal, but nevertheless are successful in a sufficient number of cases to be useful.

In some cases, plausible arguments to justify the presence of a fractal brightness dimension can be advanced. One such case arises in radiography of bones to monitor the progress of osteoporosis, the progressive thinning of bones that occurs in women, for instance, after a hysterectomy. The variation of image brightness recorded in the X-ray film is a function of density in the interior of the bone (trabecular bone). This in turn is a measure of the clustering or clumping of the cells and protein present. This clustering has many of the same characteristics as the kind of particle clustering discussed in earlier sections. If the clustering is fractal, the image brightness will also be fractal (Jacquet, Ohley et al. 1990). This has also been shown (Lynch, Hawkes et al. 1991) to have diagnostic consequences for estimating the degree of osteoarthritis.

Figure 29 shows two point-projection X-ray images of leg bones from experimental animals, one a control and the other five weeks after removal of the ovaries. Of course, the dimension of these patterns does not correspond to either a physical clustering fractal or to a surface dimension, and is therefore not constrained to less than 3.0 (or to less than 2.0 for the projection). The actual numerical value has been shown not to depend on the film contrast or exposure, just as the dimension of brightness patterns of light scattered from surfaces is independent of the lighting details, but can be used comparatively between samples, and correlated to the history of disease and treatment.

Other medical imaging applications (Lundahl, Ohley et al. 1985; Lundahl, Ohley et al. 1986; Dellepiane, Serpico et al. 1987; Chen, Daponte et al. 1989; Kuklinksi, Chandra et al. 1989; Caldwell, Stapleton et al. 1990; Rigaut 1991; Fortin, Kumaresan et al. 1992; Xiao, Chu et al. 1992) of the fractal measurement of brightness patterns from radiography have been used for identification and image segmentation. It appears that fractal geometry appears with some frequency in natural situations, and consequently its use as a characterization tool has diagnostic utility. This even extends to treating the slight irregularities in heartbeat as a fractal (West and Goldberger 1987; Basingthwaighte and van Beek 1988; Goldberger 1990; West and Shlesinger 1990), and attempting to relate that through chaos theory to the health of the patient. It is considered that a perfectly regular (nonchaotic, nonfractal) heartbeat is not as robust as one with fractal variations. However, such temporal records are not our concern here.

Light scattering from particles is often associated with fractal patterns (Wilcoxon, Martin et al. 1986). The scattered intensity from particles which are small compared to the wavelength

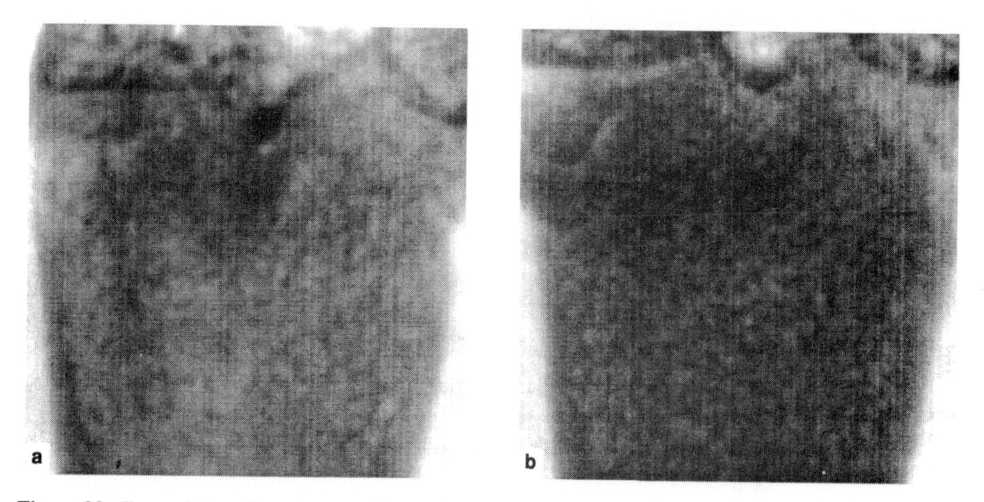

Figure 29. Transmission X-ray images of bones from leg bones: **(a)** osteoporosis five weeks after ovarectomy and **(b)** control.

of light (less than one-tenth the wavelength) is isotropic and increases with decreasing wavelength. The total scattered power is proportional to the sixth power of particle size and inversely proportional to the fourth power of wavelength. This is why light scattered from particles in the atmosphere is blue, since the small particles there scatter blue light about five times more strongly than red light.

For particles that are large compared to the wavelength, phase differences produce interference so that the light is not isotropically scattered. The total power scattered in all directions is nearly independent of wavelength, and the spatial distribution is a complex function of the particle size and shape. If the particle size distribution also follows a power law of number vs. size, the result is that the wavelength distribution becomes a power law that includes the size distribution. The larger the particle, the more light intensity is scattered in the forward direction. Consequently the distribution of scattered light includes information on the particle size distribution. For a cluster of particles, all of the clusters, subclusters, and individual particles act as scattering sites. Measurement of fractal cluster geometry by scattering of light or other electromagnetic radiation of appropriate wavelength is beyond the intended scope of this book, dealing with fractal surfaces.

Also beyond the intended scope, but interesting in that it reveals much about our human assumptions about natural scenes and surfaces, is another use of the brightness fractal dimension. Bhatt, Munshi et al. (1991) report using the fractal dimension of grey scale images reconstructed by computed tomography (CT) as a quality measure for the reconstruction. The higher the dimension, or the greater the "roughness" in the image, the more noise is present in the reconstruction.

Range Measurement Methods

Of course, the most widely used methods for measuring surfaces directly determine the elevation of points. This may be done by physical contact, for instance using a contact profilometer to scan the elevation profile along a linear traverse, or using noncontact methods, for instance using an interferometer to determine the phase shift of reflected light and from that calculate the elevation. Church (1982) has compared these two methods for precision machined surfaces and showed that they agree, producing very similar power spectra and other statistics. The measurement may be done at very small scales, for instance using the scanning tunneling microscope, or at very large ones, for instance using stereoscopy with aerial photographs. It is worthwhile to briefly review some of these methods, recognizing that this list will be incomplete and that any technique that provides a series of elevation measurements can be used. The order of entries in the list is arbitrary.

Contact profilometry is typically used to scan along a single linear traverse across a surface. The shape of the tip controls the lateral resolution and limits the steepest slope that can be followed. Typical systems have lateral resolution of tens or hundreds of micrometers, and vertical resolution of about 1 nm. Many authors have used such elevation profiles for fractal, as well as conventional, analysis of surface roughness (Church, Jenkinson et al. 1979; Church 1982; Banerji and Underwood 1984; Kaye 1984; Church, Vorburger et al. 1985; Takacs and Church 1986; Chermant, Chermant et al. 1987; Brown and Savary 1988; Stupak and Donovan 1990; Scott 1991; Sasajima and Tsukada 1992).

The atomic force microscope (AFM) is in some respects an extension of the profilometer to smaller dimensions. The probe tip is smaller and can therefore record smaller details, in the range of nm. Unlike the profilometer, in which the tip is held in contact with the surface by mechanical forces, the AFM relies on the interaction of the electron clouds around atoms in the surface and the tip. Of course, at the scale of measurement of the AFM many surfaces are

contaminated by oxidation, dirt, etc. This means that special attention to the environment of the surface may be required, even extending to maintaining the surface in high vacuum from its creation. The AFM is most often used to scan a complete raster over a surface to produce a range image instead of a single elevation profile.

The scanning tunneling microscope (STM) uses a similar mechanism to the AFM but moves the scanning tip to maintain a constant electron tunneling current between the tip and specimen. Of course, this requires that the specimen be electrically conducting. For semiconductors, the "surface" which is determined may not be the physical surface of the specimen but instead may depend on electronic states in the material (Chapman 1992). Surface contamination is an even greater concern for the STM than for the AFM.

The scanning electron microscope (SEM) also relies on traversing the specimen with a raster pattern. However, the various signals which are generated when the electron beam strikes the specimen do not directly provide elevation data. (As noted above, some of them produce brightness patterns which may reveal fractal surface behavior.) SEM images can be used to obtain elevation information using stereoscopy. Two images recorded from different viewpoints, either by tilting the specimen or shifting it laterally, can be used. Locating matching points in the two images and measuring their relative displacement or parallax provides the elevation of the surface at that location. This approach usually produces a few discrete elevation values, or points along a linear traverse. Matching enough points to produce a range image for the entire surface is difficult, and for the images of many real surfaces cannot be done because of the lack of unique detail.

Stereoscopy can be applied at other scales as well. Matching of aerial photographs taken a known distance apart (based on the speed of the airplane) is used to produce most of the topographic maps of the earth, particularly for the less accessible regions. The matching is typically performed using large, parallel computers which apply cross-correlation methods. The range image which is produced can have lateral resolution as great as that of the images (on the order of tens of meters), and elevation resolution a few times poorer. The "noise" in these images due to errors in the matching process is often removed by application of a median filter. This clears up isolated high or low points, but also affects the fractal dimension calculated from the data.

Photometric stereo is not stereoscopy, in spite of the name. Instead of two points of view which are fused just as human vision uses two eyes, in photometric stereo a single viewpoint is used, and two (or more) images are recorded with different light source locations. Based on equations (generally called "shape from shading" models) for the scattering of light from locally smooth surfaces, the local orientation of points in the image is calculated. These are then collected together to map the entire surface and convert it to elevation. This method is primarily applicable to structures with very large relief, and does not measure small scale irregularities or texture well.

Structured light is another method that controls the light falling onto the object. A series of parallel apertures, similar to the common venetian blinds in house windows, produces an array of stripes which fall onto the surface of interest at an angle which may vary from nearly horizontal to about 45 degrees. Usually, these stripes (sometimes called "fringes") are then viewed from above, perpendicular to the average surface orientation. If the surface is flat, the result is simply a set of straight, parallel lines. If there is relief in the surface, the lines are displaced. Measurement of the displacement gives the change in elevation, provided the angle of the light source is known. This method is widely used in tool making to measure the dimensions of parts, as well as in medical applications such as determining the curvature of the eye before corrective surgery, or the configuration of the spine in a subject's back. The measurement accuracy depends on the angle of the light, of course, as well as the precision

with which the location of the stripes can be measured in the viewed or digitized image. This is rarely good enough to characterize small-scale deviations associated with surface roughness.

Interferometry uses the reflection of light from the specimen as one leg of an interferometer to combine with light in a reference beam. Small changes in phase due to variations in the distance the light traveled produce interference which reduces the light intensity, producing a series of fringes across the surface. Measuring the change in light intensity can determine elevation differences to a fraction of the wavelength of the light used, for an overall measurement precision of better than 1 nm. The method is comparatively fast since an entire surface can be measured at once, with a lateral precision corresponding to the resolution of a conventional light microscope, or better than 1 μm. Changes in surface reflectivity due to changes in composition or local contamination present problems, but the greatest difficulty arises from very steep slopes on the surface which do not reflect light to the detector. For the measurement of fairly flat and smooth surfaces, this is a very rapid method which is capable of revealing the fine-scale texture.

The conventional light microscope has a comparatively large depth of field; depending on the objective lens aperture this is typically on the order of 1 μm. Because the light intensity which is reflected from the surface and focused through the optics falls off very gradually with defocus, this has not proved useful for precise depth measurements. However, the comparatively new development of the confocal scanning light microscope (CSLM) overcomes some of these problems. The depth of field is reduced to about 0.2–0.4 μm but the out-of-focus light is strongly rejected and so it becomes practical to perform measurements on surfaces.

The instrument is generally used to collect light from the entire image plane, either by scanning the specimen or the optical path. This is repeated as the specimen is raised or lowered relative to the optics. At each location in the image plane, the brightest light is recorded when the corresponding point on the sample reaches the plane of focus. Keeping track of the intensity at every point to determine the brightest value, and storing the corresponding elevation value in an image, are straightforward tasks for the computer. For surfaces with somewhat greater relief than can be conveniently studied with interferometry or AFM, the confocal microscope offers an efficient tool. Metrology of surfaces with dimensions of the order of 1 μm, such as modern microelectronics, makes extensive use of CSLM.

The acoustic microscope is also a confocal instrument, in that the source and detector for the high frequency (gigahertz) sound waves are located at the same point, and the optics focus the sound onto a corresponding point on the sample. It is also an interferometric tool in that the destructive interference of the sound waves reflected from surfaces on or in the specimen is measured to determine the distance to a fraction of a wavelength. Since the wavelength of high-frequency sound is similar to that of light, the lateral and depth resolution of the technique is potentially similar to the interference light microscope. Practical difficulties such as electronic noise limitations in the transducers generally make it poorer, but the greatest problem arises from the multiple interferences that result from surfaces or specimens which contain internal flaws, rough surfaces, variations in density or composition, etc. While it has been used for some metrology purposes, this is a comparatively fussy technology with few commercially available instruments.

On the other hand, sonar is a mature and widely used technique. Lower-frequency sound waves have a correspondingly longer wavelength which penetrates through water with little attenuation. Given the velocity of sound (which depends on temperature, salinity, etc.), the echo time for a pulse can be converted to a range value. This makes it possible to scan the sea floor to construct elevation maps with a vertical precision considerably less than 1 meter, and a lateral resolution of a few meters. There has been some use of such maps to study the texture of the sea floor. Sideband sonar, which uses interference techniques to obtain higher precision data, has also been employed for this purpose (Linnett, Clarke et al. 1991).

Radar is also a range-measuring tool. Conventional radar uses the delay time of the electromagnetic radiation pulse, just like conventional sonar. Synthetic aperture radar (SAR) is different. It collects the echoes from the surface as the radar dish moves (in an airplane or satellite) along a known path, and measures the phase shift and reflectivity as a function of position to construct a surface elevation map. The current mapping project of Magellan at Venus is a good example of the current state of the art, which can produce lateral and vertical resolution of tens of meters. Use of the data for the measurement of fractal dimensions has not yet been reported, but they appear to be quite suitable for it.

Range Images and Surface Parameters

All of these measurement methods produce elevation or range maps, typically in the form of a pixel array whose values represent surface elevation. In the ideal case, the pixels are square, indicating that the measurements have the same horizontal resolution in both x and y directions. This is typically true for interferometry, confocal microscopy, AFM, and other similar raster-scan methods. It may not be true for techniques like structured light, stereoscopy, SAR, etc. And for most of these methods, the vertical resolution (in the z direction) is usually much better than that in the horizontal direction(s).

Analysis of such a set of data, either for an entire two-dimensional pixel array from a range image, or along selected rows or columns of such an array (which form an elevation profile), or for a data set which only covers one traverse line to begin with (such as a profilometer trace), is intended to reveal important information about the surface. In most cases, a surface that has been created by machining, deposition, fracturing, or some other process will only be approximately described by a simple Euclidean shape. In many cases, it is a useful simplification to express the elevation values as differences from a plane, cylinder, etc. This is most commonly done by using least-squares fitting to the elevation data, which are initially based on some arbitrary "zero" plane. The fit is often applied to make the average elevation (or the average slope, or the average curvature) zero. If the surface has been produced by machining, grinding, or polishing, it is possible to fit the elevation data to the intended surface shape or "figure," which will leave just the deviations and texture information.

Once the best plane or other surface has been fit to the data, the remaining values represent the local elevation of each point from that figure. Now we wish to summarize the information in a compact way that will describe the surface. The traditional parameter which is widely used for this purpose is the standard deviation sigma (σ), of the elevation values. This is also known as the rms or root-mean-square roughness. It is calculated as a standard statistical parameter by summing the squares of the individual elevations, dividing by the number of points, and taking the square root. In normal descriptive statistics, it would be proper to divide by $(n - 1)$ instead of n (the number of points), but since that number is fairly large even for a linear traverse, and much greater for an entire image, the difference is inconsequential.

Sigma describes only a part of the surface roughness, but is at least better than the peak-to-peak roughness, which is the difference between the highest and lowest elevation values. This latter value depends critically on the number of points measured in the line or area, and generally varies significantly from one measurement to another. Also, it tells nothing about how frequently these extreme points are encountered. Two surfaces, one almost everywhere perfectly smooth with only a single high point and the other composed of alternating points at the two extremes, would have the same peak-to-peak distance but obviously very different values for sigma, and quite different roughnesses as well.

For any distribution of values, the standard tools of traditional descriptive statistics do not end with the standard deviation, of course. The mean value is zero once the best-fit plane

or other shape has been subtracted. But only for a Gaussian or normal distribution of elevation values does the standard deviation fully describe the data. There is no *a priori* reason to expect this distribution of elevations, regardless of whether the surface is machined, fractured, deposited, or created in some other way. The usual statistical description of an arbitrary distribution includes the skew and kurtosis.

These are the higher moments of the distribution. The mean is the first moment, and the variance (square of the standard deviation σ) is the second. Skew is the third moment and kurtosis the fourth. For a Gaussian distribution, or any other symmetrical distribution, the skew is zero; if it is nonzero, the sign describes the direction in which the data are spread out. Kurtosis is exactly 3 for a Gaussian distribution; smaller values correspond to a distribution which is flatter than the classic bell curve and higher values to one that is narrower and has fewer values on the extremes. For distributions which have arbitrary shapes, such as multimodal or uniform sets of values, none of these descriptive statistical parameters provides very much information. Plotting a histogram of the actual elevation values may be performed, but this does not provide a compact description or a single number that can be compared to other surfaces.

Of course, looking only at the elevation data without taking into account their possible organization in the lateral directions parallel to the surface is not enough to describe a rough surface fully. It would seem appropriate to select some transverse dimension which could at least serve as a first-order characterization, like sigma. Historically, it has been more difficult to measure lateral roughness parameters and there has not been a descriptive model which made it clear what should be chosen for this role. When most surface measurements were obtained as elevation profiles along linear traverses, the correlation length was often used. This may be defined in several ways. Two of the most straightforward are the average distance between points at the same elevation (typically whatever is defined as zero), or the average distance between successive maxima. Both of these definitions have a variety of modifications, but they do not alter the basic concept.

Plausible as they may sound, these parameters are in fact not appropriate for describing rough surfaces, and have been shown to be artifacts of the measurement procedure. The standard deviation of the elevation values varies with the area of the surface examined. In fact, it was shown in previous chapters that this variation offers a way to measure the surface fractal dimension. The correlation length depends on the lateral resolution of the surface measurement (O'Neill and Walther 1977; Freniere, O'Neill et al. 1979; Church 1986).

To understand this, and to present the link between such parameters and the fractal dimension of the surface, it will be helpful to consider again the power spectrum of the surface elevation profile introduced in the preceding chapter. Beginning with the elevation profile $Z(x)$, we can define a structure function $S(T)$ (Sayles and Thomas 1977) which is the average or expected value of the difference in elevation of two points as a function of their separation, or $S(T) = \langle |Z(x) - Z(x+T)|^2 \rangle$. This is closely related to the autocorrelation function $G(T) = \langle Z(x) \cdot Z(x+T) \rangle - \langle Z(x)^2 \rangle$ since $S(T) = 2 \cdot (G(0) - G(T))$, where $G(0)$ is just $\langle Z(x)^2 \rangle$ or σ^2, the variance. The Fourier transform of $G(T)$ is $P(k)$, the power spectrum of the profile.

There are a few relationships between parameters already described and the power spectrum which may be interesting (Church and Berry 1982). First, the variance of the elevation data σ^2, the variance of the slope of the profile m^2, and the variance of the curvature of the profile c^2 are given by integrals of the power spectrum over all frequencies:

$$\sigma^2 = \int_0^\infty S(f)\,df$$

$$m^2 = (2\pi)^2 \int_0^\infty S(f) f^2 df$$

$$c^2 = (2\pi)^4 \int_0^\infty S(f) f^4 df$$

For the case of a random surface (no fractal structure and each point an independent random sample of a Gaussian elevation distribution), the average distance between zero crossings will be $\pi\sigma/m$, and so the correlation length or surface wavelength in the horizontal direction will depend on the displacements in the vertical direction. Likewise, the average distance between successive maxima is $2\pi m/c$. A "generalized" autocorrelation length is sometimes defined as the shortest distance T which satisfies the relationship $S(T) = q\ S(0)$, where q may be either zero (which corresponds to the nearest zero crossing) or $1/e$. Figure 30 illustrates the latter case. None of these parameters is very easily generalized to measurements which cover the entire surface to produce a range image.

A recent review of the conventional approach to surface roughness measurement (April, Bouchard et al. 1993) summarizes the use of profilometer traces to determine the standard deviation of the elevation values, the correlation function, and the $1/e$ correlation length. Light scattering (diffraction) is also used to measure the same surfaces. The normalized value of the specular peak intensity is shown to be related to the rms roughness value for the case of very smooth surfaces, and to be strongly dependent on the correlation length for rough surfaces. In fact, for a fractal surface there is more information in the scattered light profile as shown above in Figure 28, while these two classic parameters are not unique descriptors of the surface but rather depend strongly on the measurement system.

We saw in the preceding chapter that a fractal profile produces a power spectrum of the form $P(k) = B \cdot k^{-\beta}$, where the exponent β is related to the fractal dimension D as $D = (4 + \beta)/2$. It is immediately apparent that the integrals given above for the variance of elevation, slope, and curvature are unbounded for the power spectrum of a fractal. The actual value which would be determined from a measurement of an elevation profile and the construction of its correlation function and power spectrum depends, therefore, on the range of frequencies over which the data actually extend. The lowest frequency (which dominates the integral for σ^2) is a function of the length of the profile which is traversed. The highest frequency (which dominates the integral for m^2 and c^2) is a function of the finest spacing between points, or the lateral resolution of the instrument.

Consequently, the traditional measures of surface "roughness" are artifacts of the measurement tool and procedure. Comparison of values determined under similar conditions may be useful as an ad hoc criterion of surface quality, but the numbers do not really describe the

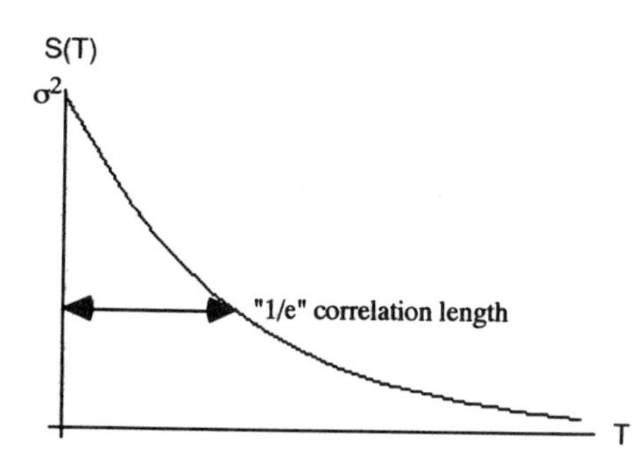

Figure 30. Correlation plot for the structure function.

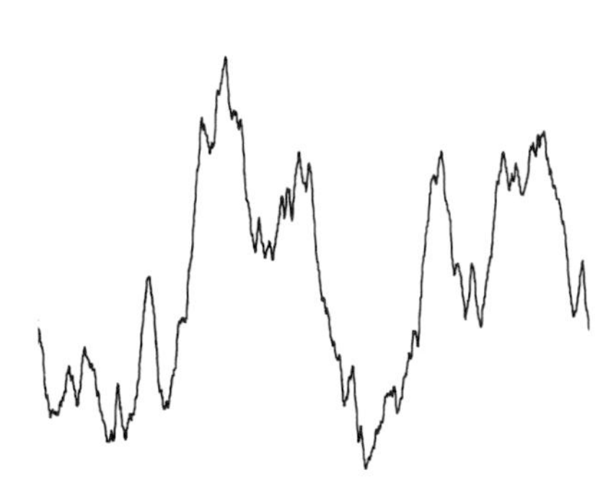

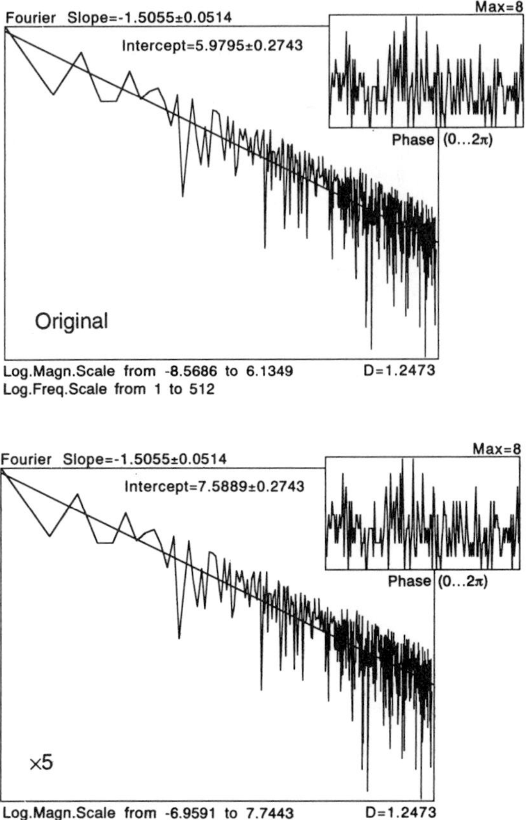

Figure 31. FT power spectra for a fractal profile and for the same profile with each value multiplied by 5. The slope and fractal dimensions are unchanged, but the intercept changes.

surface in any absolute sense. Church (1986) discusses the effect of measurement bandwidth on the observed values of these and other common surface description parameters.

For surfaces which are fractal, there are better descriptive parameters available which can be determined with instruments that cover a finite range of distances from the unit of resolution to the size of the sampled region. It is equivalent to describe this as an instrument with a finite band pass. The real test of these methods is whether the parameters which are determined by one instrument agree with those of another whose band pass may be quite different. Comparison of visible light interferometry with atomic force microscopy (Russ 1991a) on machined surfaces does produce consistent results, but much more work is needed to verify that this is true for all of the various measurement tools which are available. For one thing, many surfaces may not be ideal fractals in the mathematical sense, but may have a mixed fractal behavior that varies with scale. This is discussed in Chapter 7.

For an elevation profile of a fractal surface, the power spectrum is simply $P(k) = B \cdot k^{-\beta}$, where $1 < \beta < 3$. For a surface which is isotropic, the two-dimensional power spectrum is $P(k) = B \cdot \Gamma[(1 - \beta)/2]/\Gamma(1/2) \cdot \Gamma(-\beta/2) \cdot k^{1-\beta}$. Note that this function drops in magnitude one order of magnitude faster with frequency k than does the one-dimensional case. Two parameters, B and β, are needed to specify this power spectrum fully. The fractal dimension D is related simply to β, as $D = (4 + \beta)/2$. This is the parameter we have been primarily discussing in the prior sections on fractals, which is determined by plotting various kinds of data on log–log plots.

It is easy to construct two profiles or surfaces which have the same value of D (or β) but a different value of B. Figure 31 shows a simple example. The same profile had all of the elevation values multiplied by 5, to create a greater magnitude of vertical displacements of the points. The slope (and hence the fractal dimension) is unchanged, but the intercept value is shifted.

The vertical scale for the data has a dimension of length. There are several ways this length dimension can be specified. The one most commonly used is the topothesy Λ introduced in Chapter 4. This is defined as the distance over which the slope between two surface points is 1 radian. These dimensions are typically very short, often much less than any instrument resolution or even less than the dimensions of atom spacing in the specimen. It is not clear that there is any physical significance which can be assigned to the topothesy. The topothesy is defined as $S(\Lambda)/\Lambda^2 = 1$. It can be calculated from both B and β as

$$\Lambda^{3+\beta} = \frac{B \cdot (2\pi)^{-\beta}}{2\Gamma(-\beta)\cos(-\beta\pi/2)}$$

Church (1988) has derived the relationships between the conventional standard deviation of elevation values (σ) and the correlation length (ℓ).

$$\sigma = \left[\frac{B \cdot L^{-(\beta+1)}}{-(\beta+1)}\right]^{1/2}$$

$$\ell = \frac{(\beta+1)^2 \cdot L}{-2(2\beta+1)}$$

These depend on the total length of the measured elevation profile L, as was noted before. This confirms the need to use the fractal parameters (either B and β, or D and Λ) to meaningfully characterize surfaces, instead of the traditional σ and ℓ. Any of the measurement techniques described in earlier chapters may be efficiently used to calculate the fractal dimension D, for either elevation profiles or range images. The B or Λ value is most straightforwardly determined using the power spectrum, which can be calculated for either an elevation profile or an isotropic surface. The interpretation of these values for anisotropic surfaces is more complicated, and is discussed in Chapter 7.

6

Modeling Fractal Profiles and Surfaces

In order to test and understand the characterization tools, and to compare real surfaces, it is important to be able to generate fractal profiles and surfaces using mathematical models. The description in Chapter 1 of a patterned Koch profile shows one way to do this for a line profile. A similar effect for surfaces can be achieved by adding and removing blocks; as smaller and smaller blocks are added and removed from the surface, the volume beneath it is not changed but the surface area increases without bound. The pattern of blocks added and removed may either be strict (as in the classic Koch fractals) or randomized.

There are many different kinds of structured fractals. Mandelbrot's 1982 book *The Fractal Geometry of Nature* shows Menger sponges in which blocks are removed from within a solid in various sizes to create a porous material, and branching fractals in which each branch of a tree has a self-similar set of smaller limbs, and so on. This may serve as a useful model in two dimensions for river systems and in three dimensions for blood vessels and the air passageways in the lung. All of these (and more) can be made random rather than structured, and when this is done the appearance becomes more natural and the visual similarity to real structures more striking. It is this latter characteristic that has led to the increasing use of computer-generated fractal scenes and objects in movies, as a less expensive and more versatile alternative to natural or constructed sets, to create realistic and yet slightly strange visual effects.

A very complete treatment of branching patterns generated using an iterative technique known as "L-systems" is presented by (Prusinkiewicz and Lindenmayer 1990). This approach, described in Chapter 1, uses a series of characters which are ultimately interpreted as lines and branches, with rules to repetitively substitute sets of characters for each existing character. The results are capable of modeling the three-dimensional growth patterns of plants with amazing realism, using astonishingly few rules.

Nor is the generation of fractals limited to dusts (distributions of points), profiles, or surfaces. Higher dimensions can be used as well. Producing a distribution of density as a fractal function of x, y, and z produces a useful model of clouds, for instance (Voss 1985a). Thresholding this three-dimensional function, which is equivalent to taking the zero-crossings of a surface or line profile, corresponds physically to the density of water vapor needed for condensation. This produces three-dimensional "islands" that appear as clouds, and very realistic images.

Fractal Profiles

For generating fractal surfaces, there are a variety of models available (Saupe 1989; Russ 1992b). Many of them are generalizations of models for generating profiles. Fractal profiles

have been defined in preceding chapters operationally, based on a measurement method, of which there are several. It is also possible to define a fractal profile based on a generating model, and again several methods are available. It is possible to generate a line profile that is fractal, with a known fractal dimension. One simple way to accomplish this is to use the method proposed by von Koch (Koch 1904). The technique is most commonly shown for closed outlines, called Koch islands, but the method will work for profiles as well. This method was shown in Chapter 1.

There are a few problems with the resulting profile as a model for elevation profile data as would be measured from a surface. For one thing, it doesn't look very natural. This kind of fractal is known as a structured or patterned fractal because the generating shape is iterated exactly at ever-smaller dimensions. This can be overcome by using random numbers to vary the local details of the pattern to better match observed natural lines. The second problem is that while the Koch patterns shown work pretty well for the kind of boundary line that might occur for a shoreline, they produce elevation profiles that are not single-valued. The irregularities produce an endlessly finer detail of undercuts and under-undercuts so that most points along the x axis have more than one, and perhaps infinitely more, points in the z direction. This is not the way most elevation profiles behave, and even if the surface being examined may have such undercuts, most measuring instruments will not detect them and will report only the uppermost value. There are other methods for generating either systematic or random fractals which avoid this problem of multiply defined points, and we will now look at one which also introduces the use of random numbers.

Consider another profile consisting initially of a straight line segment. Displace its midpoint up or down (with equal probability) by some distance h which is obtained from a Gaussian random number generator with mean value $\langle h \rangle$. As shown in the example in Figure 1, this will produce a profile consisting of two straight line segments (Figures 1–3 repeat illustrations in Chapter 1). Now repeat the procedure for each of the midpoints of these segments. As the length of the line segments has been halved (or, more exactly, the horizontal distance w between the end points has been halved), the magnitude of $\langle h \rangle$ must also be reduced. This is done by setting $\langle h \rangle = w^{\alpha}$ where the exponent α is between 0 and 1. Then, as for the Koch patterns above, the procedure is repeated for each of the shorter line segments. The procedure can be repeated mathematically, at least, without limit. In practice, a profile such as the one shown in the figure consisting of elevation values at more than 1000 discrete points can be completed in 10 iterations ($2^{10} = 1024$).

Depending on the magnitude of the exponent, the perceived roughness of the line can be varied as shown in Figure 2. These profiles are satisfyingly natural-looking to the human eye, and are single-valued. The fractal dimension of each profile can be calculated, and the rougher values corresponding to smaller values of α have higher fractal dimensions. The quantitative relationship is $D = 2 - \alpha$. We will later see other methods for generating profiles with fractal

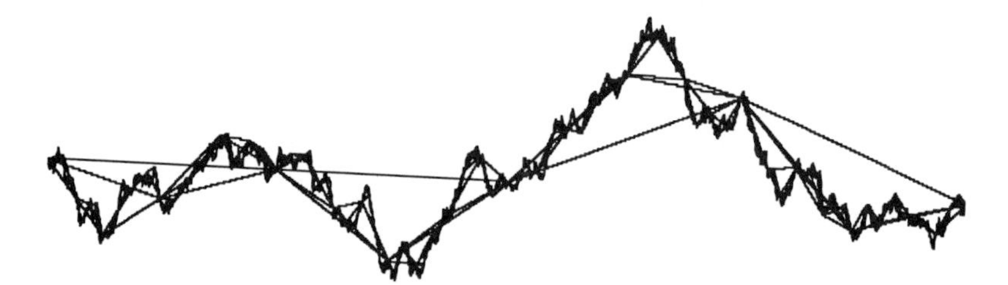

Figure 1. Step-by-step process of producing a fractal profile by random midpoint displacement.

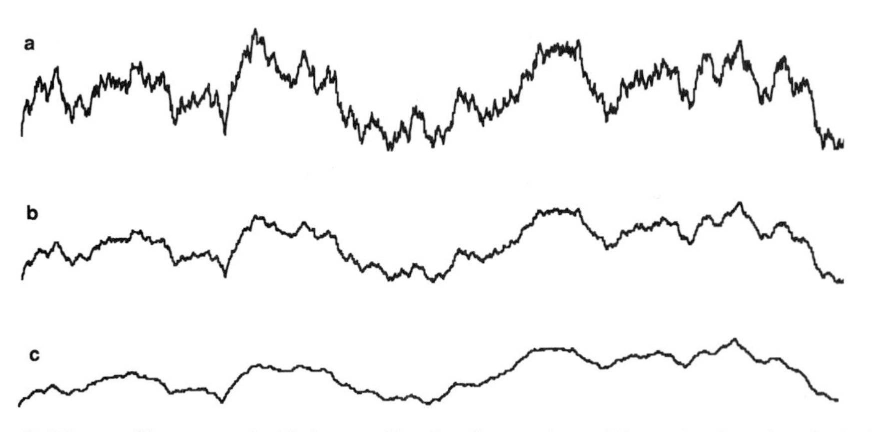

Figure 2. Three profiles generated with the same list of random numbers while varying the value of α and the fractal dimension: midpoint displacement fractals generated with α = **(a)** 0.5, **(b)** 0.7, and **(c)** 0.9.

shapes and adjustable dimensions, but this one is among the most efficient and most easily understood.

The justification for using a Gaussian random number generator in this method is based on an interpretation of the central limit theorem. If a large number of individual events, controlled by many independently random parameters, is sampled, the result is expected to produce a single Gaussian peak. Specifying the mean value $\langle h \rangle$ of the displacement overlooks the other parameter of a Gaussian distribution, the standard deviation. It is known that while changing this parameter affects the appearance of the plots, they are all fractal with the same formal dimension. Figure 3a shows a plot constructed with the same sequence of random numbers and the same value of $\alpha = 0.7$, as Figure 2b, but changing the standard deviation of the Gaussian random number generator to double the average displacement magnitude. Mandelbrot has proposed a parameter called the "lacunarity" of profiles which takes this into

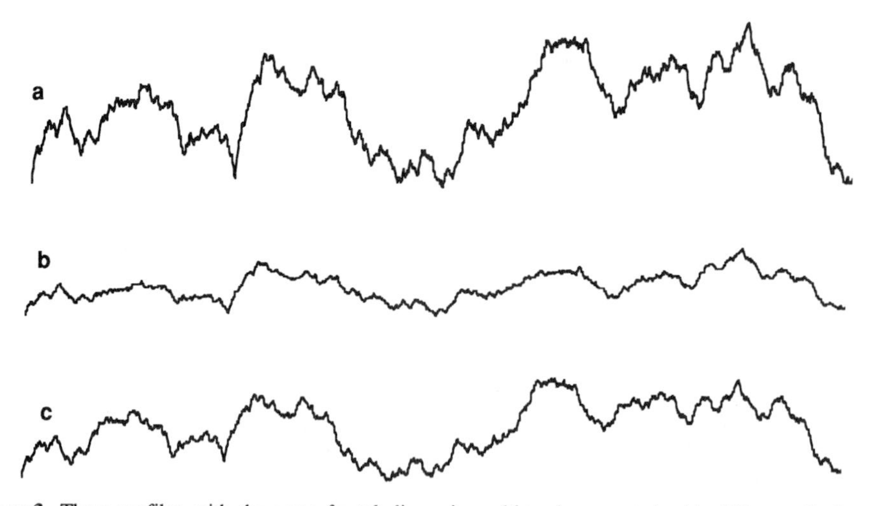

Figure 3. Three profiles with the same fractal dimension, although generated with different displacement magnitudes or with displacement values taken from distributions which are not Gaussian (compare to Figure 2b). Midpoint displacement fractal generated with $\alpha = 0.7$: **(a)** magnitude × 2; **(b)** squared distribution; **(c)** square-root distribution.

account, but no interpretation has yet been developed. The topothesy, discussed in previous chapters, also provides a measure of this value.

Other changes in the distribution of displacement values can also be made without changing the fractal dimension of the profile. These include biasing the results so that only displacements to one side of the original line can occur. It is even possible to use an arbitrary distribution function; Figures 3b and 3c show an example in which two different distributions of values are used, the square and square root of the Gaussian in Figure 2b. Since the relative scaling of the mean displacement is controlled by the α value, the fractal dimension of the profile is not altered.

There are a variety of other techniques available for generating fractal profiles. These include the addition of random steps to simulate Brownian motion, and addition of summations of regular functions such as sine curves or triangles. Like the iterated midpoint displacement technique, each of these methods can be generalized to deal with surfaces. Some of the methods that work moderately well for profiles (e.g., the Mandelbrot–Weierstrass function discussed below) are not very satisfactory for surfaces. Others (e.g., the Takagi and inverse Fourier methods) can be easily generalized to produce surfaces with controlled kinds of anisotropy. There are a few methods (such as extrusion) that do not have a corresponding equivalent for producing a fractal line profile.

Since the major interest in this book is with surfaces, each of these methods will be discussed in those terms. Where appropriate, the generation of a line profile using the same technique will be shown as well. It should be understood, however, that while these profiles could be used to model many natural one-dimensional phenomena, such as noise in electrical circuits, stock market prices, or river floods, our intent here is that they represent elevation profiles across a physical surface. As noted above, the scale of such profiles may vary over many orders of magnitude, and the measurement resolution of any particular instrumental technique is likely only to cover a much smaller range, perhaps a factor of 100 to 1000 times from the largest traverse to the smallest resolved horizontal distance. Also, the vertical elevation resolution may be equal to that in the horizontal direction, or considerably better than it, depending on the measurement technique.

A single-valued function such as an elevation profile can only be self-affine and not self-similar (Mandelbrot 1985a; Mandelbrot 1985b; Voss 1985b; Mandelbrot 1986; Thomas and Thomas 1988; Voss 1988; Voss 1989), because a small feature can only appear on a larger one if the differential is larger. In many applications, even if the actual surface is not single-valued but has undercuts, the vertical elevation profile may provide the relevant description. This would apply, for instance, in surface contacts or wear. This is fortunate since most measurement methods provide single-valued vertical elevations. Elevation profiles have been measured on real surfaces, both man-made and naturally occurring, for many years and in many fields.

Self-affinity also occurs when the vertical and horizontal axes of the profile are not the same, for instance voltage and time. When true self-similarity does not hold, fractal behavior in terms of a continual refinement of detail may still be present. A self-affine curve will have a linear Richardson plot, but the slope need not fall within the range of 0…1. A least-squares fit on log–log axes may be calculated by rescaling each of the measures. Distances along the horizontal axis can be expressed as a fraction of the total length, and those in the vertical direction either as a fraction of the maximum range or the standard deviation of the data. In either case, the values become dimensionless ratios so that the units disappear. This method allows calculating a slope which is sometimes interpreted as a fractal dimension. However, Mandelbrot has shown that for a self-affine rather than self-similar curve, the slope H of a Richardson plot is related to the dimension of the boundary as $D = 1/H$ instead of $D = 2 - H$.

This means that the calculated dimension can exceed 2, which clearly does not have the usual geometric meaning in terms of boundary roughness.

Chapter 8 will discuss in detail some of the particular kinds of real surfaces that have been studied in some detail and found to exhibit fractal behavior. These include (at least under some circumstances) machining, fracture, corrosion, deposition of particulates, solidification, and other processes. Few of the mathematical models for generating surfaces consciously attempt to mimic any of these physical processes. However, there are models for particle deposition by diffusion or by ballistic motion which seem to capture the essence of the physics to produce fractal structures. These have adjustable parameters that make physical sense and also modify the structures in important ways.

Particle Aggregation

Surfaces which are formed or grown by chemical or physical deposition include many structures deposited either by chemical or electrochemical means, or by the physical deposition of particulates by diffusion or ballistic agglomeration. They may also include some instances of surface corrosion. Studies of such surfaces and modeling of their formation processes (Kaye 1984; Kardar, Parisi et al. 1986; Stanley and Ostrowsky 1986; Kaye 1989a) indicate that they are fractal.

The classic and simplest description of such a deposition process is diffusion-limited aggregation (DLA) in which individual particles random walk until they encounter the surface or another particle that has already adhered to it. Wherever the particle touches, it has a certain probability of sticking (which in principle may vary with direction, etc.). When performed in a two-dimensional grid starting with a single particle or point as a seed, this gives rise to very open structures with dendritic arms, because the likelihood of a particle finding its way down between the growing arms to lodge near the original center is small. The resulting cluster is fractal, and can be observed in such macroscopic cases as the growth of ice crystals on an automobile windshield. It is also important in the formation of agglomerates in soot particles, the deposition of catalytic surfaces, and even the formation of electrical discharges.

Aggregates such as this are often characterized by a fractal dimension, called the mass dimension. The mass within a distance λ of the surface, or for a three-dimensional particle, within a radius λ of the center of gravity, increases as λ^E for a solid object, where E is the Euclidean dimension of the space (2 for a surface, 3 for a solid). For a fractal structure, the exponent is less. For a classic diffusion-limited aggregate as defined above, the relationship is mass $\propto \lambda^{1.73}$ in 2D, or mass $\propto \lambda^{2.5}$ in 3D.

Changes in the rules for motion of the particles or their sticking probabilities alter the appearance and dimension of the clusters. Figure 14 in Chapter 1 shows the effect of altering the sticking probabilities from 100% to 25%, 5%, and 1%, respectively. The cluster becomes progressively more compact, and the mass fractal dimension increases. If the same procedure is carried out in three dimensions, as shown in Figure 4, the same trend is observed. However, since the fractal dimension of a cluster in 3D is greater than 2.0, the projected image of the cluster should be solid. For a finite size cluster, this limit is not reached and the projected density of the cluster is a function of size.

All of the clusters in Figure 4 contain the same total number of particles (30,000). The mass fractal dimension of the cluster in 3D depends on the sticking probability, but always exceeds 2 so that the projected cluster should always have the limiting dimension of 2.0. However, as shown in the figures, this is not the case. Mass fractal plots for the projected clusters are straight lines with slopes that also vary with sticking probability and have a dimension less than 2.

Measuring the real fractal dimension of the cluster, or of the surface of the cluster, is a difficult problem. It has been proposed (Vicsek 1992) that as the sticking probability becomes low, the projection of such a cluster will have a boundary that is fractal. Figure 5 shows an example. A 3D agglomerate containing 100,000 particles, formed with a 1% sticking probability, is shown in projection. Because of the finite size of the cluster, it is not entirely dense. The exterior boundary of the cluster was used to obtain the Euclidean distance maps of both the cluster interior and the surroundings. From these, a Minkowski plot of the dimension of the boundary can be formed as shown. The plot suggests that the outline is described by fractal geometry, but no rules are known to relate the dimension of the outline to the sticking probability, the mass fractal dimension of the cluster, or the number of particles in the cluster.

There are other parameters for the agglomeration that modify the geometry of the cluster. One is to bias the direction of particle motion, as for instance to model formation of clusters in a moving air stream. A second is to make the sticking probability vary as a function of direction, simulating crystallographic effects. A third is to eliminate the grid which controls the motion of the particles and allow them to move freely in 2D or 3D space, until they touch. This is computationally more demanding but produces more realistic structures.

Another modification of the basic technique is to use weighting (Vicsek 1992). Each time a random-walking particle reaches a site adjacent to a particle that is already part of the cluster,

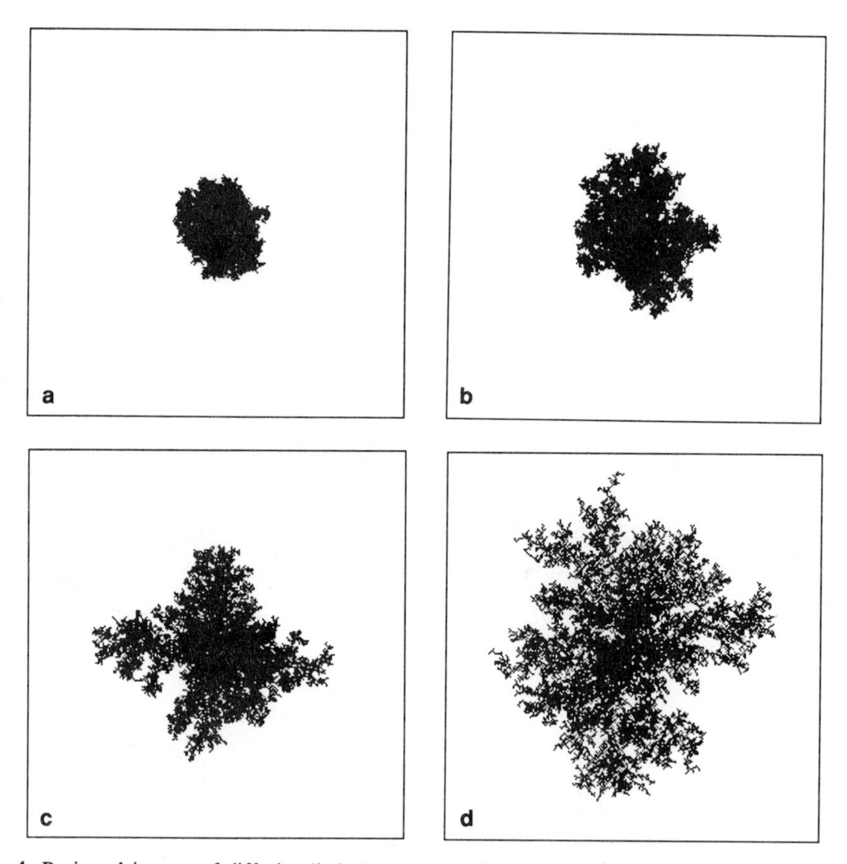

Figure 4. Projected images of diffusion-limited aggregates formed with 30,000 particles on a grid in three dimensions, while varying the sticking probability for the particles. **(a)** 1% sticking probability; **(b)** 5% sticking probability; **(c)** 20% sticking probability; **(d)** 100% sticking probability.

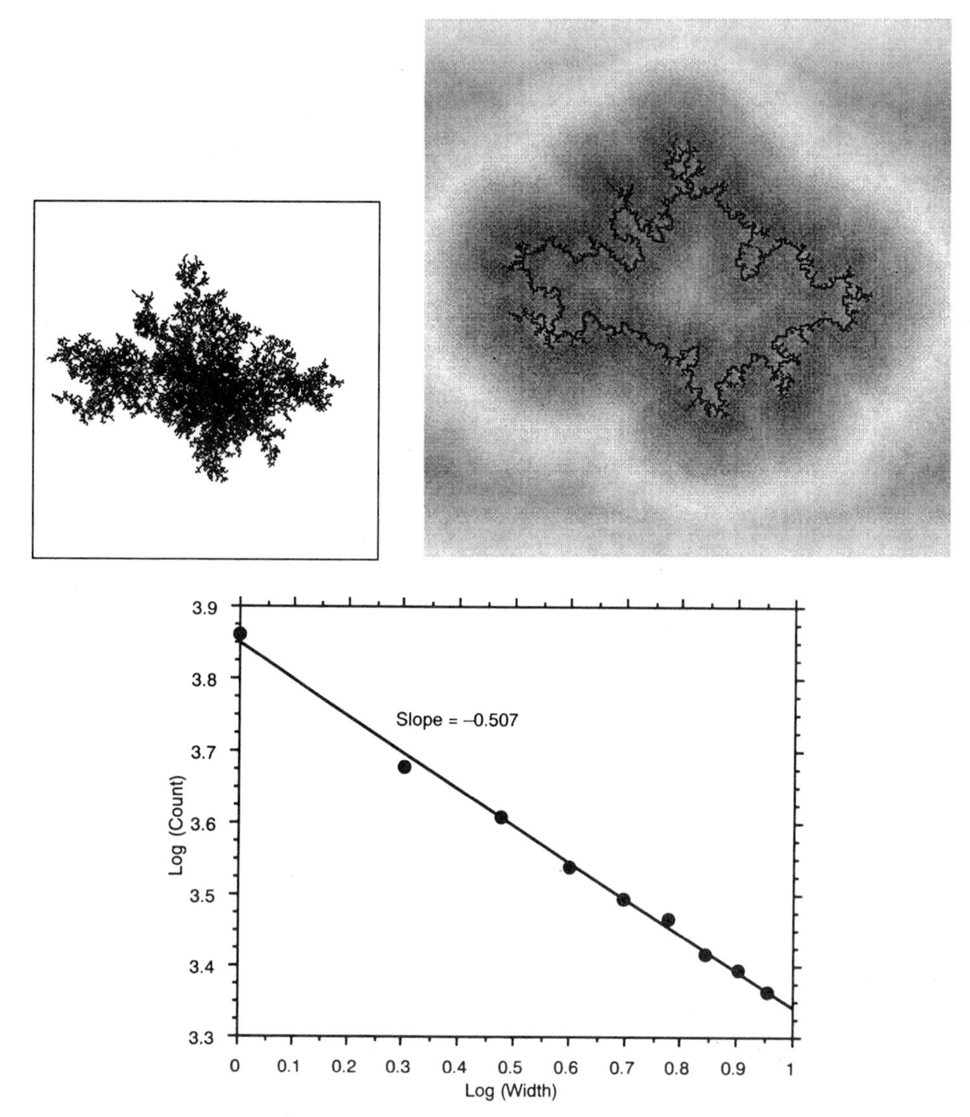

Figure 5. Measurement of the fractal dimension of the exterior projected boundary around a diffusion-limited aggregate (100,000 particles, 1% sticking probability). The Euclidean distance map of the cluster and the background is used to construct the Minkowski plot showing the number of points as a function of distance from the boundary.

it is counted but not necessarily added to the cluster. The number of times a particle must reach a site before one is allowed to stick there is a weighting value that can be adjusted to alter the appearance and dimension of the growing cluster as shown in Figure 6. Larger weight values produce increasingly needle-like dendritic shapes.

Clusters are important in many fields, including the formation of soot particles and the packing of food into containers. Whenever particles stick together, the size of the clusters increases faster than the cube root of the weight (which would be the case for dense Euclidean

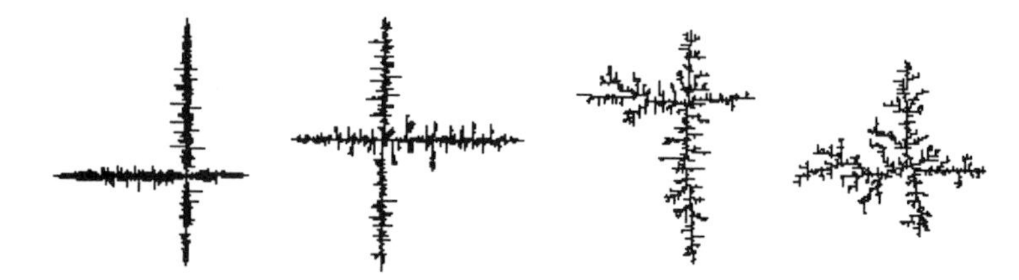

Figure 6. Cluster formation in 2D using Vicsek's point weighting method, with weights of 100, 20, 4, and 1. Each contains 100 points.

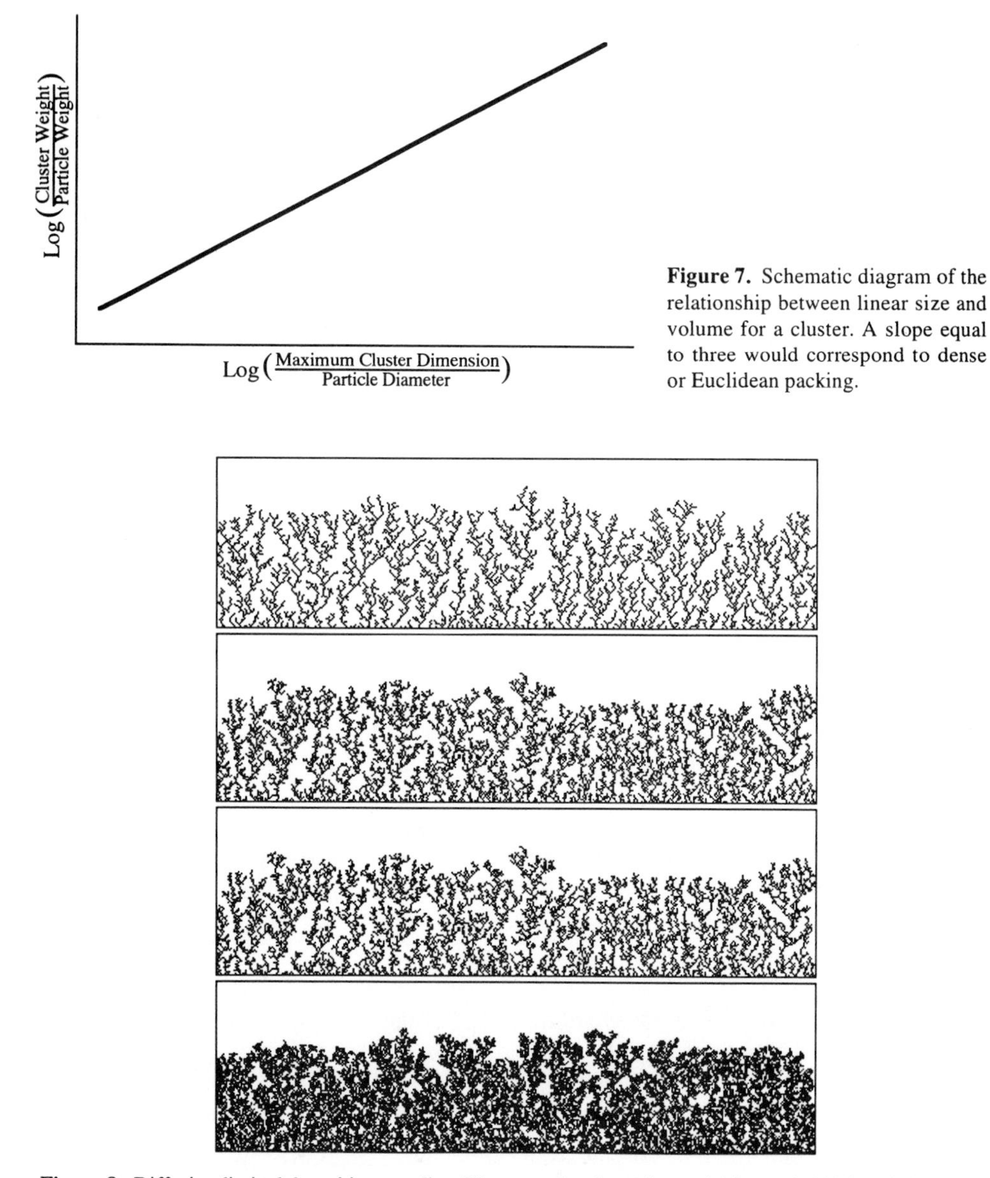

Figure 7. Schematic diagram of the relationship between linear size and volume for a cluster. A slope equal to three would correspond to dense or Euclidean packing.

Figure 8. Diffusion-limited deposition on a line. The examples shown have sticking probabilities of 100, 20, 5, and 1% respectively. The mass fractal dimension in the vertical direction varies from 0.918 to 0.956.

packing). Figure 7 shows schematically the relationship between the linear dimension of a cluster, for instance the maximum caliper dimension, and the number of particles in the cluster.

Deposited Surfaces

Diffusion-limited aggregation onto surfaces can also be modeled and may offer insights into deposition of materials, corrosion, and other processes. Changing the boundary conditions so that sticking occurs on a line or surface instead of a point does not alter the fractal dimension of a growing cluster, nor the effect of altering the rules. Figure 8 shows examples of deposition onto a line, allowing particles to random walk on a two-dimensional grid. Varying the sticking probability increases the density of the deposit.

When the simulation is extended to a three-dimensional grid, the deposit on the surface becomes even more open and dendritic. Figures 9–11 show examples of deposits produced with sticking probabilities of 100%, 20%, 5%, and 1%. Three different views are presented. In Figure 9, the view is a projection in the horizontal direction, across the surface. The superposition of points in the deposit from front to back makes the structure appear much more dense than it really is. Looking down on the surface from a point above it, in Figure 10, shows

Figure 9. Horizontal projections of diffusion-limited aggregation of particles on a surface, with varying sticking probabilities. The mass fractal plots show the variation in the density of the coating. These are the same depositions shown from the top down in Figure 10. **(a)** 100,000 particles, 1% sticking probability; **(b)** 100,000 particles, 5% sticking probability; **(c)** 100,000 particles, 20% sticking probability; **(d)** 50,000 particles, 100% sticking probability.

only the highest point at each x, y location, and reveals that even with 100,000 points deposited on a 256×256 grid, there are a great many points at which the surface is still bare. The deposit that extends into the z direction is very open and feathery. The cross-sections in Figure 11 emphasize this even further. These show the occupancy of grid points in a single plane. The points tend to occur in clusters, as is expected for any zeroset through a fractal, but they are separate and isolated from each other. There is no hint of continuity to the structure as a whole.

The upper surface of such a deposit is not fractal in the sense of fractal surfaces explored in this book. It can be seen from the images in Figure 10 that the lower sticking probability simulations produce surfaces with less dendritic character, and in the limit would produce a fully dense structure. The surface of this may be fractal.

Another important modification of the deposition rules in the simulation is to bias the direction of motion of the particles. In the limit, this becomes ballistic aggregation in which the particles move in a specified direction until they strike the surface or another particle. Ballistic deposition on surfaces is different from diffusion-limited aggregation, producing a

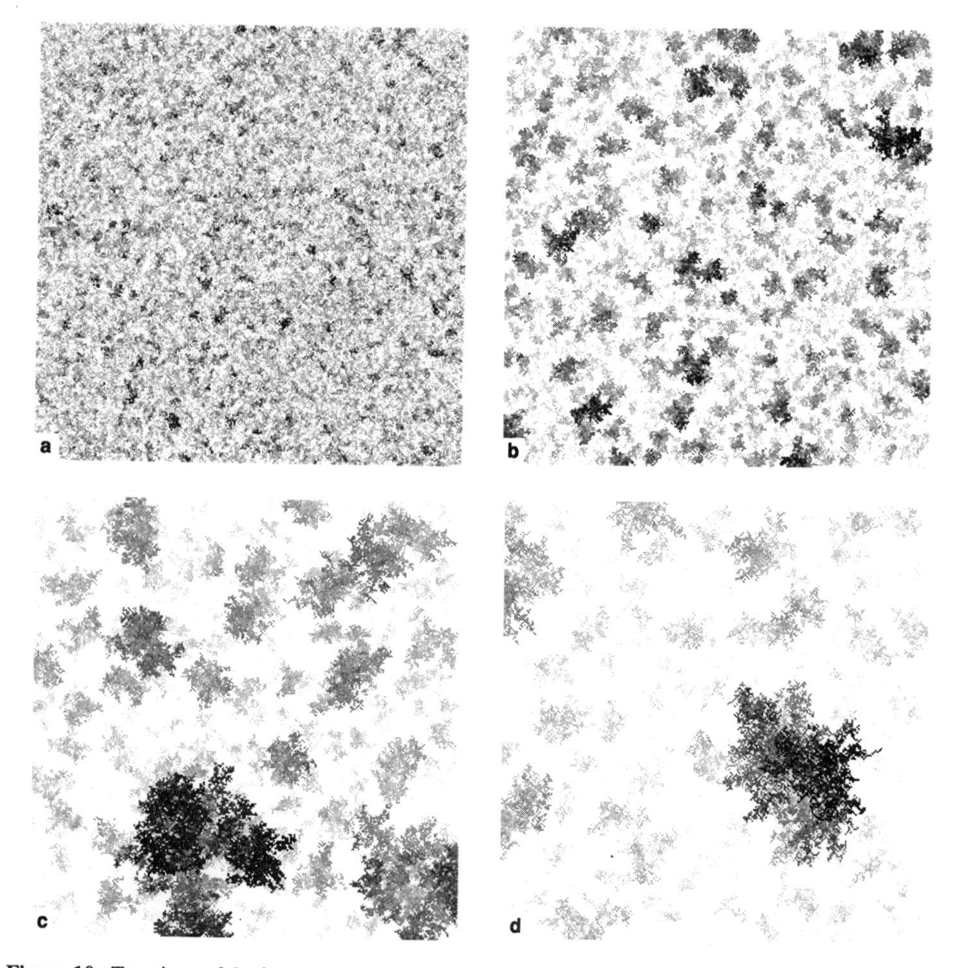

Figure 10. Top views of the four deposited surfaces shown in Figure 9. Each range image has grey scale values proportional to the highest elevation at each point, and each image is autoscaled to cover the total range of heights for the surface. **(a)** 1% sticking probability; **(b)** 5% sticking probability; **(c)** 20% sticking probability; **(d)** 100% sticking probability.

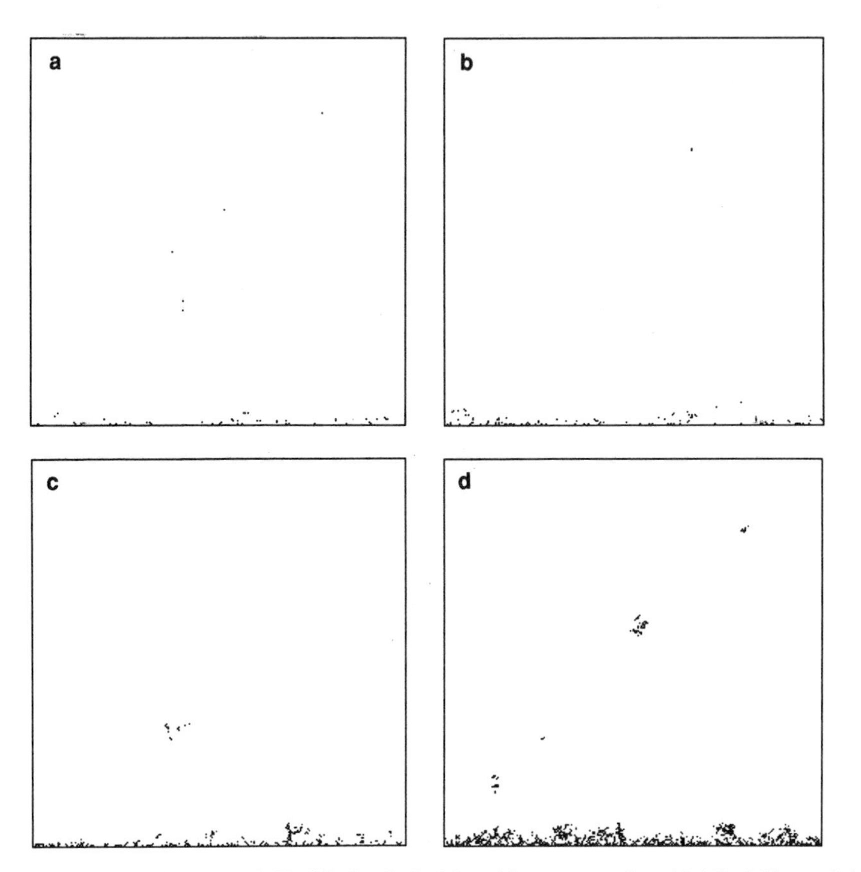

Figure 11. Vertical sections through 3D diffusion-limited deposition onto a surface. **(a)** 1% sticking probability; **(b)** 5% sticking probability; **(c)** 20% sticking probability; **(d)** 100% sticking probability.

characteristic columnar structure or a superposition of conical features, each with its point down. The larger cones eventually shadow, and hence cut off, smaller ones. So as the structure grows it becomes progressively coarser, with larger and larger cones whose bases appear on the surface of the coating as "cauliflower-like" structures. Figure 12 shows an example (with 100% sticking probability on an inclined surface), and Figure 13 shows an isolated cone (Vicsek 1992), with its characteristic growth angle. As shown in Figure 14, this kind of surface is not strictly self-similar, but is instead a self-affine fractal (Sander 1986a; Sander 1986b; Messier and Yehoda 1986; Aharony, Gefen et al. 1984; Aharony 1985; Aharony 1986; Meakin, Ramanlal et al. 1987; Meakin 1989a; Meakin 1989b; Meakin and Tolman 1990).

Modeling a Fractal Surface

For profiles and surfaces, the random displacement method shown earlier provides one of the simplest algorithms. For analysis purposes, it is particularly useful to have an algorithm that creates the same kind of data arrays as most scanning acquisition methods for real range images, namely an x, y array of z (elevation) values. As shown in Figure 15, each point at one corner of a square grid is initially set to the same value. Then the points in the center of each grid square are randomly displaced upwards or downwards from the average of the four corners

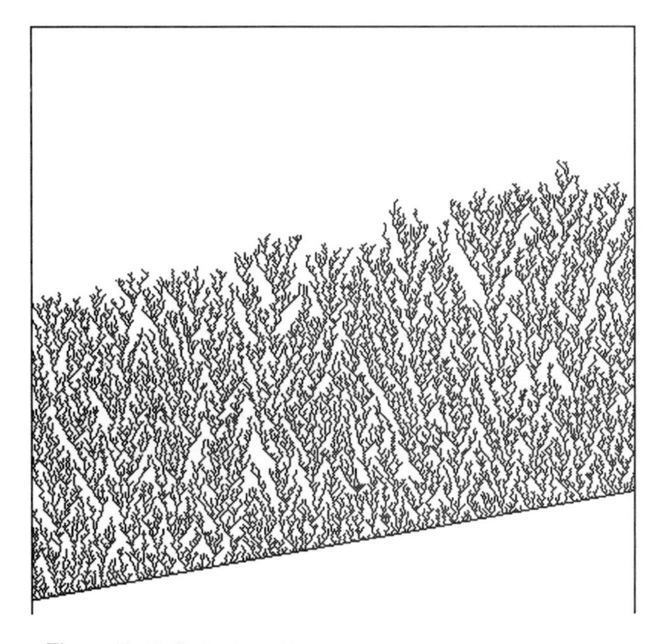

Figure 12. Ballistic deposition (100% sticking probablity) on an inclined surface.

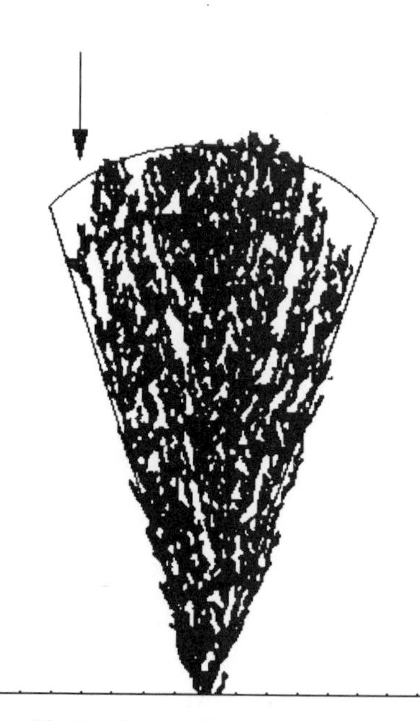

Figure 13. Simulated ballistic deposition, showing a single growing cone and its characteristic growth angle.

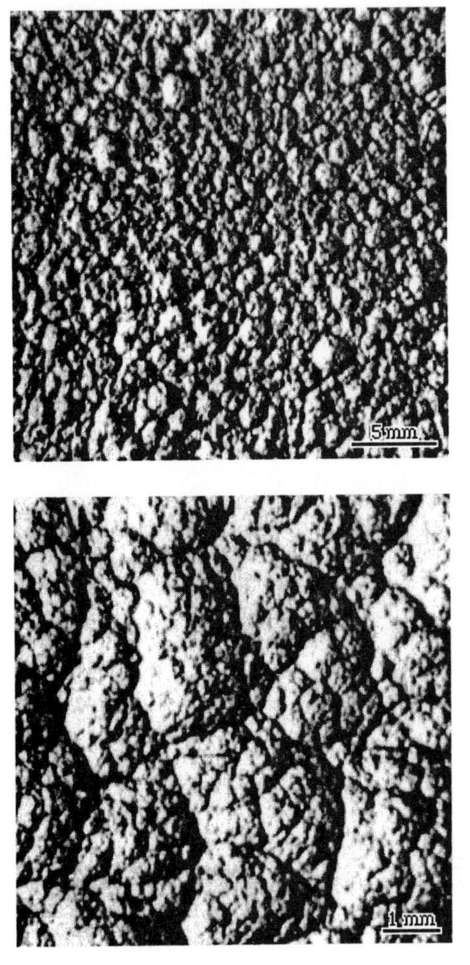

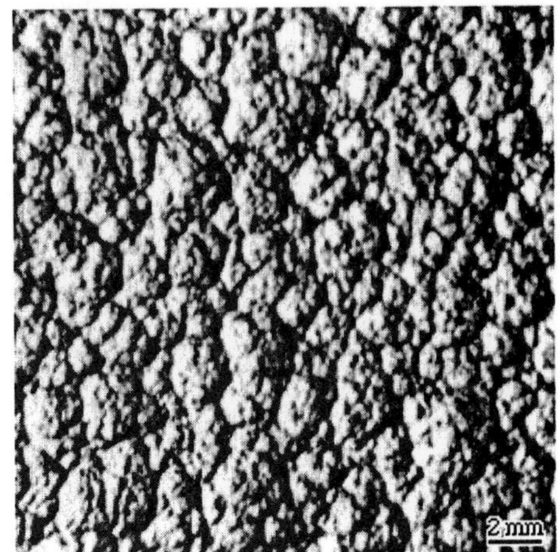

Figure 14. SEM images of a ballistically deposited surface showing the characteristic "cauliflower" appearance. However, note that the scale of the visible structure changes with magnification, which indicates that the structure is not strictly self-similar.

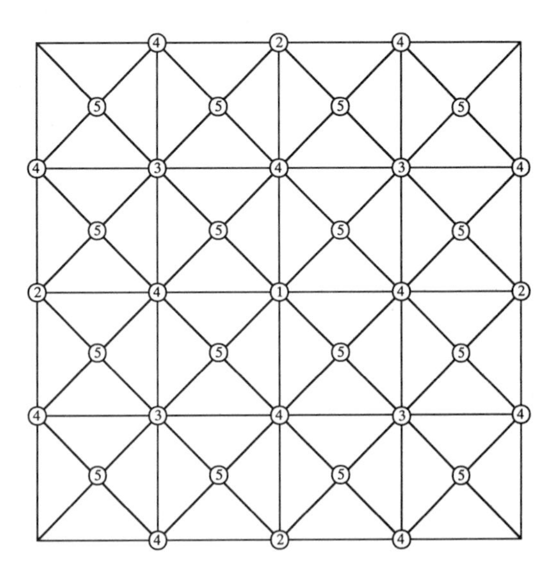

Figure 15. Diagram of midpoint displacement for a surface. Keeping the corner points fixed, points are displaced vertically up or down in the numerical order shown, with a decreasing magnitude.

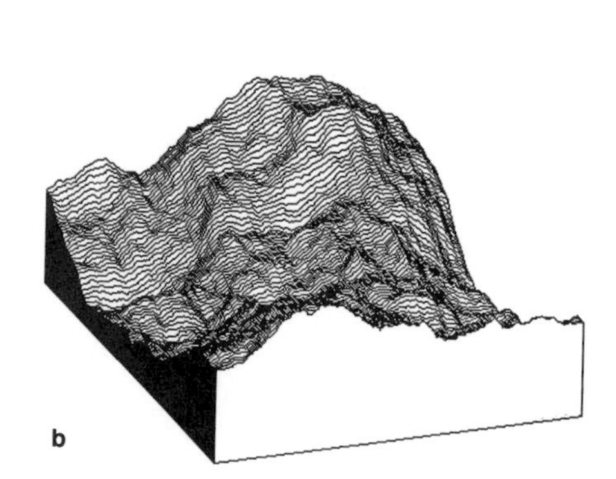

Figure 16. Some presentation modes for surface elevation data, applied to a fractal surface generated by iterated midpoint displacement: **(a)** range image (grey value proportional to elevation), **(b)** pseudo-isometric display (series of line profiles), **(c)** pseudo-isometric display (points displaced vertically and shaded), **(d)** shaded grid display, **(e)** iso-elevation contours, and **(f)** rendered surface (note fold lines from generation method).

with equal probability, by an amount taken from a Gaussian random number distribution with mean value $\langle h \rangle$.

This produces a new grid of points arranged in a square, lying at 45 degrees to the original and with a spacing $\sqrt{2}$ of the original. Displacing the central points in these squares, as described above, moves those points which were at the midpoints of the edges in the original grid. This produces a new grid with each side having half the original dimension. The algorithm is iterated to create new displacements along the midpoints of each of the resulting smaller squares. The mean value of the displacement magnitude is reduced according to $\langle h \rangle = w^{\alpha}$, where w is the width of the square and α is a coefficient between 0 and 1. The surface that results will have a fractal dimension of $3 - \alpha$. Figure 16 shows several views of an example

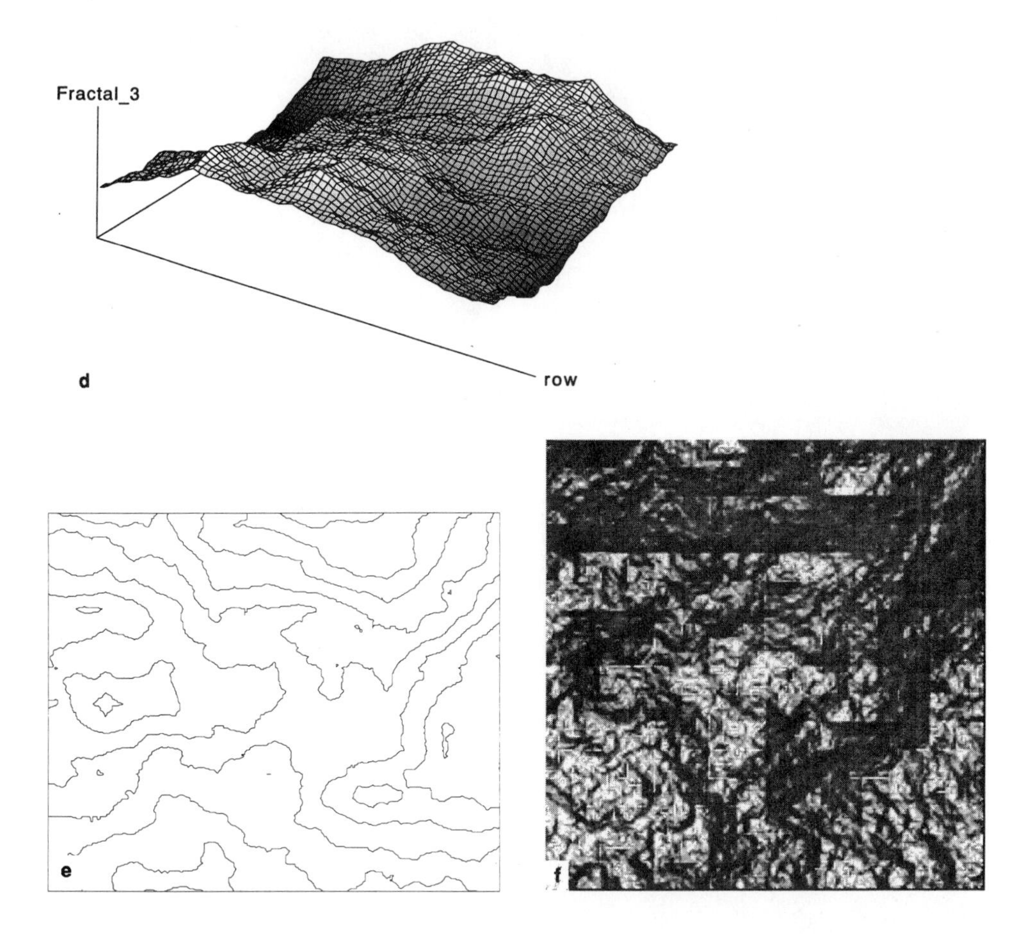

of a fractal surface range image produced in this way. Figure 2 in Chapter 3 shows the step-by-step generation of the surface.

Notice that most of the presentation modes for the generated fractal surface shown in Figure 16 look quite natural. Indeed, this method is often used to generate surfaces for artistic purposes. It is only in the rendered surface image (Figure 16f) that the defect in this surface becomes evident. The folds which appear when light is reflected from a shiny surface result from the process by which the points are displaced. Such folds are not seen in natural fractal surfaces. They indicate that this kind of model should not be used for most technical purposes.

It should be noted that some displays of fractal surfaces may be superimposed onto a Euclidean shape. For instance, if the elevation values are mapped onto a sphere by adding or subtracting elevations from the mean radius, a fractal globe or planet can be formed (Peitgen and Saupe 1988). Wrapping the range image around a cylinder adds a visual texture that may be useful for graphic arts, and for a suitable anisotropic surface may model tree bark or the surface of extruded wire. Similarly, in measuring fractal surfaces it may be necessary to subtract away the gross details of an underlying Euclidean shape to reveal the fractal, or to have enough precision in the range values to measure it. This is particularly a problem with the Fourier methods of analysis since at low frequencies, the magnitude plot is often dominated by the nonfractal large-scale features of the surface.

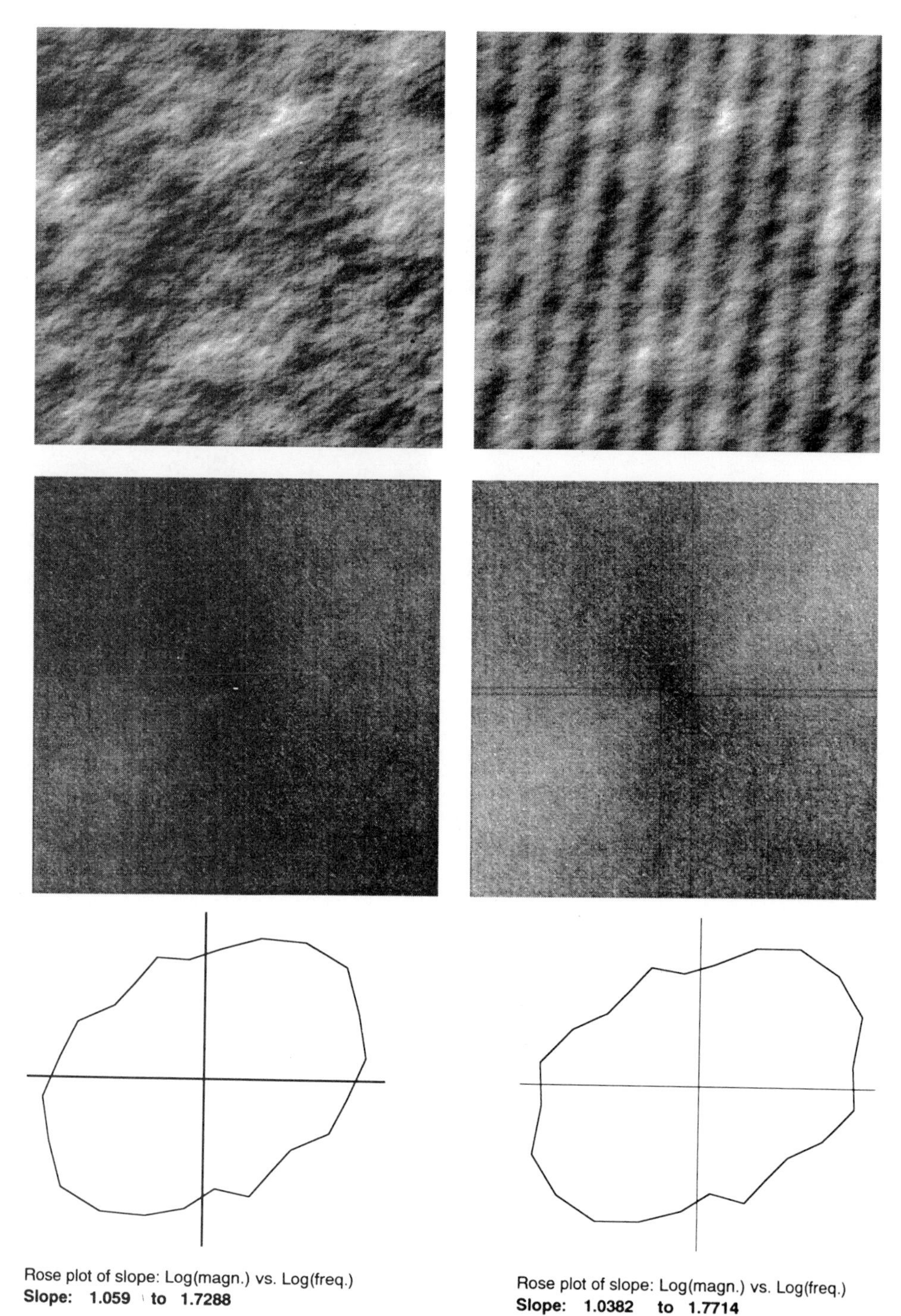

Figure 17. Superposition of an anisotropic fractal surface onto a sinusoidal pattern **(top)**, with the 2DFT power spectrum display **(center)**, the rose plot of slope vs. direction from the 2DFT **(bottom)**, the HOT **(top, next page)**, and the rose plot of slope vs. direction from the HOT **(bottom, next page)**, for the surface with **(right)** and without **(left)** the sine pattern. The "spike" produced in the 2DFT power spectrum by the sinusoid does not affect its rose pattern.

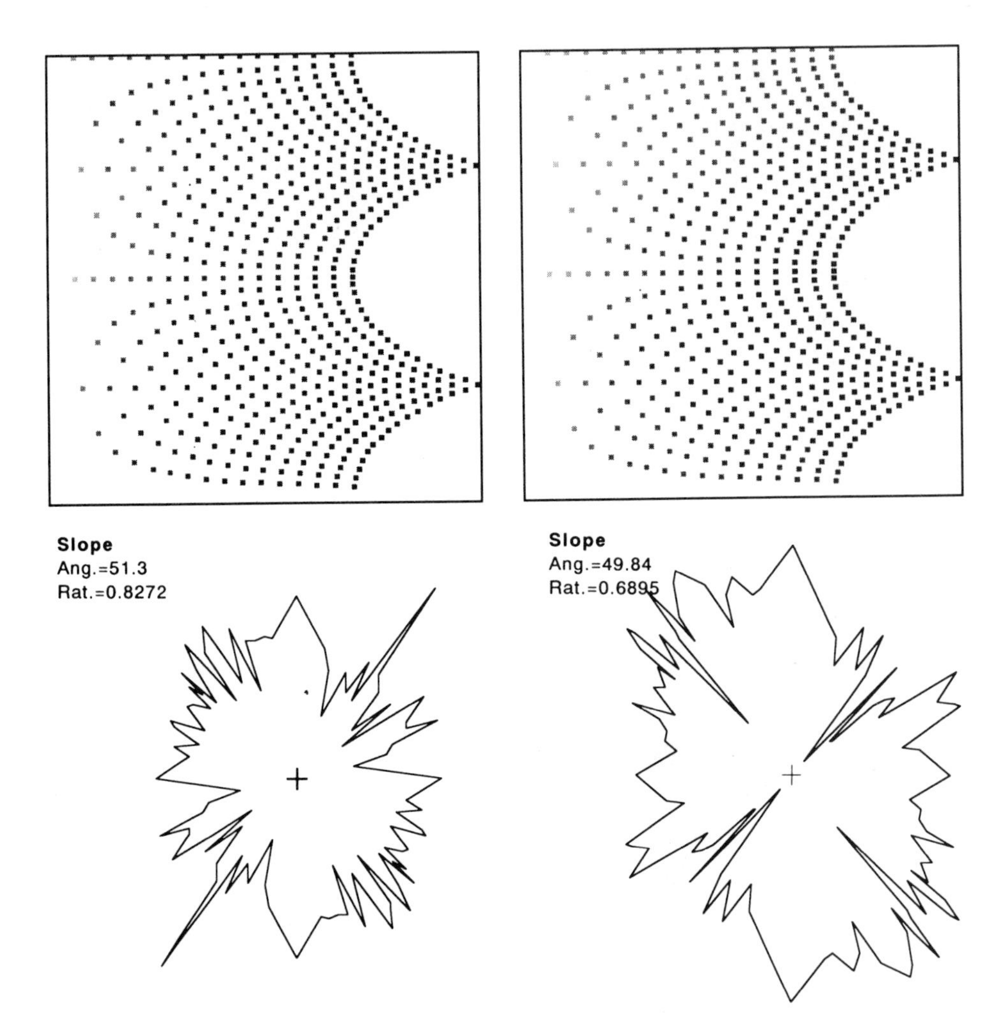

Figure 17 shows a fractal surface superimposed on a sine wave or corrugation, and the resulting 2DFT and HOT analyses. Visual separation of the sine wave from the texture is straightforward, but the large peak in the frequency space representation may influence attempts to determine the slope. With care, for instance by ignoring the low frequency terms in a Fourier transform, these features can be separated, with enough precision remaining for useful measurement and characterization.

Fractal Brownian Surfaces

There are other algorithms for generating fractal profiles, and by extension fractal surfaces. One is simple summing of steps. A single step is a function that can be written as $z = z_0$ for $x \le x_0$; $z = z_0 + h_0$ for $x > x_0$. This is clearly not a fractal, but a simple Euclidean shape. It might seem surprising that any sum of Euclidean shapes could produce a fractal shape, but this is so. Adding together a series of step functions in which the magnitude of the step h decreases with the number of additions according to the rule $h = h_0 \cdot n^{\alpha}$, where n is the number

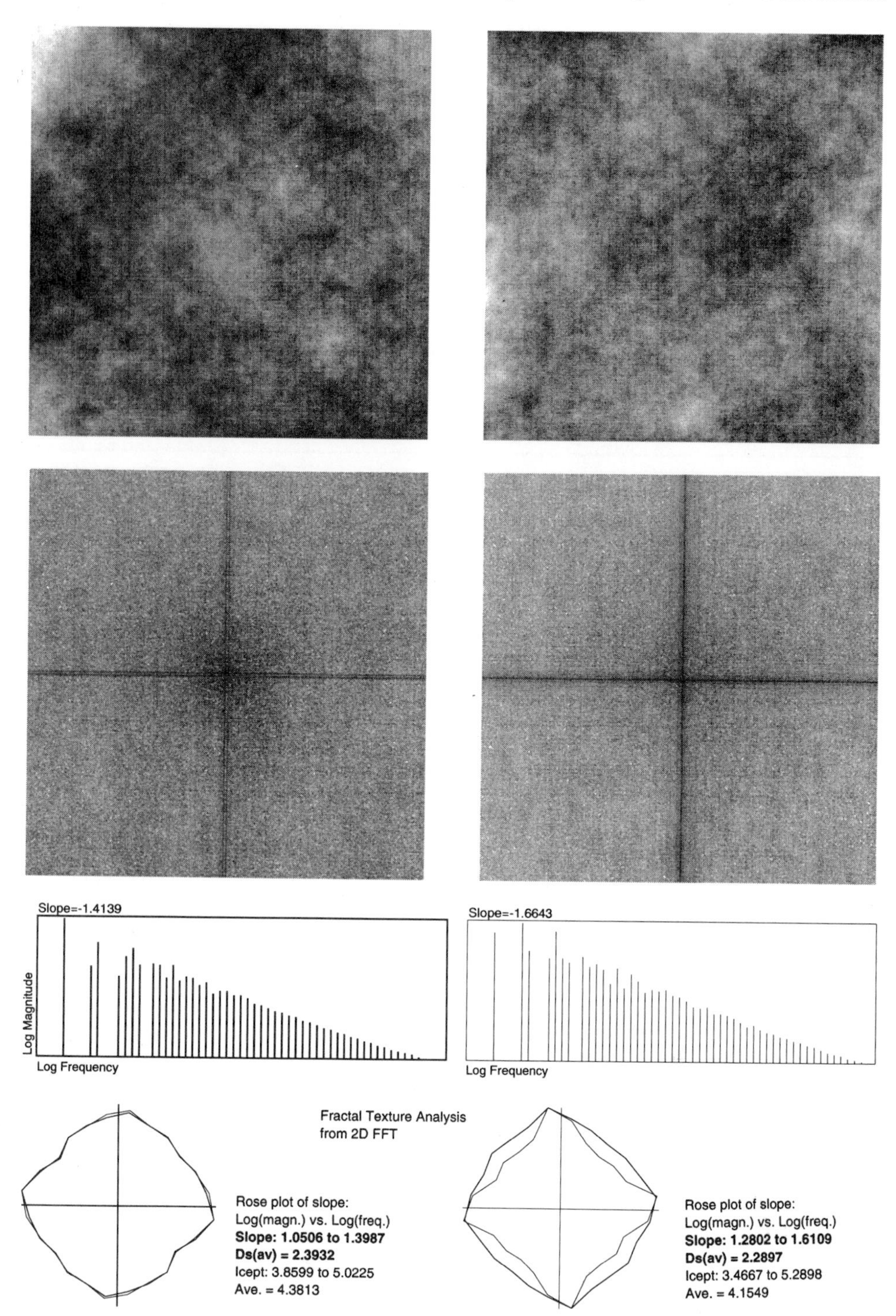

Figure 18. Surfaces generated by fractal Brownian model (step summation); left side image is generated with steps decreasing in magnitude by a factor of 0.5, and the right side by a factor of 0.9. Below are shown the slope of the log (Magnitude^2) vs. log (Frequency) plots from the 2DFT of each image. Notice that although the same α value was used for both images, the slopes of the power spectra and the fractal dimensions are different.

of steps (and where as usual α varies between 0 and 1) and the location of the step x_0 is uniformly random over the range of the profile, produces a random fractal profile with dimension $2 - \alpha$.

This method is generalized to two dimensions (surfaces) by adding together step functions in which the height h decreases with the number of steps as above, and the step position and orientation are both uniformly randomized. For both the profile and the surface method, the results are indistinguishable from the random displacement method but much more work is required to achieve those results. The displacement method works by iteratively reducing the scale width w at each step. For a range image 1024 points on a side, reduction by a factor of 2 requires only 10 steps to reach the scale of individual points ($2^{10} = 1024$), regardless of the magnitude of α. The summation of steps must be large enough that the individual step size becomes negligibly small, which does depend on the coefficient α. For $\alpha = 0.5$ and a surface in which three significant digits are used for elevations, this requires 10^6 steps, and as α varies so that the steps decrease in magnitude and the surface becomes rougher, even more additions are needed. However, as shown in Figure 18, the surface generated with a large number of steps ($\alpha = 0.9$) is quite isotropic and realistic in appearance.

This summation of steps is an implementation of Brownian motion, closely related to the generalization of fractal Brownian motion from a one-dimensional case of distance vs. time, to a two-dimensional function. This is not simply the case of Brownian motion in which the motion takes place on a two-dimensional surface. The displacement of the moving particle is random in both the x and y directions but the displacement is a function of a single dimension, time. The vector distance of the point from the origin increases as the square root of time, just as for the one-dimensional case for motion along a line. The same thing is true for Brownian motion in a three-dimensional space.

Mathematically, it is possible to define a function z (elevation) which depends separately on two variables, x and y, so that the elevation difference between any two points x_1,y_1 and x_2,y_2 has a mean or expected value which increases as the square root of the vector distance between the two points. This is a two-dimensional Brownian surface. If the exponent of increase is different from 1/2, it is called a fractal Brownian surface.

Mandelbrot–Weierstrass Functions

Another summation operation in which nonfractal shapes are added together to produce a fractal result is the Mandelbrot–Weierstrass function (Berry and Lewis 1980; Mandelbrot 1982; Majumdar and Bhushan 1990). The method has specifically been applied to modeling elevation profiles of surfaces (Roques-Carmes, Wehbi et al. 1987). This algorithm adds together a series of sine functions, but unlike the Fourier expansion theorem, the frequencies do not form a linear sequence. Instead, they increase geometrically as $f_j = c \cdot f_{j-1}$. The values of c are greater than 1, typically in the range from 1.1 to 3. The amplitudes of the terms decrease as $f^{-\beta}$, and the phases are uniformly randomized. The fractal dimension of a profile created by this summation is $D = 2 - \beta$. Changing the value of c changes the appearance of the generated profile, as shown in Figure 19. While it is fairly common to use values of c near 2, and the resulting profiles may look realistic, it is only with much smaller values of c that enough terms are added together to produce a profile whose Fourier transform is reasonably continuous and free from major spikes, as shown in the figure.

While this is the most common formulation of the Mandelbrot–Weierstrass method, other selections of frequencies are possible. Feder (1988) shows an example in which the frequencies increase as $f_j = f_{j-1}^{\gamma}$ for values of γ in the range 0.7…1.4, and it is also possible to use Fourier

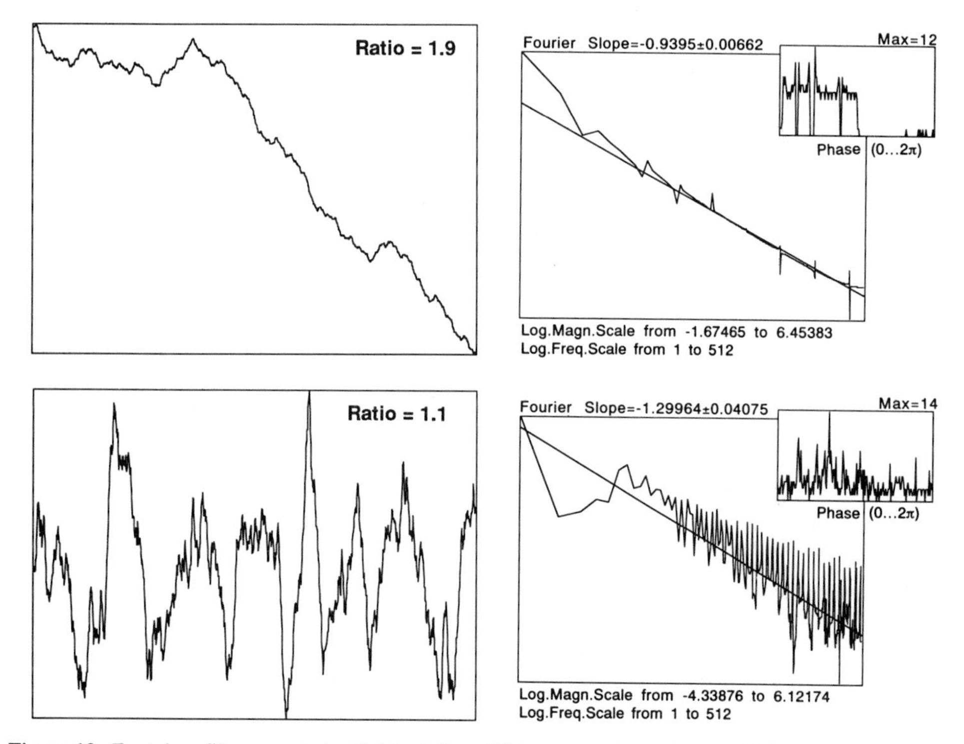

Figure 19. Fractal profiles generated with Mandelbrot–Weierstrass series, using $\alpha = 0.75$ and frequency ratios of 1.9 and 1.1. The Fourier power and phase spectra of each are shown at right.

series in which the frequencies increase linearly. It is the decrease of magnitude with frequency that determines the fractal dimension.

In principle, all of these summations should be an infinite series. In practice, the lowest frequency of interest corresponds to a wavelength of the order of the width of the profile, and the highest to the order of the point spacing along the profile. Higher and lower frequencies do not contribute anything to the observed profile, which is only an approximation to the theoretical fractal shape which continues its self-similarity to all scales. Enough terms must be used, of course, to thoroughly hide the individual sinusoidal components and to randomize the phase. For nontechnical purposes, a manageable number of terms is sufficient, but as the Fourier spectra in Figure 19 indicate, even a large number of terms (many frequencies) does not eliminate artifacts.

Unfortunately, the use of the Mandelbrot–Weierstrass formulation, or any of the other summations of sine terms, generalizes very poorly to surfaces (Kumar and Bodvarsson 1990). The result of superimposing a series of terms of increasing frequency and decreasing magnitude, in which the orientation as well as the phase is randomized, does not usually include enough terms to become truly fractal, and even more important does not generate an isotropic result. While this approach is commonly used as a tool to generate realistic-looking surfaces to represent "scenery," quantitatively it is limited by having too few phases and frequencies in each direction. The fractal dimension varies with direction in an unpredictable and uncontrollable way, so that neither isotropic nor specific anisotropic surface modeling is practical. Figure 20 shows two surface models produced using Mandelbrot–Weierstrass series. As for the case of the profiles shown above, using a small ratio of

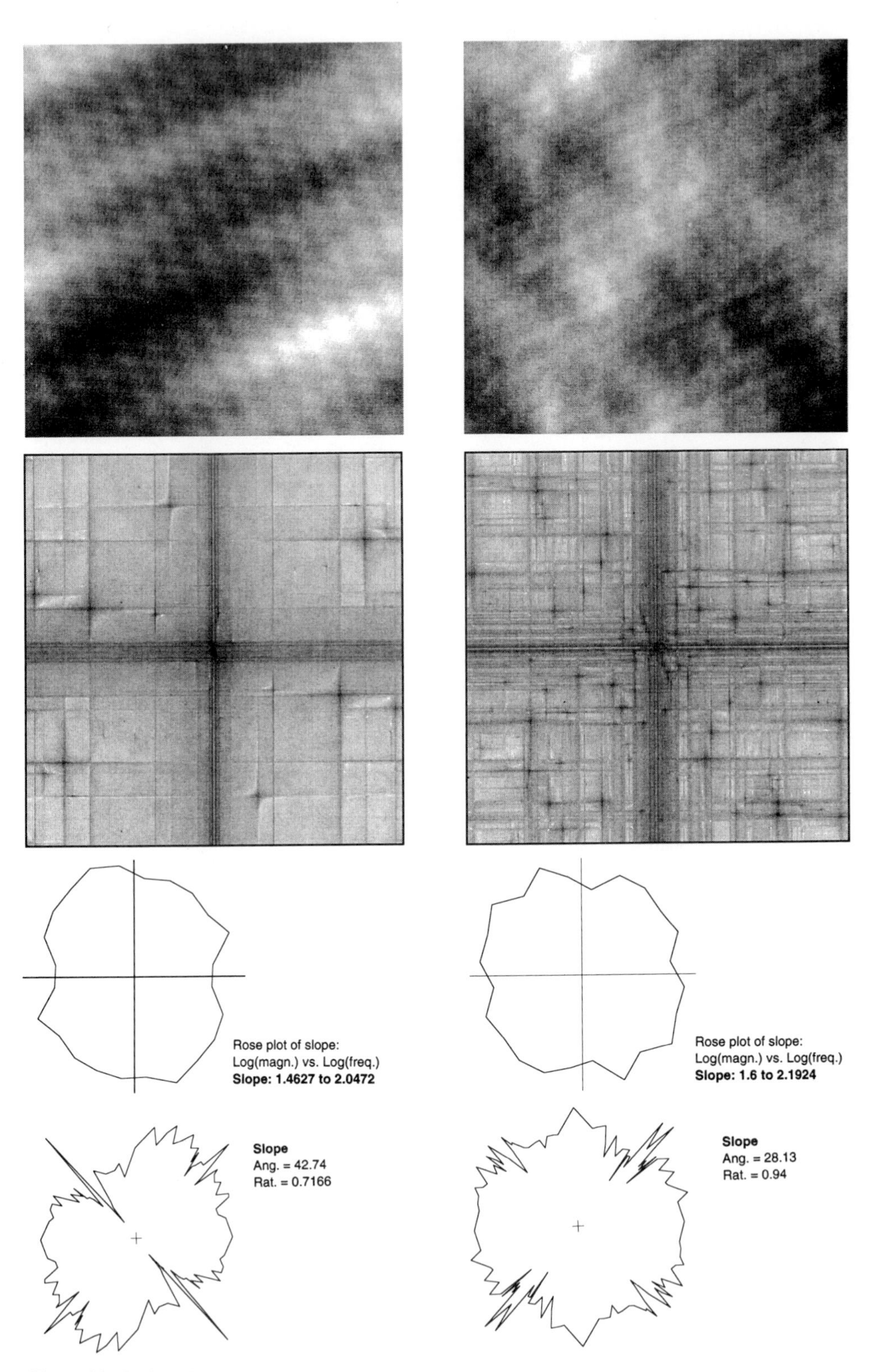

Figure 20. Surfaces images generated with Mandelbrot–Weierstrass series, using $\alpha = 0.75$ and frequency ratios of 1.9 **(left)** and 1.1 **(right)**. The 2DFT images and plots of log (Magnitude^2) vs. log (Frequency) are shown for each. Even with the smaller ratio, the spikes are evident and frequency space is sparsely filled, although the directional uniformity appears fairly good. The rose plot of the HOT results is also shown **(bottom)**.

Figure 21. Anisotropic fractal surface generated by inverse Fourier series.

frequencies introduces more terms (and takes longer to calculate), but produces a somewhat more uniform result as shown by the 2DFT and HOT measurements.

Using other sequences than the geometrical one used in the M–W series, which include enough terms to provide phase randomness and frequency isotropy, is computationally demanding and allows only very gradual variations of dimension with direction. The fractal surface shown in Figure 21 was generated by summing all of the terms in a 2D Fourier series (Osborne and Provenzale 1989). The phases were randomized and the magnitudes generated to fit a 3:1 ellipse, using a Gaussian random number generator. This pattern was then inverted using a two-dimensional FFT to produce a surface range image which is anisotropic. The method is also exceptionally good for producing isotropic surfaces. However, it is difficult to generate realistic surfaces which have larger ratios of anisotropy or abrupt directional variations.

The power spectrum averaged over all directions produces a straight line fit for log (Magnitude^2) vs. log (Frequency), with a slope that averages the dimension values used in the generation. It also shows an ellipsoidal variation in slope (but not in intercept) as expected (Figure 22). But measurement of this generated surface by other techniques produces some

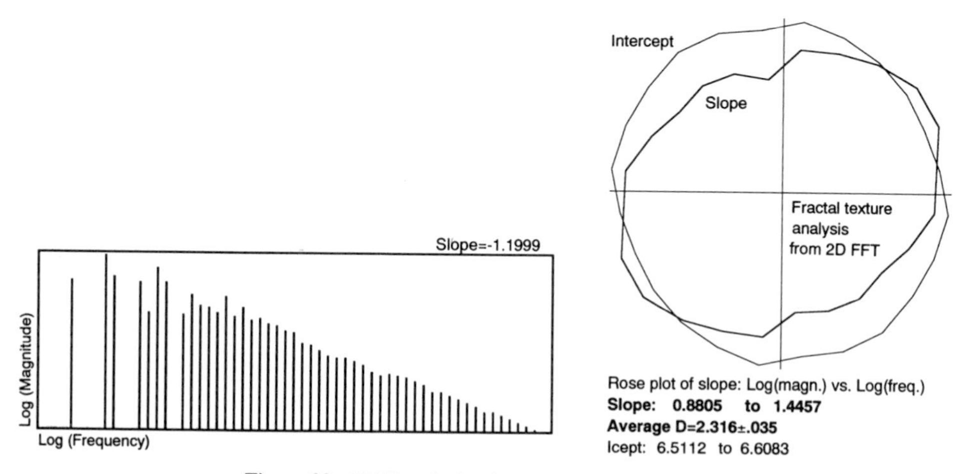

Figure 22. 2DFT analysis of the image in Figure 21.

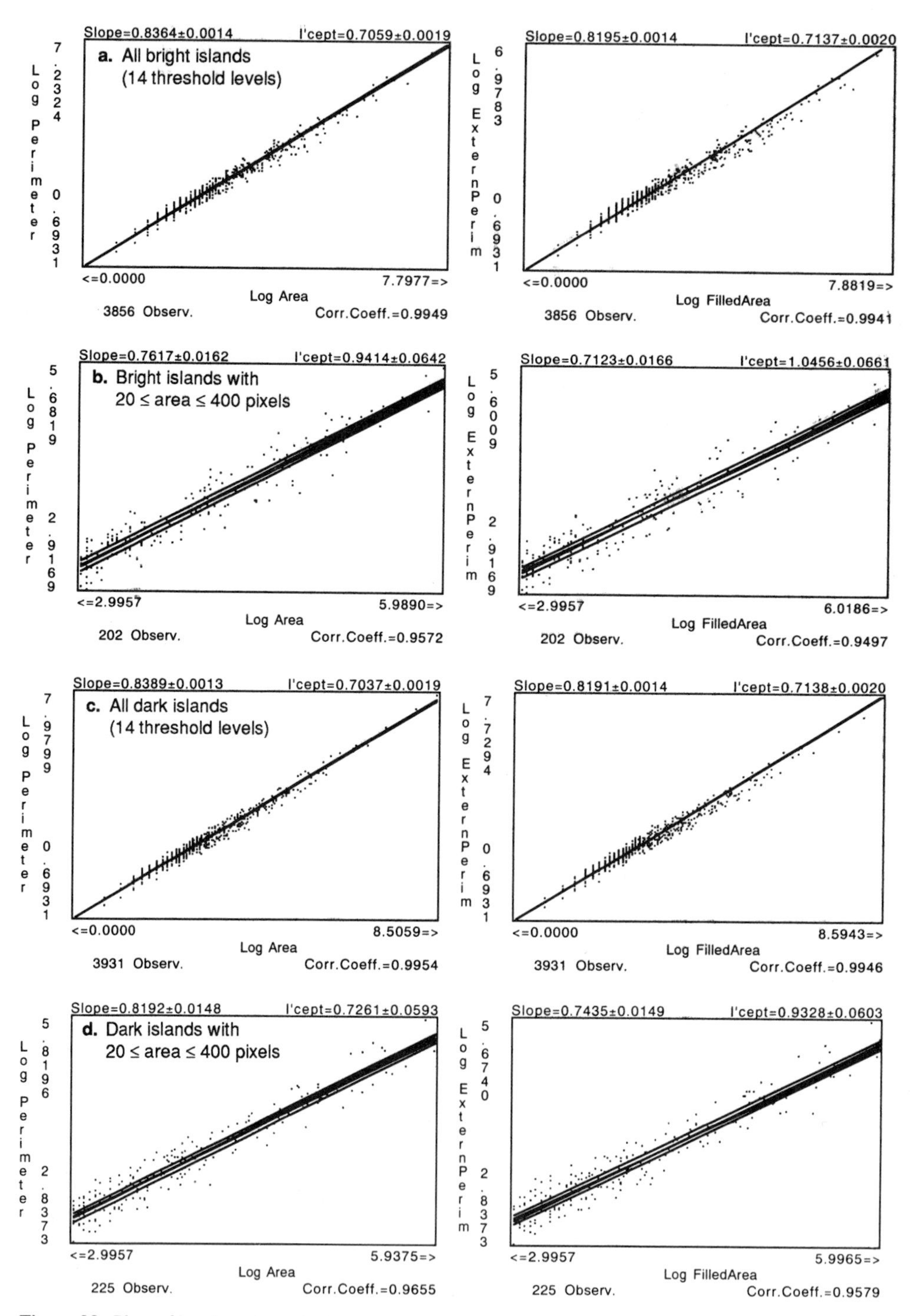

Figure 23. Plots of log P vs. log A for islands from Figure 21. Both dark and light islands are shown, for all of the data and for just those islands with areas between 20 and 400 pixels. Plots are shown both for the net area and perimeter, and for the external perimeter and filled area.

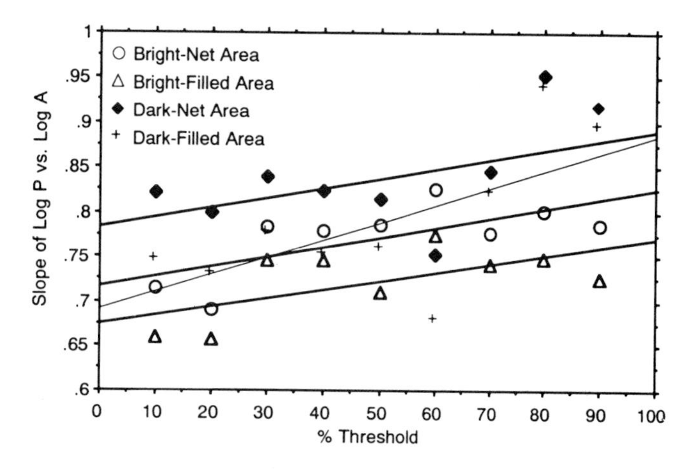

Figure 24. Variation of slope of log P vs. log A for bright and dark islands as a function of threshold level on the anisotropic surface. All of the slopes are significant ($p = .05$).

anomalies. Figure 23 shows plots of the perimeter and area of islands (dimensional analysis) from a series of thresholds at different elevation values. The individual plots appear normal, although the slope does not give the expected dimension value. Furthermore, plots from the individual thresholding levels show a significant trend as a function of elevation value (Figure 24) which is unexplained.

Other measures that use a slit island approach, such as measurement of the fractal dimension of the boundaries using a Kolmogorov technique, are also affected by the directional anisotropy as shown in Figure 25. Although the horizontal section method produces a zeroset of an isotropic fractal, which can then be measured conventionally, for a directionally anisotropic surface this is not the case. Since many real-world fractal surfaces may be anisotropic, this places some constraints on the use of measurement methods.

Comparing Models and Measurements

It is interesting to compare several of the models available for surface generation by using some of the measurement techniques discussed in earlier chapters. It might be imagined that all of these model and measurement procedures would produce equivalent results, perhaps

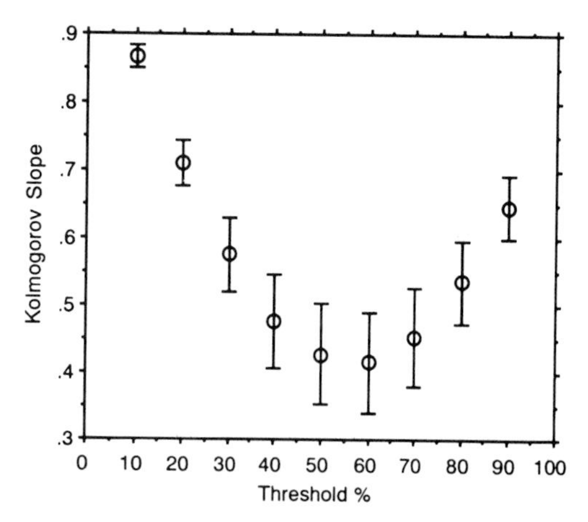

Figure 25. Inconsistent results from measuring Kolmogorov dimension on slit island contours at different threshold levels (large error bars indicate poor straight line fits for the box counts).

differing in some minor but consistent ways. However, the situation is much more complex than that.

Beginning with simple profiles, Figures 26–29 summarize a series of measurements performed on ninety profiles generated by each of several methods. The generating methods used were midpoint displacement with Voss' modification (Voss 1985a) to remove some artifacts, fractal Brownian motion (addition of steps), the Mandelbrot–Weierstrass function (with randomized phase values), and the inverse Fourier transform. In all cases, nominal fractal dimensions from 1.1 to 1.9 were produced by varying the alpha parameter in each model, and ten profiles with each alpha were generated using a Gaussian random number generator to produce the required values.

Each of these profiles was then measured using several of the techniques discussed earlier: the Minkowski (variational or covering method); the Korcak method (cumulative plot of intercept lengths); a plot of rms (root-mean-square differences) vs. window width; and both Hurst and Fourier plots. Although both the generating procedures and measurement methods have some biases, and there is an inherent variation in the generated profiles because of the random numbers used, it is still possible to draw some generalizations from the plots.

Most of the plots show a rough correlation between the measured and the nominal profile dimensions. However, in all cases the Minkowski measured dimension is the lowest value, and drops farther below the identity line as the dimension increases. Likewise, the rms method tends to produce the highest measured values, while the Korcak and Hurst values are in between and generally give the best agreement with the nominal value.

The Fourier measurement method produces excellent agreement with the inverse Fourier generation procedure, but of course it should. It is interesting that it does not do so well with

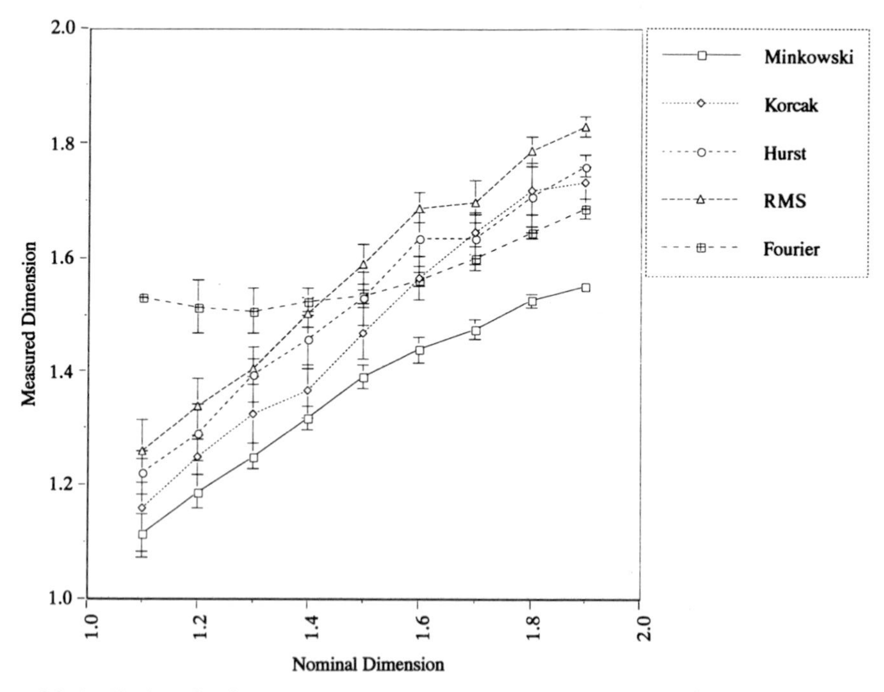

Figure 26. Application of various measurement methods to profiles produced with iterated midpoint displacement.

the other generation procedures. Even the Mandelbrot–Weierstrass summation of sinusoids produces deviations in the Fourier measurement at low dimensions, probably because the power spectrum is too sparse for a good linear fit. The midpoint displacement and Brownian addition procedures produce profiles with nominal dimensions below about 1.5 that the Fourier method does not measure properly. The steps and folds introduced by these methods apparently introduce high-frequency terms in the power spectrum that distort the result.

These results reinforce the idea that comparison of measurements on profiles using the same measurement method may be useful and valid even if the exact numeric value of the dimension is not necessarily very accurate. They also suggest that some of the generating models are less suitable for some purposes. For instance, the inverse Fourier transform method produces results that are well behaved by all of the measurement procedures, which is not the case for the midpoint displacement or Brownian step addition procedure.

Figures 30, 31, and 32 show the results of similar tests for surfaces. There is no clear difference evident between the use of midpoint displacement and inverse Fourier modelling, as there was for profiles. The fractal Brownian addition method produces surfaces whose measured values differ widely and (as shown by the missing points in the plots) cannot always be measured. Again, the various measurement methods produce somewhat different results for these surfaces. The Minkowski or covering dimension is lowest, especially at high dimension values. The Korcak results from cumulative plots of island areas, dimensional analysis from plots of log (Perimeter) vs. log (Area), and slit island results from a Kolmogorov or box-counting method applied to island boundaries, all use horizontal sectioning planes through the surface, and all three methods show rather similar results for both the midpoint displacement and inverse Fourier generated surfaces.

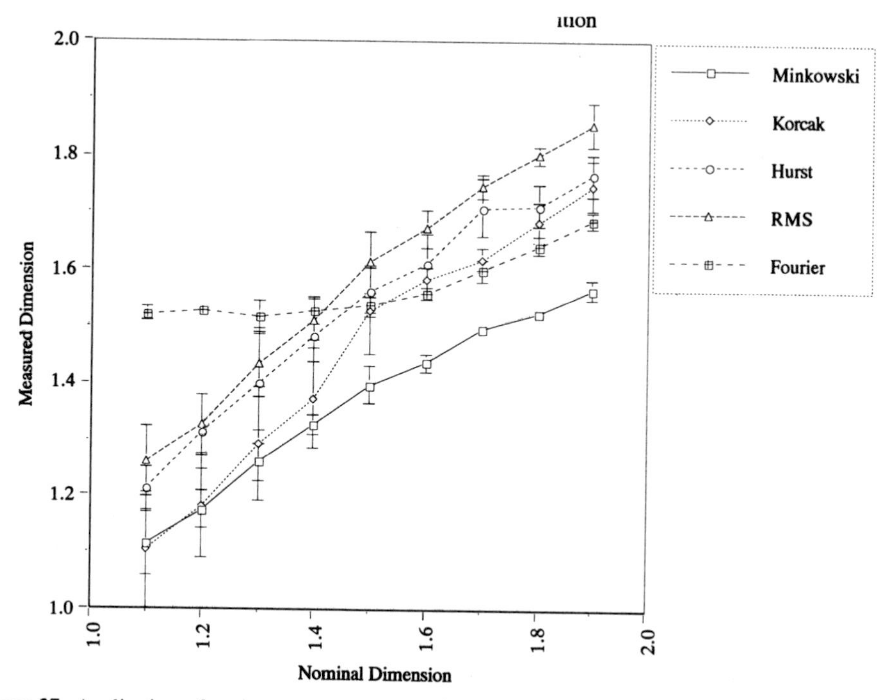

Figure 27. Application of various measurement methods to profiles produced with fractal Brownian step addition.

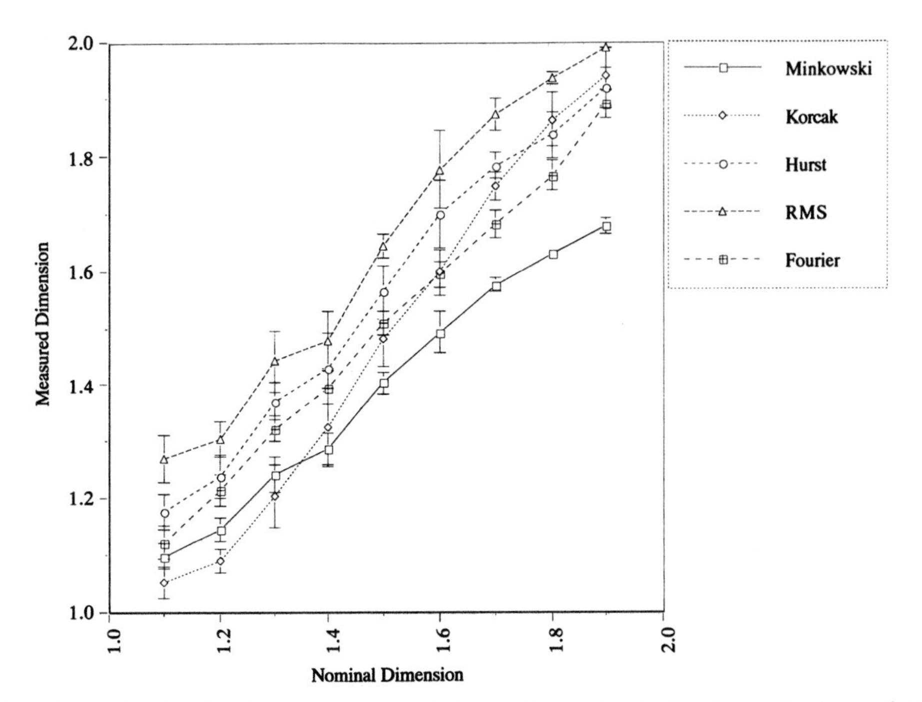

Figure 28. Application of various measurement methods to profiles produced with an inverse Fourier transform.

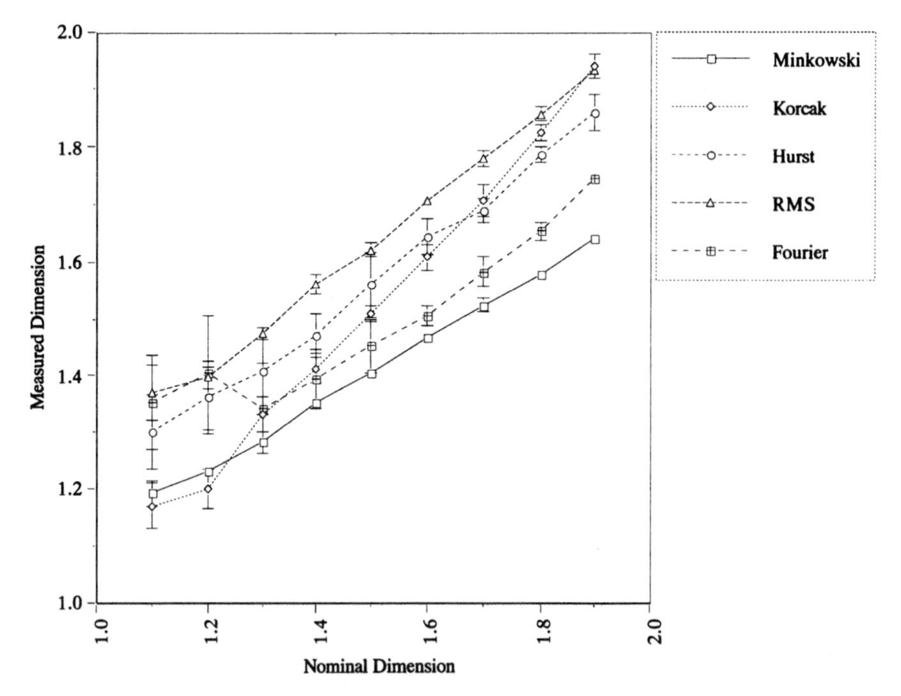

Figure 29. Application of various measurement methods to profiles produced with a Mandelbrot–Weierstrass summation.

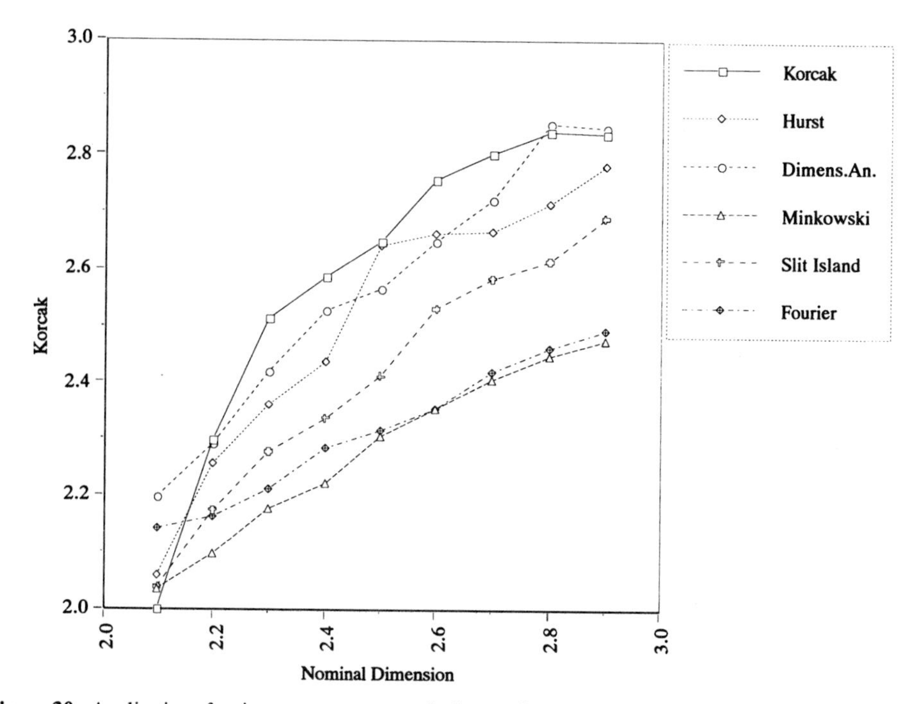

Figure 30. Application of various measurement methods to surfaces produced by iterated midpoint displacement.

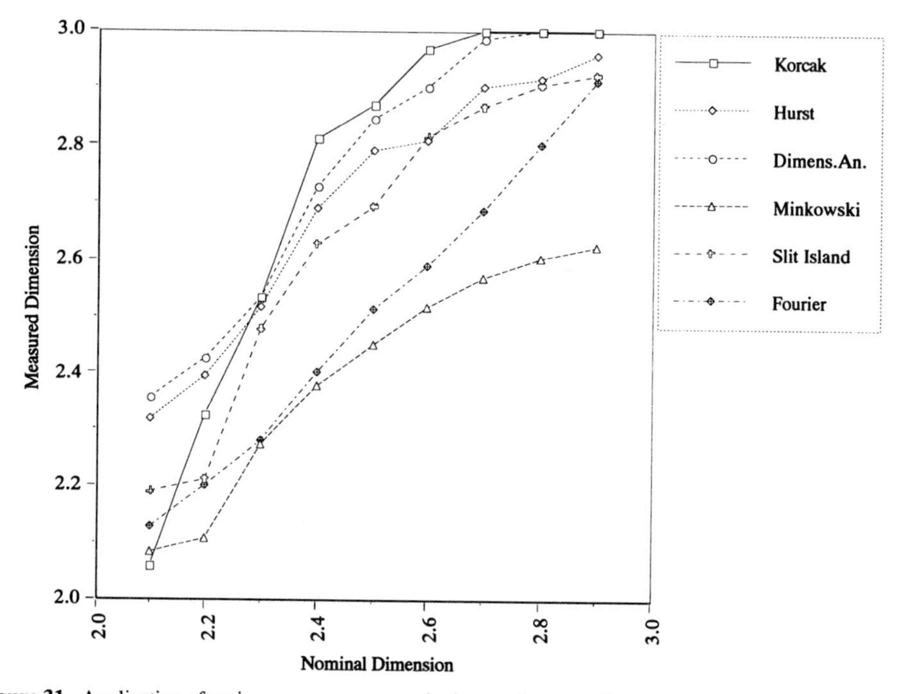

Figure 31. Application of various measurement methods to surfaces produced by an inverse Fourier transform.

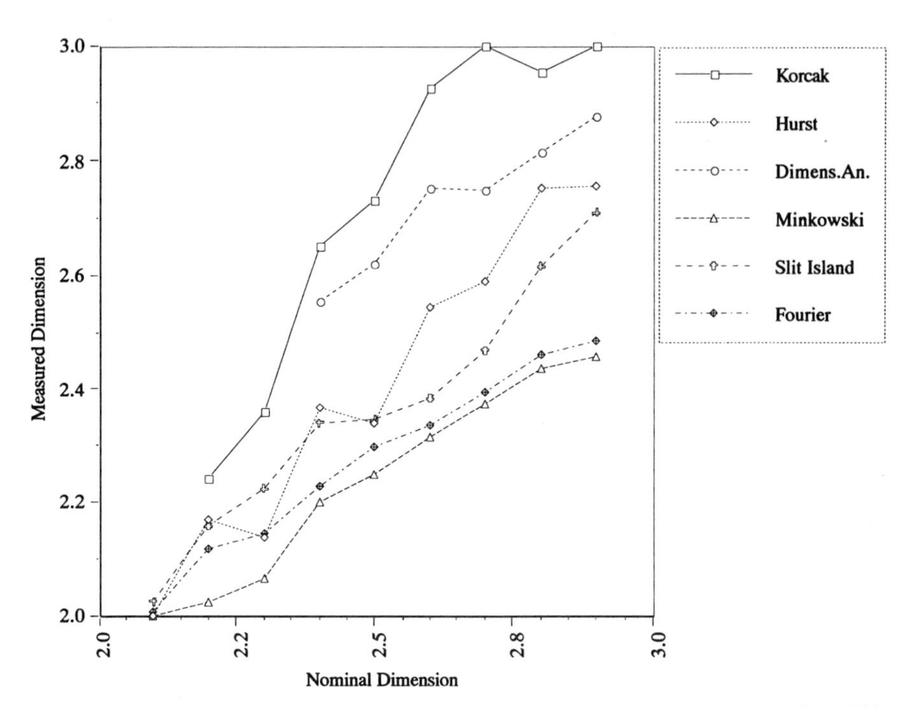

Figure 32. Application of various measurement methods to surfaces produced by fractal Brownian addition.

Takagi Functions

Of course, other functions than sinusoids can be superimposed to produce profiles or surfaces. The Koch structured fractals shown above use functions which are quite simple, although they cannot be described as compactly as a sine function. One set of functions that can be used are Takagi functions (Takagi 1903; Roques-Carmes, Wehbl et al. 1988; Dubuc 1989; Russ 1991a). Figure 33 shows these triangular shapes, which may be considered as simplified sine waves. For simple profiles, the addition of Takagi functions looks much like the summation of sinusoids in a Mandelbrot–Weierstrass function. To generate a fractal profile, the series of Takagi functions that decrease by a factor of two in width (or increase in frequency by a factor of 2) are added together. The starting location (phase) is randomized and the height decreases as $\langle h \rangle = w^{\alpha}$, where w is the width of the triangle base and α is a coefficient between 0 and 1. This produces a fractal profile with dimension $2 - \alpha$.

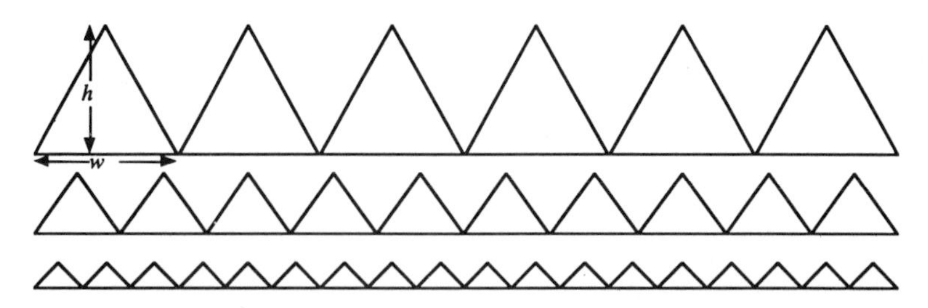

Figure 33. Takagi functions for generating a profile, consisting of a series of triangles whose heights h scale as a power of width w. The series of triangles are added with random phase (horizontal position).

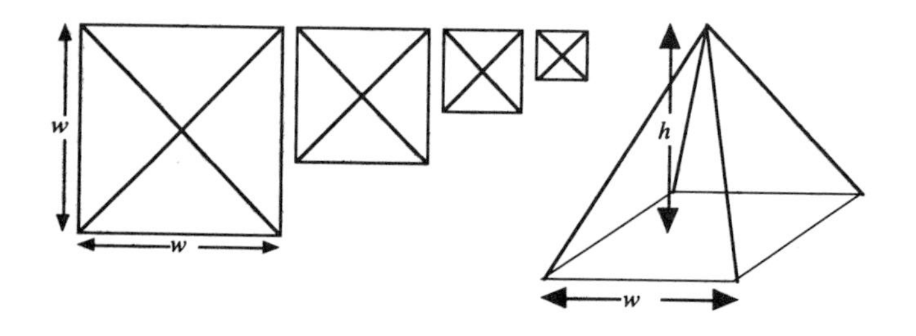

Figure 34. Generalization of Takagi functions to surface generation. The triangles become square pyramids.

This may seem to offer no advantage over the superposition of sine waves, beyond the elimination of the calculation of the value of the function itself, which is trivial using a computer. But in fact, the Takagi method can be generalized to surfaces much more easily, both to generate isotropic and controlled anisotropic fractals. An isotropic fractal surface can be created by adding together two-dimensional Takagi functions, which are simple square-based pyramids as shown in Figure 34. An array of pyramids of the largest size (perhaps a single pyramid as large as the image) is placed first. Then another array of pyramids with a base width w equal to half the first is superimposed. The placement must be randomized by shifting the starting point (randomizing the phase). The altitude of the pyramids is reduced according to the usual rule $\langle h \rangle = w^{\alpha}$, where w is the width of the triangle and α is a coefficient between 0 and 1. Continuing this process until the pyramid base width is close to the point spacing in the array produces an isotropic fractal surface with dimension $3 - \alpha$. A modification of this scaling rule will be introduced below to generate anisotropic fractal surfaces.

Takagi functions properly consist of a series of arrays of triangles (for an elevation profile) or pyramids (for a surface) whose base width for each step is reduced as 0.5^{α}. Hence, Takagi functions are an iterative midpoint displacement method in which the base width for each step is reduced by a factor of two. A more general formulation of this method for profiles is the Takagi–Landisberg function, in which the base width for step n is w^n. Takagi functions have $w = 0.5$, while Landisberg functions have $0.5 < w < 1$. For profiles, both produce a fractal dimension $D = 2 - |\log_2(w)|$. If the steps are randomized in height and can be either up or down, this is equivalent to a midpoint displacement technique. And, of course, if the steps are uniform in magnitude and alternate up and down, the result is a Koch structured fractal. Varying the standard deviation of the midpoint displacement, or using a non-Gaussian or even nonsymmetric distribution for the displacements, produces no change in D but a large change in the perceived surface roughness. No comprehensive analysis of these effects exists, but such "tricks" are sometimes used in generating landscape models with this approach.

It is possible to generalize these functions further to produce anisotropic surfaces. Figure 35 shows the principle, in which the pyramids have rectangular bases whose dimensions w_1 and w_2 scale independently with the height h. Figure 36 shows two surfaces, one isotropic and one anisotropic, generated with Takagi pyramids. The generated surface is most realistic and shows the fewest visual artifacts when w is close to 1 so that many terms are superimposed.

It is not entirely clear how an anisotropic fractal surface should be characterized. The fractal dimension of a line profile in the direction parallel to the sides of the pyramid base is $2 - \alpha_1$ (or $2 - \alpha_2$) and varies smoothly from one direction to the other. The surface fractal dimension determined by the Minkowski covering function or variational method is neither $3 - (\alpha_1 \cdot \alpha_2)^{1/2}$ nor $3 - \min(\alpha_1, \alpha_2)$.

The practical limit to the degree of anisotropy which can be produced by these generalized Takagi functions is set by the resolution of the surface range image being generated. If the pyramids shrink to less than the point spacing (or pixel width) in one

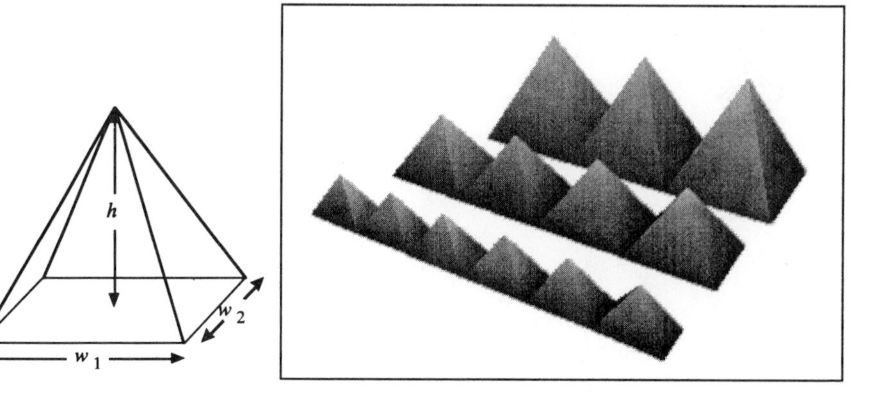

Figure 35. Takagi pyramids used to generate anisotropic surfaces. The two sides of the rectangular base scale independently with height.

direction while they are still many pixels wide in the other, or if the elevation becomes reduced to negligible proportions while the pyramid width in one direction is still significant, the resulting surface will not be completely fractal in one direction. This will be revealed by a Hurst or Minkowski plot that levels off below a certain range of neighbor distances. In effect this means that the generated images must have fine point spacing (and many points), and a grey-scale range used to represent the surface elevation of at least 256 levels (more are better). Within these limitations, the generalized Takagi functions can model surfaces whose directional fractal dimension varies smoothly. However, the generated anisotropic surfaces do not have the same appearance as a typical machined surface, in spite of the agreement of the 2DFT and HOT analysis results.

Note both the similarity and the differences between this method and the structured fractal and random midpoint displacement fractal. In the structured fractal, the phase or alignment of subsequent structures is not randomized, which produces results that do not look natural. Also, the scaling rule does not have the flexibility of selecting a coefficient α to produce varying fractal dimensions. The random midpoint displacement method requires the generation of a large quantity of random numbers, whereas the Takagi method uses a single height value for all of the pyramids of a particular size and iteration step. But the random midpoint displacement method does not randomize the phase or positioning of each step. Both the random midpoint displacement and Takagi methods produce isotropic and natural-looking fractal surfaces, while those generated by a structured shape are neither.

It is interesting to consider a kind of surface which corresponds in the physical mechanism of its formation to the generalized Takagi model and for which it might be a suitable model. Normal cratering of a surface by particles is well known to produce a fractal surface. If the incidence is not normal, the effect is to make all of the craters into ellipses, but this results only in a weakly anisotropic fractal and is equivalent to stretching the surface produced by normal impact. If the pyramids are imagined to be craters (subtracting them instead of adding them works just as well for generating the surface model), then changing the aspect ratio with size is equivalent to changing the angle of incidence with size, which could be the effect of an atmosphere above a planet, or gravity effects in particle bombardment. Similar effects may be present in agglomerated or deposited surfaces due to anisotropy in the deposition, electrostatic attraction, or other effects.

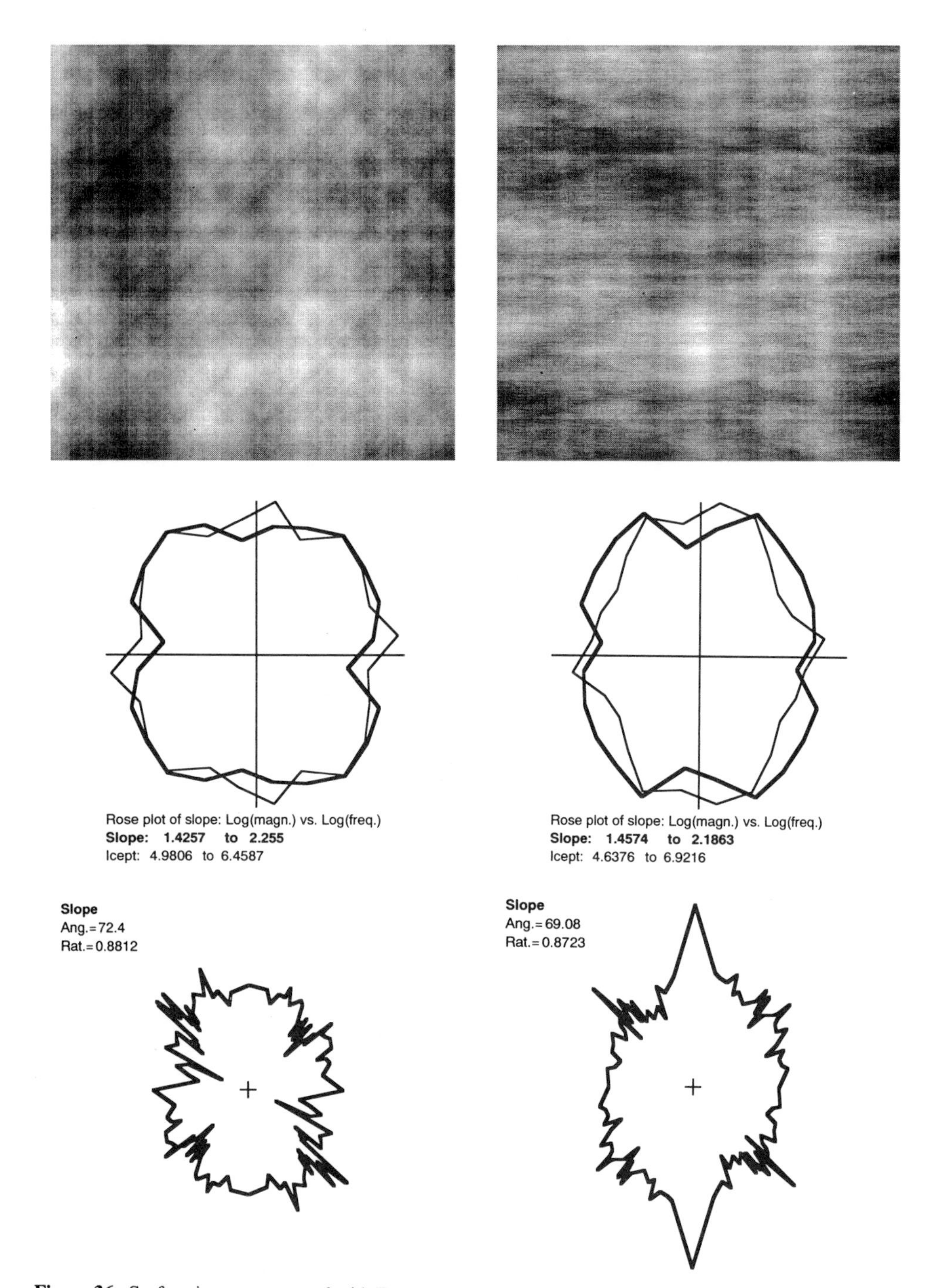

Figure 36. Surface images generated with Takagi functions: **Top:** left: square-based pyramids, $\alpha = 0.75$; right: anisotropic pyramids, $\alpha_1 = 0.5$, $\alpha_2 = 0.9$. In both images the sequence of terms uses a dimension ratio of 0.95. **Middle**: Rose plots of the slope and intercept of the log (Magnitude^2) vs. log (Frequency) plot from the 2DFT of the image. **Bottom:** Rose plots of the slope of the Hurst Orientation Transform.

Weak and Strong Anisotropy

The random fractal surfaces generated by the methods described above are good models for some real surfaces. These include natural terrain, corrosion surfaces, agglomerates and some catalysts, etc. They are not good models for many other surfaces which are fractal but not isotropic. These include many fracture and wear surfaces, and especially machined surfaces. Attempts to produce anisotropic fractals in which the degree of anisotropy can be controlled can be separated into weak and strong anisotropy. The difference is not necessarily in the visual appearance of the surface, as will be shown. It instead has to do with whether the fractal dimension itself varies with direction.

The anisotropy of some engineering surfaces, such as those produced by machining (Thomas and Thomas 1986), wear, etc., is an important characteristic of the history and properties of the surface. It is clearly inadequate as well as impractical simply to measure profiles in many different directions. In addition, it is not clear that there is any unique relationship between the surface fractal dimension and those of the profiles.

A weakly anisotropic fractal surface can be created by stretching an isotropic fractal surface (Bush, Gibson et al. 1978; Thomas and Thomas 1988). Such surfaces are visually anisotropic, as shown in the example of Figure 37. However, the fractal dimension measured in any direction is the same as shown by the rose plots from 2DFT and HOT analysis. This seems at first to present a paradox, but in fact it is inherent in the very definition of fractal shapes. It was pointed out before that a fractal surface might be self-affine rather than strictly self-similar. An isotropic self-affine fractal would have the same scaling and units in the x and y directions, but different in the z direction. It is only the way that the magnitude of z differences between points varies as a function of point spacing, as represented by the α coefficient shown in many formulations above, which specifies the fractal dimension. Stretching the isotropic fractal surface simply makes it self-affine in all three directions. The dimension is not changed.

This is true regardless of what method is used to perform the measurement. Either a Minkowski (variational) plot of area/width vs. measurement scale or a Hurst plot of maximum difference vs. distance will be shifted vertically by the stretching operation, but the slope will not change, and it is the slope which is related to the fractal dimension. To characterize these weakly anisotropic fractals, the vertical displacement of the line must be considered.

This is related to the topothesy of the surface, discussed in Chapter 4. The mathematical precision of the determination and the dominance of the low frequency terms in the Fourier expansion make it difficult to measure the intercept or the topothesy with precision. Furthermore, the measure of topothesy for a self-affine profile rather than a strictly self-similar one is suspect since the units of measure in the x and z directions may have no relationship, just as the scaling exponent may be different. Also it is difficult to generalize this to the case of a 2D range image since the central value in the 2DFT power spectrum is the intercept in all directions, but clearly the magnitude of the roughness may vary with direction.

For these reasons, the variation in the intercept from the HOT method remains a very sensitive tool for examining the degree of weak anisotropy present. Since the lateral scale within a single image is uniform, the comparison of values in different directions can be performed without concern for the actual measurement scale. Measurements performed at different magnifications or with different instruments will not generally produce intercept values which agree quantitatively, but the ratios of the values as a function of direction (the axial ratio of an ellipse fitted to the HOT data) are dimensionless and can be used for this purpose independent of scale. For a weakly anisotropic surface, the slope of the HOT transform is uniform with direction but the intercept is not, and its variation can be used to characterize the anisotropy.

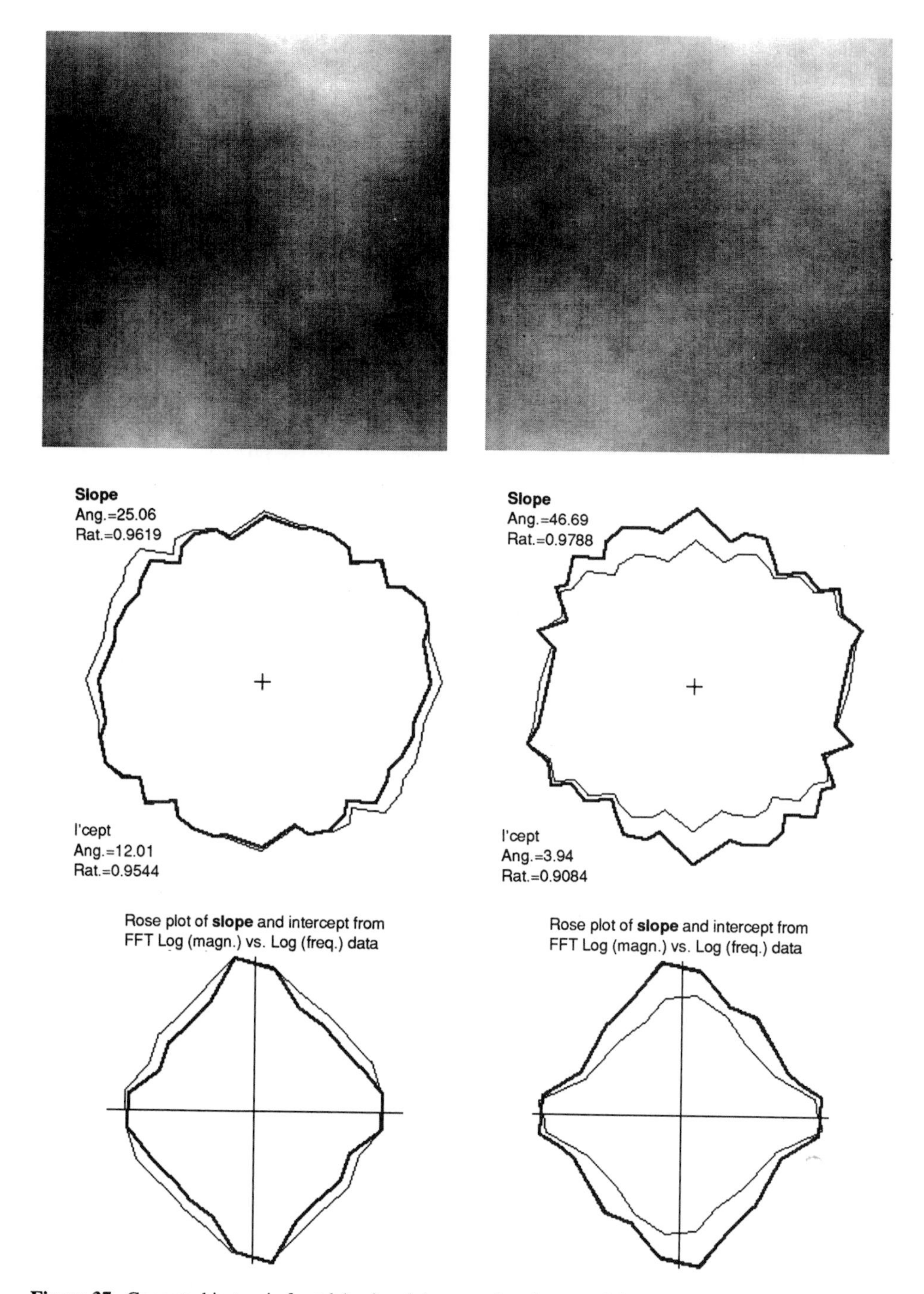

Figure 37. Generated isotropic fractal **(top)** and the same data (image originally shown as Figure 17) stretched horizontally by 50%. Rose plots of the slope and intercept of the HOT data **(middle)**, and of the slope and intercept of log (Magnitude^2) vs. log (Frequency) data from the 2DFT **(bottom)** show that the slope (and hence the fractal dimension) remains isotropic, but the intercept varies.

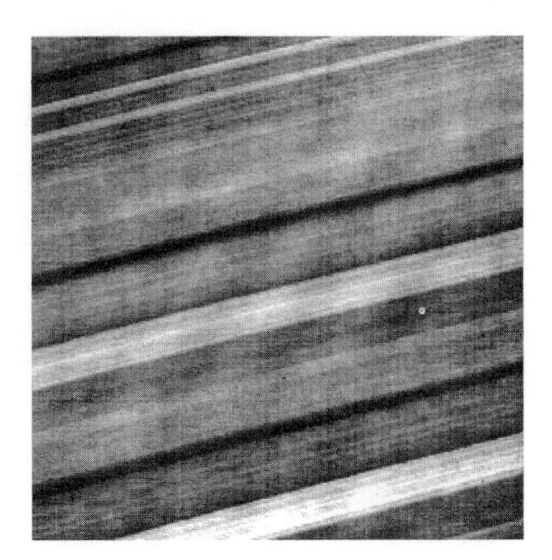

Figure 38. Atomic Force Microscope (AFM) image of machined surface of nickel.

Strongly anisotropic surfaces have a fractal dimension which varies with direction. This is observed for surfaces eroded by particle impact at an angle, for some fracture surfaces, and especially for machined surfaces. Scott (1990) has discussed the physics of machining in terms of chaotic dynamics, and it seems reasonable to expect different fractal dimensions along the direction of travel of the tool and perpendicular to that direction. Different sets of chaotic oscillations present in those two very different time domains, involving the tool tip, the local atomic and dislocation motions in the workpiece, and stress in the support machinery, should each contribute to these in different ways.

Observations of real machined surfaces lend support to this, as shown in Figure 38 and discussed further in Chapter 8. The obvious ridges which run in the machining direction arise largely from the shape of the tool tip, while the much smaller and more irregular variations along those ridges reveal oscillation frequencies too high to be anything but very local motions in the material as it is fractured at the moving tip.

Modeling an Anisotropic Surface

It might seem that a useful model for machined surfaces would be a technique known as extrusion. Consider a fractal elevation profile translated at right angles to the x direction. This produces a corrugated surface as shown in Figure 39, which bears some superficial similarity to a machined surface. If the extrusion is performed not along a straight line, but following another curve, a cross-extruded surface is produced. If this second curve is another fractal elevation profile, the result is as shown in the figure. Since the two elevation profiles used in this process may have different fractal dimensions, it might be expected that the surface roughness could be controlled to produce useful simulations. Of course, it makes no difference which profile is extruded along the other; the results are identical. And it is possible to add the extrusions at orientations other than right angles, or to add more than two.

First, the question arises as to what the fractal dimension of such a cross-extruded surface is, and how it varies with direction. Mandelbrot (1982) and Feder (1988) both suggest the "rule of thumb" that any superposition of fractals with dimensions D_1 and D_2 should produce a result whose fractal dimension is their sum, $D = D_1 + D_2$, unless the sum exceeds the dimension of the embedding space (3.0). For the case of the straight-line extrusion, in which $D_2 = 1.0$, the result is to predict that the surface would have a dimension whose fractional value is the same

Figure 39. Surface images generated by extruding a profile (generated by midpoint displacement with $\alpha = 0.6$) along a straight line, and along another fractal profile ($\alpha = 0.8$).

as the generating profile, only increased by 1 to take into account the topological difference between a line and a surface.

This somewhat nonintuitive result is quite correct. To verify it, imagine constructing a Minkowski or Hurst plot along lines running in different directions across an extruded or corrugated surface. In any direction except exactly along the extrusion direction (where the lines are straight and Euclidean), the profile which is traced out is exactly the same, only stretched to different degrees. As discussed before, this is simply a self-affine transformation of the original data and produces no change in the fractal dimension. Consequently, in (almost) any direction the line profile dimension is equal to that of the original line, and the singularity in the extrusion direction becomes lost in averaging. A direct two-dimensional measurement of the surface fractal dimension using the Minkowski variational or covering method confirms that the surface fractal dimension is 1.0 greater than the extruded profile.

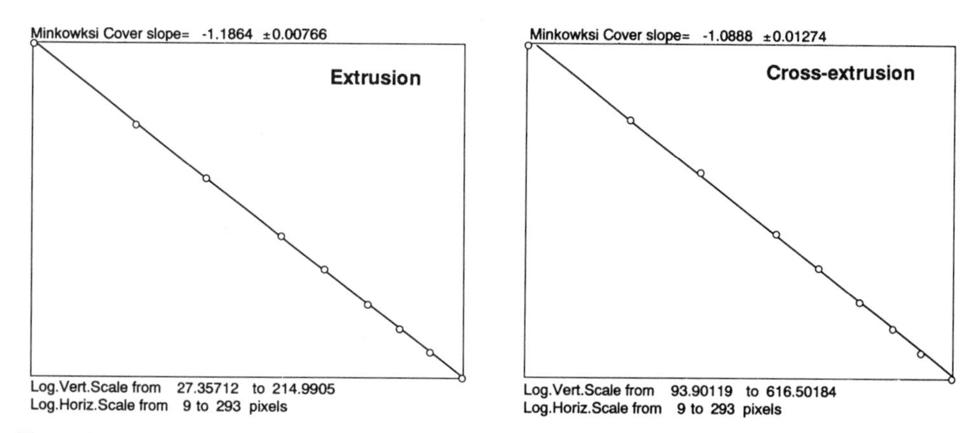

Figure 40. Minkowski (variational or cover) dimension analysis for the extruded and cross-extruded surfaces from Figure 39.

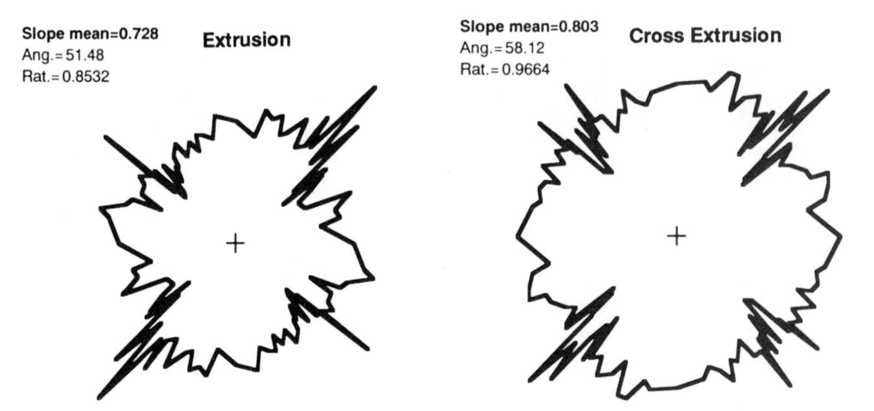

Figure 41. HOT analysis of extruded and cross-extruded surfaces from Figure 39.

But it is not clear from intuition that averaging is a correct procedure to determine the dimension of the surface (compared, say, to adding 1.0 to the larger profile value), or if it is, whether arithmetic or geometric averaging is appropriate. For a singly extruded surface, either will produce the same, correct result. For the case of cross-extrusion, the Minkowski variational method gives a result that is not easily interpreted (Figure 40). In both cases, the slope of the plot is greater than 1.0 which would correspond to a dimension for the surface that is less than 2, which is impossible. In fact, the results from a Minkowski plot applied to an anisotropic surface often exhibit this behavior. It is not clear that any of the measurement methods that assume isotropy give useful results for anisotropic surfaces. A further discussion of such mixed fractals is presented in Chapter 7.

Analysis of the directionality of the dimension using the HOT method (Figure 41) shows a uniform polar plot within the available precision (the relatively small size of the image and the large degree of stretch in directions close to the extrusion axis make those measurements less precise). Analysis of the slope of the plot of log (Magnitude^2) vs. log (Frequency) in a 2DFT image is also in agreement (Figure 42), although the small size of the image also limits the precision of this result.

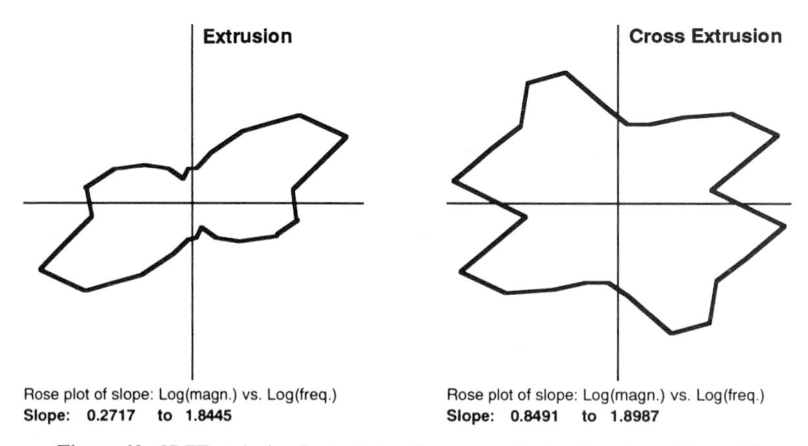

Figure 42. 2DFT analysis of extruded and cross-extruded surfaces from Figure 39.

If the two elevation profiles used in the cross-extrusion do not have the same fractal dimension, the resulting surface does not conform to simple expectations. First, while the measured dimension along the x and y axes corresponds to those of the original profiles, in other directions the value that is measured varies with direction and does not equal $D_1 + D_2 - 1.0$, which might be expected from the rule of thumb quoted above. The resulting surface is clearly anisotropic, but is "rougher" in a diagonal direction than in either of the extrusion directions. A two-dimensional measurement of the surface fractal dimension using the covering or variational method produces a value that is not simply $D_1 + D_2$. This is true even if the two extrusion directions are not orthogonal. If more than two extrusions are combined, the result is not simply the sum of the individual profiles, but rather 2.0 plus the sum of the fractional parts of the individual profile dimensions, until the sum reaches 3.0 (Russ 1991b).

If the two generating profiles have different fractal dimensions, as illustrated in Figures 39–42, the resulting 2DFT and HOT analyses show that the surface has a dimension that does not vary in a smooth elliptical manner (and is in fact quite difficult to interpret). Likewise, the two-dimensional measurement of the covering dimension does not yield $D_1 + D_2$. The difficulty of analyzing and controlling the surface roughness using cross extrusion, the impossibility of getting a smoothly anisotropic result as one possible outcome, and the paradox presented by the failure of the rule of thumb, make this method a poor choice for the modeling of complex, strongly anisotropic surfaces.

There is another paradox which arises in considering strongly anisotropic surfaces. The intercept value or topothesy needed to interpret weakly anisotropic surfaces has not been mentioned in connection with the strongly anisotropic case. It is not clear what, if any, physical meaning this would have, but it does give rise to a thought experiment. If the topothesy or intercept value is not the same in the two directions (as is typically the observed situation), then plots of the structure function in the two directions would have some distance value for which they would be equal (Thomas and Thomas 1988). For smaller distances, the magnitude of the vertical roughness would be greater for one direction, and vice versa for larger distances. The direction with the smaller fractal dimension would have the greater vertical displacements at smaller dimensions. This is certainly counter-intuitive, and indicates a need for further investigation.

A practical demonstration of this limitation of the use of fractal dimension as the characterizing parameter for surface roughness is shown in Figure 43. Two perpendicular extrusions with different fractal dimension (generated with $\alpha = 0.9$ and $\alpha = 0.5$) were combined by simple addition, but using different scaling constants of 1:1, 10:1, and 100:1. The surfaces are quite different in appearance but have the same fractal dimensions as a function of direction as shown by the slope plots for the 2DFT. It is only the intercept plots which show any difference, and the meaning of this parameter (or the related idea of topothesy) is not entirely clear for anisotropic surfaces.

Analysis of the machined surface shown above in Figure 38 is presented in Figure 44 using the 2DFT and HOT methods. Both show a strongly anisotropic surface in which the fractal dimension for an elevation profile varies more or less smoothly as a function of direction, between two extremes in the direction parallel to and perpendicular to the machining striations. It is also noteworthy that the low-frequency periodicity imposed by the machining tool shape is not confined to a straight line transverse to the machining direction. Unless these particular frequency terms are ignored in performing the fit, the results for the fractal dimension will be biased. In most cases, the spacings must be smaller and the frequencies higher than characteristic scales such as the tool dimensions, feed distance, etc. Specific vibrational modes of the machine must also be ignored.

Realistic model images for a machined surface, one of the most anisotropic normally encountered, can be constructed by adding together the unidirectional grooves produced by a

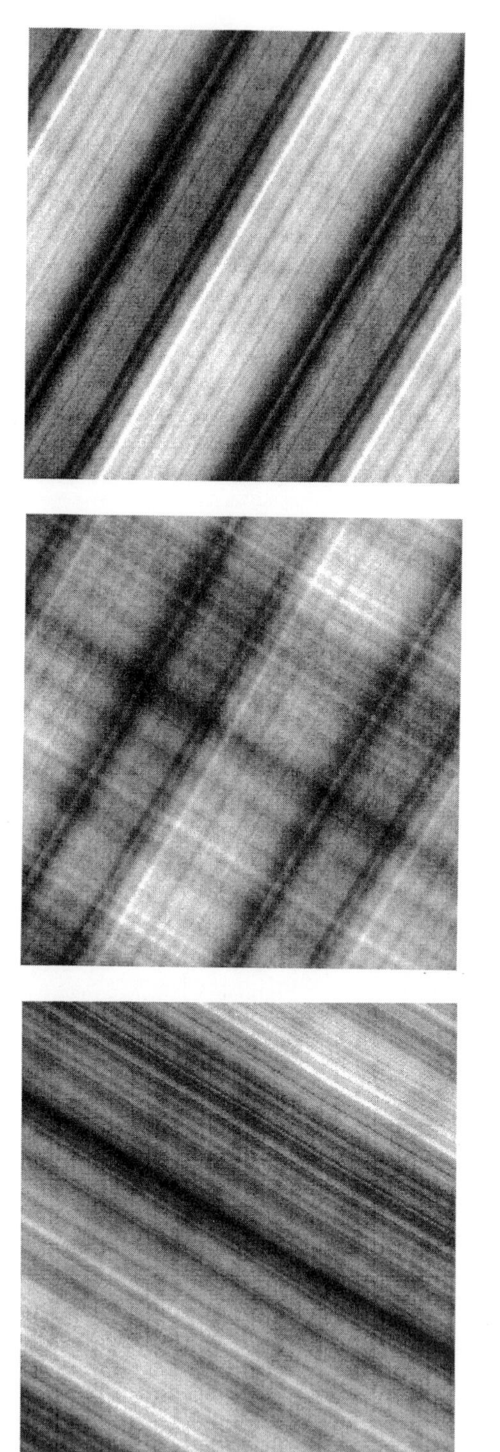

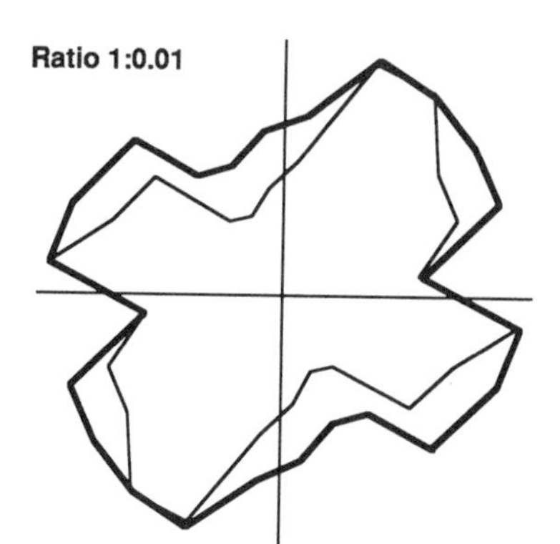

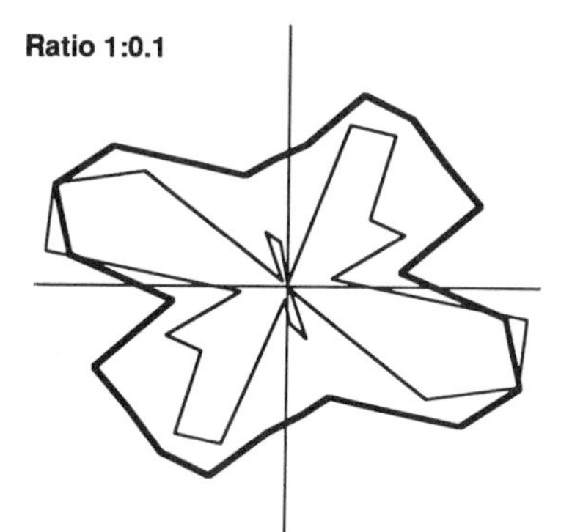

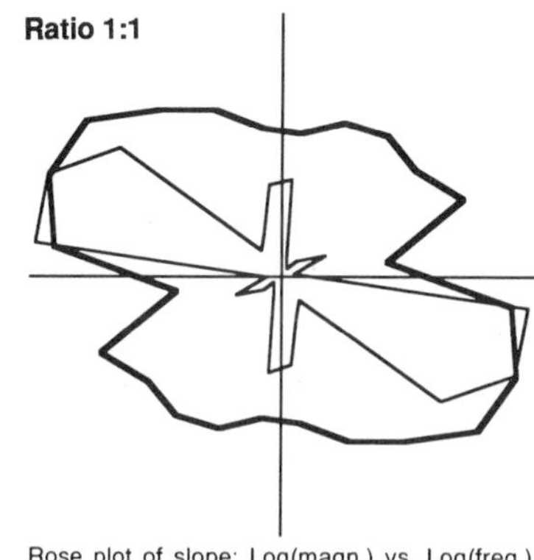

Figure 43. Sum of extrusions with $\alpha = 0.5$ and 0.9 with relative scaling changing by factors of ten, with the rose plot of the slope and intercept of the 2DFT. The slope plots are all similar, while the intercept values vary widely.

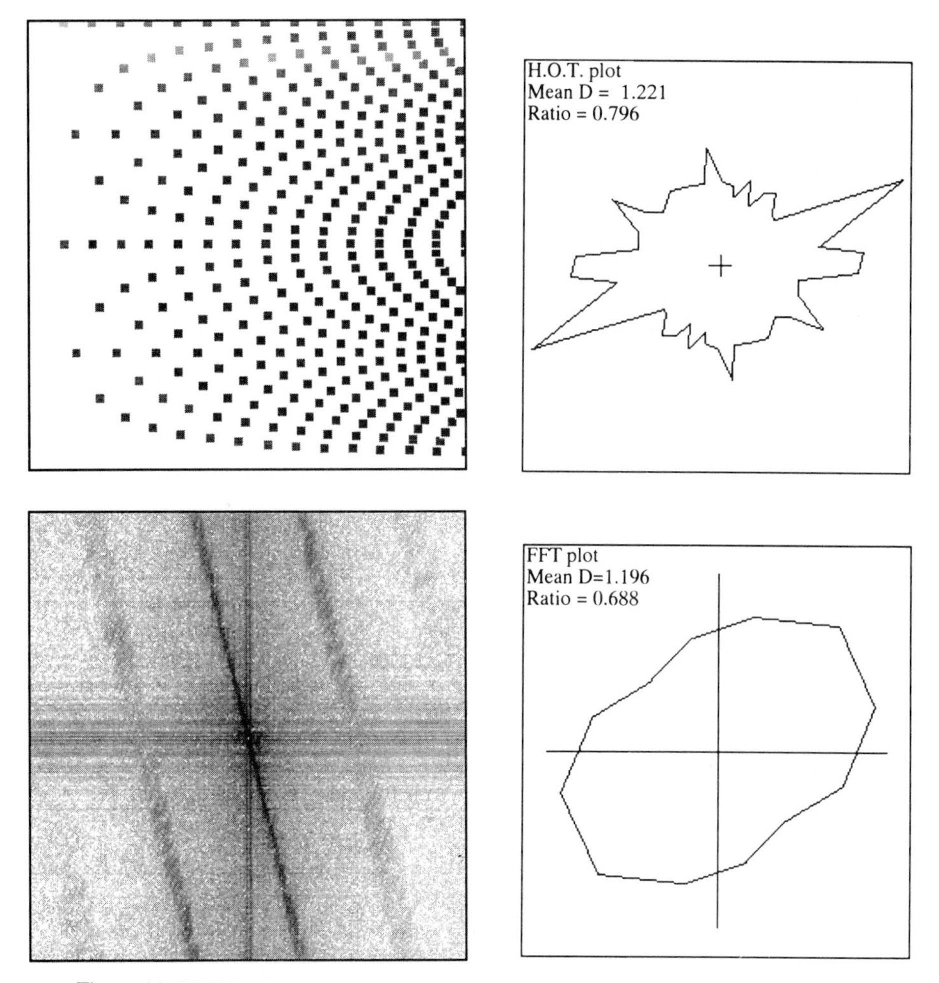

Figure 44. HOT **(top)** and 2DFT **(bottom)** measurements on surface image from Figure 38.

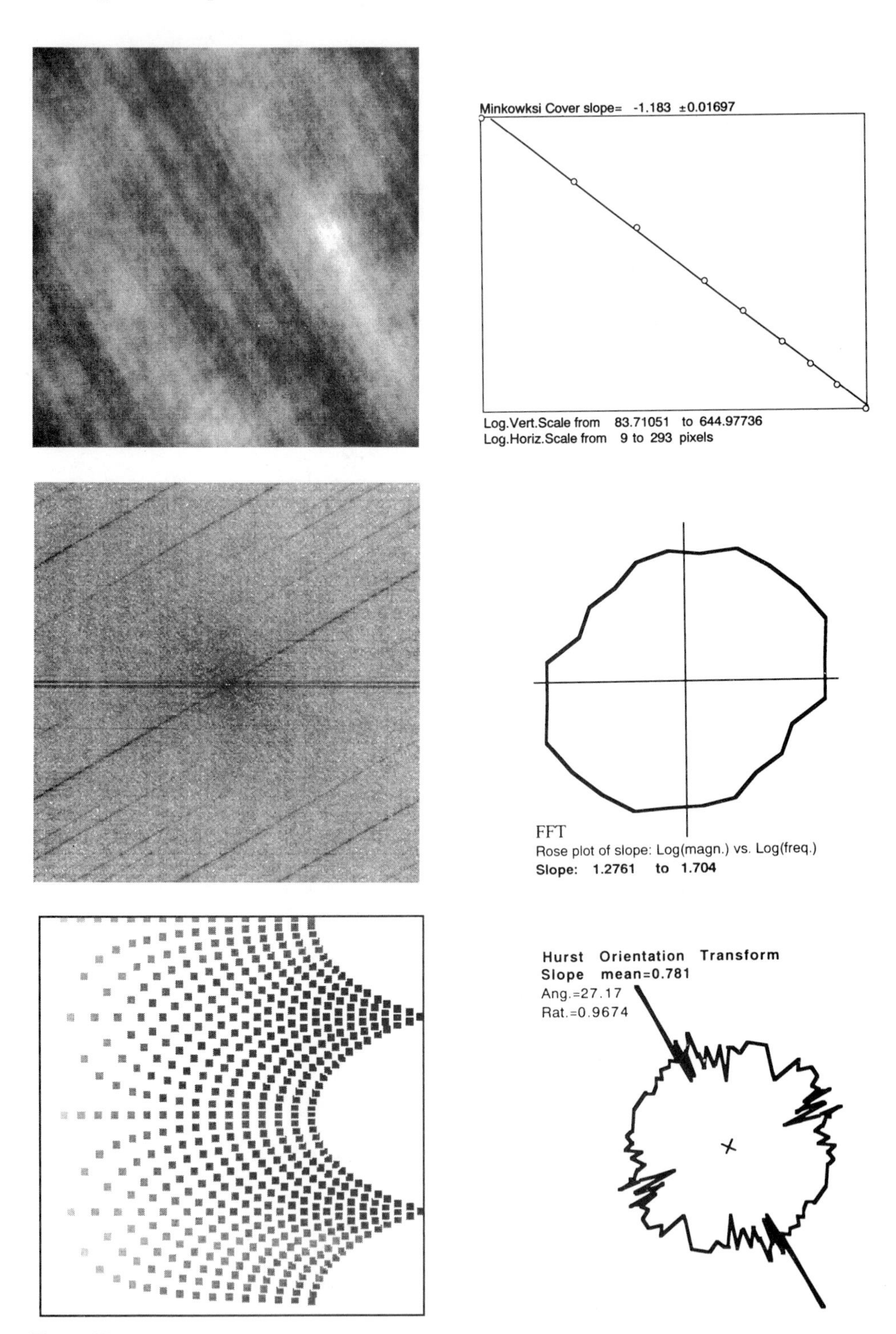

Figure 45. Composite surface image produced by adding extrusion ($\alpha = 0.6$) from Figure 39 and an anisotropic inverse Fourier transform ($\alpha = 0.6$ and $\alpha = 0.8$), with Minkowski (variational) plot **(top, right)**, 2DFT **(center),** and HOT **(bottom)** analyses.

single extrusion, with the result of an inverse Fourier transform that is either isotropic or anisotropic. Figure 45 shows an example, with its 2DFT and HOT analyses. To apply this method to match a particular machined surface, several steps are required. First, the fractal dimension in a direction perpendicular to the principal grooves is determined by averaging the measurement on several vertical elevation profiles. Then the 2DFT of the surface range image is determined and the anisotropy is determined using all directions except that in the transverse direction, which is dominated by the grooves. These two values together allow the model to generate a realistic mimic. It is not clear yet how these parameters relate to the surface history or properties.

The ability to model a variety of fractal surfaces allows visual and measurement comparison with real surface images. This provides a test bed for the various measurement procedures, and may add several new tools to the study of many man-made and natural surfaces.

7

Mixed Fractals

Limited Self-Similarity

From a purely mathematical point of view, a fractal curve or surface is defined as having a statistically self-similar form at any dimensional scale. In other words, the surface area or line length should increase without limit as the image magnification is increased. Some purely mathematical fractal objects, such as the now well-known Mandelbrot set, can be enlarged without limit to show ever more richness of fine detail, as illustrated in Figure 1. The only limitation is the finite numerical precision in the computer, and even with that constraint it is possible to find a tiny representation of the entire M-set magnified more than 20,000 times. Compare this to the magnification of the fern in Chapter 1, Figure 23.

We would not expect to find real objects which exhibit this behavior from the infinitely large to the infinitely small, but many real-world objects are observed to be describable as fractals over some limited range of length scales. The upper limit typically corresponds to the maximum size of the object, while the lower limit may either be set by the available image magnification (so that we might expect to find more fine detail with a microscope of higher resolution) or by some change in the physics which produces the line or surface. For instance, at very small dimensions effects such as atomic forces, crystallography, surface tension, etc., may intervene to create Euclidean behavior.

Between these understandable limits, many phenomena do show an ideal fractal relationship between the length or area and the measurement scale. For instance, plotting cloud perimeters and areas on log–log axes produces a line which is straight over five orders of magnitude from 10^1 to 10^5 m; the line's slope gives the fractal dimension (Lovejoy 1982). At smaller scales, we have found a similar constant slope for data measured on machined surfaces using interference light microscopy and atomic force microscopy and covering five orders of magnitude from 10^{-9} to 10^{-4} m (Russ 1991a).

For scattering data from clusters, it is useful to consider the results in the form of a frequency plot of log (Magnitude^2) vs. log (Frequency) as discussed in Chapter 4. As shown in Figure 2, at low frequencies the trend is flat (frequencies k below $1/R$, where R is the average cluster radius). At high frequencies (frequencies k above $1/r$, where r is the particle radius), the slope is –4.0, corresponding to Porod scattering. In between, the slope should give the fractal dimension of the cluster. For very large clusters of very small particles, this linear trend may persist over many orders of magnitude.

But it is perhaps more common to find in the literature that observed data depart from the ideal fractal relationship defined by this straight line log–log plot. One of the phrases often employed to describe this departure is to describe the behavior as "multifractal." The defini-

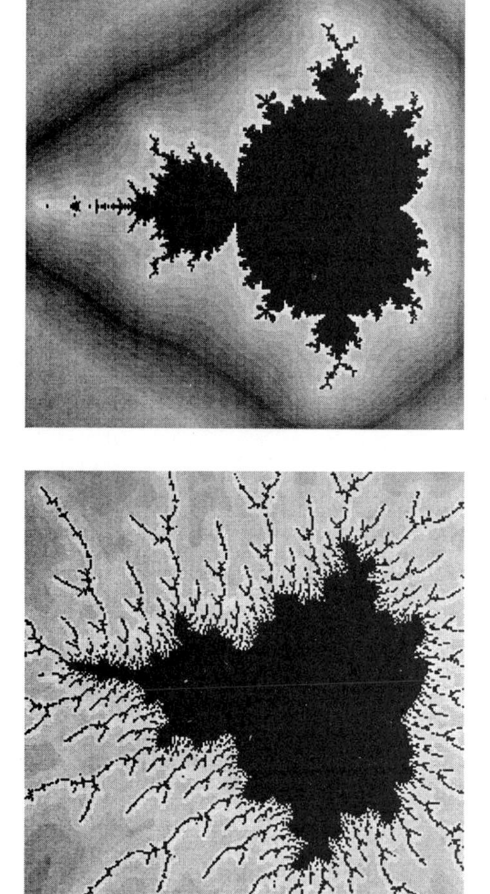

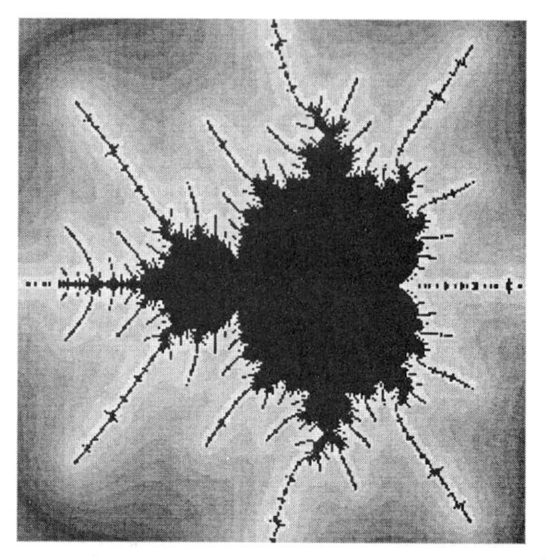

Figure 1. The Mandelbrot set, while not an ideal fractal, is endlessly self-similar. Enlargement shows detail of all kinds. One example is the infinitely many reoccurrences of the set itself, with minor rotations and distortions, at all scales. The figure shows examples of the entire set, a reproduction enlarged 45 times, and another one enlarged 20,000 times. The only limit to the regression is the mathematical precision of the program.

tions of multifractality are not all the same, and several of them have quite different consequences.

We will be discussing here the mixing together of two or more discrete fractals (Russ 1991b). This use of the word "multifractal" is quite different from another meaning also encountered in the literature (Barabsi 1991; Falconer 1992). From a mathematical point of view, multifractal behavior can describe the behavior of a population of points which are distributed on a support which is itself fractal (Rammal 1984). Imagine a two-dimensional support such as the surface of a sphere (say, the earth to a reasonable approximation) and a set of points such as all of the individual human beings present. It might turn out that these points have a fractal clustering distribution. Certainly, it has been shown that some other sets such as the distribution of oil and other mineral resources are fractal. This is a Cantor set of points and its fractal dimension can be measured.

However, suppose now that the points are distributed on a support that is not Euclidean but is itself a fractal, with its own dimension. Now the intertwining of the two fractals produces a more complicated behavior. Imagine a (fractal) rough metal fracture surface with a (fractal) distribution of corrosion sites. Slice this surface with a plane, or in any other way select a subset of the points, and a fractal dimension can be determined. The dimension of each subset is

different, hence the name "multifractal" (Mandelbrot 1984; Mandelbrot 1989). Measures of the scaling exponent in these cases, the Lipshitz–Hölder exponent, produce not a single value but a function. Such an approach has been particularly useful in the description of turbulence, which is a multifractal problem (Campbell 1990; Sreenivasan 1991).

Another classic example of multifractal behavior (Aharony 1991) can be visualized in the process of growing a diffusion-limited aggregate, as discussed in earlier chapters. The cluster itself has a fractal dimension, technically a mass fractal, arising from a plot of the number of particles as a function of distance. The growth rate of each location on the cluster, which is a measure of the probability of the next particle sticking there, is different for each point in the cluster. Each of these scales with a power law, having a different exponent. The entire growth behavior of the cluster is multifractal.

This meaning of multifractal is unrelated to the subject of this chapter, another reason to avoid further use of the term for describing the various phenomena described here. Perhaps "mixed fractals" would be a better term, but we still must deal with the large number of quite different ways in which the mixing can occur.

Kaye's Definition

One of the operational definitions of mixed fractal behavior is most strikingly illustrated by the many examples published by Brian Kaye and collected and summarized in his book, *A Random Walk through Fractal Dimensions* (Kaye 1989a). Kaye has measured particle profiles, perimeters of natural coastlines, and some other phenomena using a Richardson plot to determine the fractal dimension. This technique is quite straightforward, and has been discussed at length in earlier chapters.

Kaye's plots for a variety of specimen types and magnifications show data that do not follow the ideal straight line behavior expected for a fractal (Kaye 1978b; Kaye 1984). In practically all cases, they can be satisfactorily explained with two straight line segments which have different slopes, as shown in Figure 3. It is interesting to note that the range of dimensions covered by the plots are generally small, often less than one order of magnitude. In addition, most of Kaye's research has involved particulates such as extraterrestrial particles, diesel soot, and crushed rocks (Kaye 1985a; Kaye 1986; Kaye and Clark 1986; Kaye 1987a; Kaye 1987b), which are examined in projection so that the boundary line is not the intersection of a plane with the particle surface. This means that the dimension of the line does not correspond directly to that of the surface, and may have some further consequences for the interpretation of the two-segment lines on the Richardson plots as will be explained below.

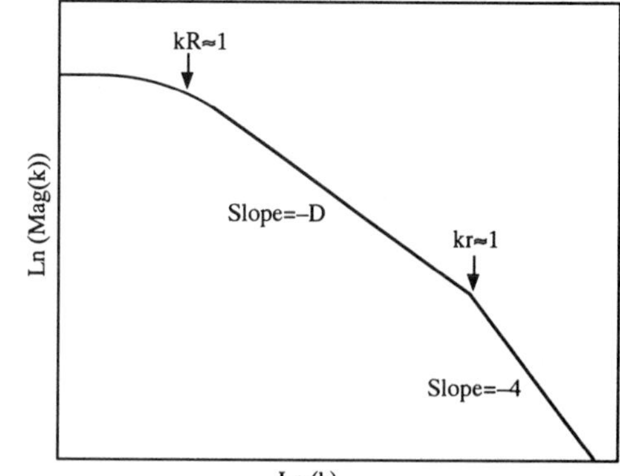

Figure 2. Schematic diagram of scattering curve showing three regimes (Vicsek 1992). R is the cluster radius, r the particle radius, and k the frequency.

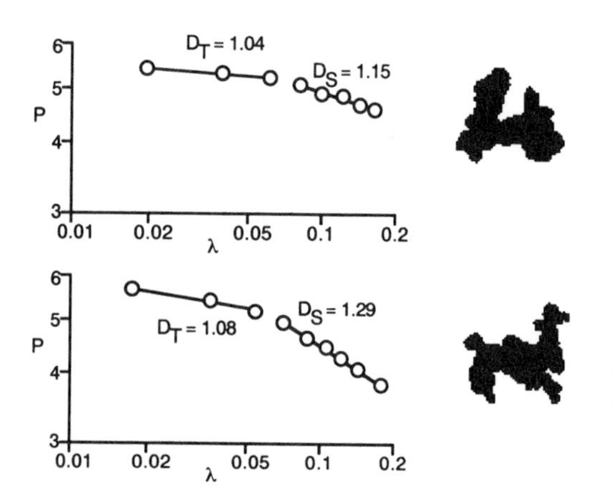

Figure 3. Examples of textural and structural fractal dimensions measured by a Richardson walk around projected profiles of particulate clusters (from Kaye 1989, p. 288).

Ideally, the projection of a classic diffusion-limited agglomerate in three dimensions onto a plane should produce a fractal dimension lower by 1.0. However, for a finite-sized cluster, the actual value is lower than this. Nelson, Crookes et al. (1990) calculate and model this effect, and offer a correction factor.

Kaye's description of the two line segments on his Richardson plots uses the terms "structural" and "textural" for the two different regions. The assumption is that at fine scales a "textural" dimension given by the slope of the line at small values of λ describes one physical process that controls the surface roughness, while at larger scales the "structural" characteristics of the object emerge. If this interpretation is true, the transition might in itself contain important information about the object. The dimension at which the transition from textural to structural behavior occurs would be significant and might be related to some underlying physical processes in the formation of the boundary or surface, or to the size of the subunits making up the structure. It is the latter interpretation that is used by Kaye, who applies the method to particle agglomerates.

It is not clear that this interpretation is meaningful for most surfaces, however. This transition point consistently lies within the comparatively narrow range of stride lengths used in Kaye's experiments, but he knows a great deal about the physics of the particles involved in forming the agglomerates and can select a size range for measurement. In the more general case, as we will see below, this transition dimension may not have a physical meaning, certainly not a direct one relating to scale. If the behavior seen in Kaye's multifractal plots is indeed the combination of two different fractal dimensions, the fact that the transition point often appears within the range of stride lengths covered by the plots may be coincidental, the result of normalization procedures, or the consequence of selecting the appropriate scales for measurement.

The preceding paragraphs should not be read as any condemnation of Kaye's data or his interpretation of it. In fact, Kaye's published graphs are used here simply as a representation of data published by quite a few other researchers, which generally show similar characteristics. Kaye presents a plausible argument for finding two different fractal dimensions at different scales when measuring two-dimensional agglomerations of monosized spherical particles, which may also apply to the more general case of 3D agglomerations and distributions of particle sizes. In private communication regarding this subject, he defends the idea that for agglomerates of particles, physical conditions often produce small subunits with a narrow range of sizes which clump together to form fractal clusters which are meaningfully described by two dimensions.

Kaye also notes that occlusion of valleys or indentations in such clusters is rare (they would not be for a surface), and when they do occur they can be recognized since they generally produce another, even lower dimension value. Kaye indicates that dimension values of about 1.3+ are typical for the structure of an agglomerate, changing to values of about 1.2 when the scale of the textural subunits is reached. However, the dimension may be lowered to about 1.1 if occlusion is present. Kaye promises to discuss this subject of occlusion in a forthcoming paper. His book is singled out here only because it presents quite a number of such plots applied to a diverse range of applications, and his terminology has been adapted, without such physical reasoning, by some other researchers.

The use of Kaye's idea of structural and textural fractals has been applied to many other sets of data with perhaps a less critical analysis. It is interesting that in every case known to this author, the Richardson plots interpreted in this way show a lower slope (lower fractal dimension) for the small stride lengths than for the larger ones. It is certainly possible to imagine real boundaries or surfaces which become "smoother" in the fractal dimension sense at fine scales and rougher at large scales. It is even possible to imagine mechanisms by which this might occur. But it is also possible to imagine the reverse. In fact, it might be expected that images obtained with any instrument having a finite noise level would show an artificial increase in roughness near the resolution limit that would appear to increase the fractal dimension. This is illustrated in Chapter 8. But the published data using Richardson plots do not show such behavior. In the modeling shown below, boundary lines which are smoother at fine scales as well as those which are smoother at coarse scales will be used to generate Richardson plots for comparison to the observed data.

Simulating the Projection

The one characteristic which all or most of the plots which show a transition from a smooth "textural" fractal to a rougher "structural" value share is that they are measured on projected boundary profiles from three-dimensional particles. It is the surface roughness of the particle surface that is of interest. One of Mandelbrot's "conjectures" (introduced in earlier chapters) is that there is a quantitative relationship between the fractal dimension of a surface and that of a boundary line produced by intersecting that surface with a plane. The relationship is simply $D_{boundary} = D_{surface} - 1.0$, subtracting 1 to account for the difference between the topological dimension of the 3D space in which the surface exists and the 2D space represented by the plane. There are some necessary caveats about this relationship, depending on whether the surface is uniform and isotropic and the fractal is random, but the method has been used in a number of cases to measure surface roughness. This can be done by mounting the object of interest in some embedding medium and polishing a plane through it to produce a boundary line for measurement (Mandelbrot, Passoja et al. 1984). It is often described as "slit island" analysis.

But it is clear that no such simple relationship can exist for the projected boundary of a rough surface. The depressions in the surface are hidden by protrusions, and the resulting projected line must be smoother than any possible intersection of the surface by a plane. In addition, it is the smaller irregularities which are systematically hidden by larger ones, so that the small-scale roughness mostly disappears while the large-scale roughness is mostly preserved. Further, the extent to which the small details are hidden by the larger ones increases with the distance in the projection direction so that the bias in any measured dimension must increase in magnitude with particle size. At least qualitatively, all of these effects should reduce the observed fractal dimension (Falconer 1990; Sreenivasan 1991) and might create the appearance of a "dual" dimension at different scales.

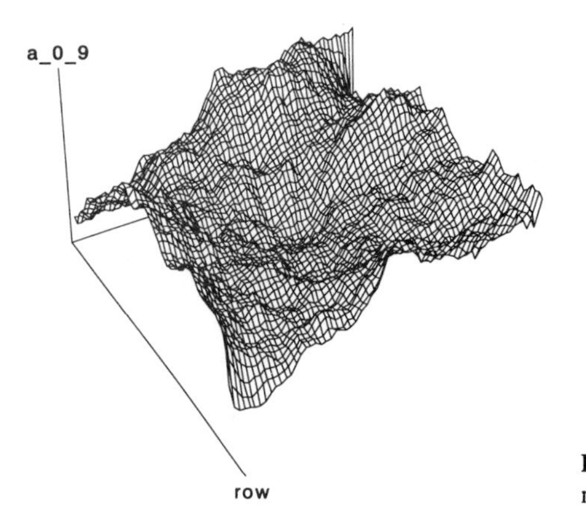

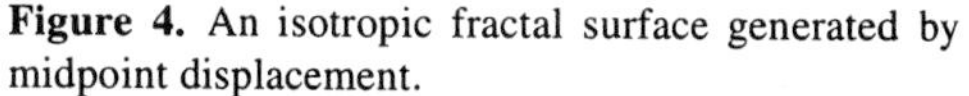

Figure 4. An isotropic fractal surface generated by midpoint displacement.

It is difficult to simulate a fractally rough surface on a spherical or other-shaped particle which can be used to evaluate this effect. As an approximation, fractal surfaces were generated using midpoint displacement as discussed in Chapter 6. This method produces displacements of points from a plane which are reasonably isotropic and have a fractal behavior that covers several orders of magnitude. The data consist of an array of 1024×1024 points, each representing elevation. Figure 4 shows one of the surfaces. For this kind of surface, the Mandelbrot ($D_S - 1$) conjecture holds. Hence, any profile across the array of values will also have a fractal character, with a dimension that is just 1.0 less than the surface. Of course, it can be argued that a Richardson plot is fundamentally inappropriate for a vertical elevation profile across a surface, which is self-affine rather than self-similar, but this is the method applied to many projection outlines and it is that method which we are studying at the moment. Applying a structured walk to this profile produces a Richardson plot (as shown in Figure 5) which is quite linear. The stride length is taken as multiples of the finest point spacing in the grid, and the plot covers a range of 50:1.

If elevation values are projected parallel to the nominal plane orientation to produce a profile in which depressions may be hidden by other points along the projection direction, the measured fractal dimension changes. Figure 6 shows the results for two such arrays, which have dimensions of about 1.15 and 1.45 for section profiles. The horizontal axis in the plot is the fraction of the array width over which the projection was made. It is clear that the numerical value of a fractal dimension obtained in this way is generally lower than the true dimension of

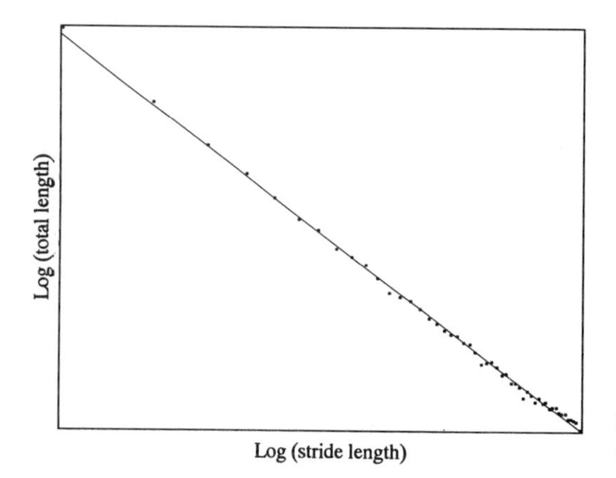

Figure 5. Richardson plot from a profile through the surface of Figure 4.

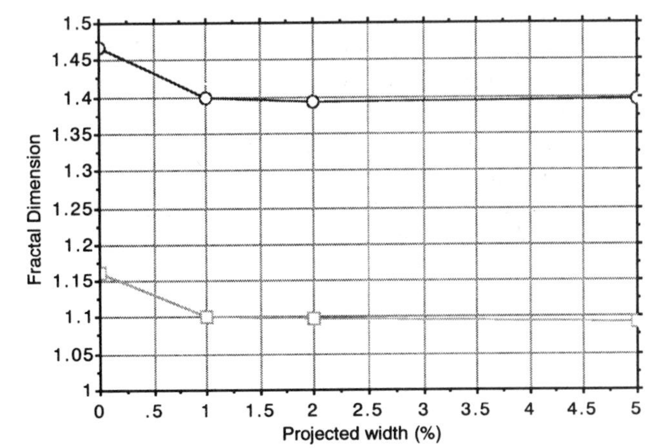

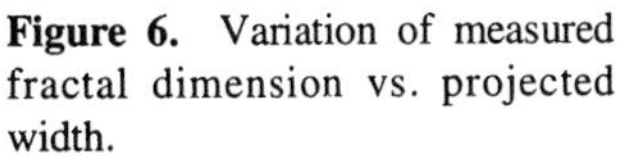
Figure 6. Variation of measured fractal dimension vs. projected width.

a profile, or of the surface. While the observed dimension is always expected to be smaller than the actual dimension, it is not clear that there is a quantitative relationship that can be used for measurement purposes.

In addition, the Richardson plot itself often shows an interesting effect, as illustrated in Figure 7 for one of the 5% projections. The data points can be quite satisfactorily represented by two straight lines, with exactly the type of behavior postulated as a distinction between structural and textural fractals. It therefore seems possible that this observed behavior may be entirely or partially an artifact of the use of projection data, and does not represent an actual characteristic of the surfaces being investigated.

This effect has been studied experimentally by Hamblin and Stachowiak (1993). Comparison of projected and sectioned outlines of actual particles produced by hacksawing ductile and brittle metals and aerosol spraying of a reagent showed a significant difference in the dimension. This was determined using a Richardson method on outlines imaged by light and scanning electron microscopy. The projected outlines had lower dimensions, but the Richardson plots did not show the predicted change in slope.

Vicsek's Fat Fractal

The formal definition of a fractal structure such as a line or surface implies that the volume occupied by the structure approaches zero as the resolution approaches zero. This is probably

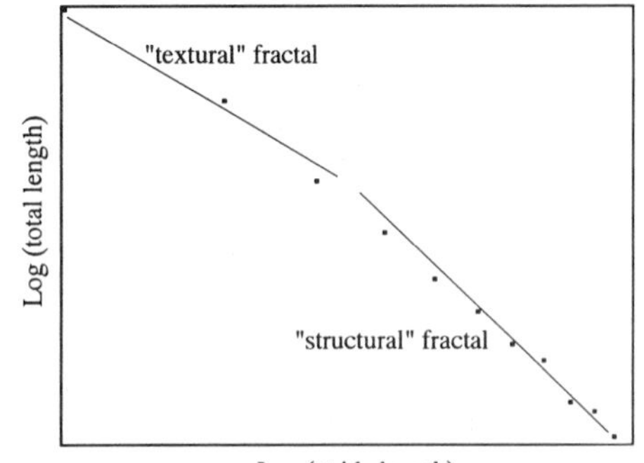

Figure 7. Result of fitting two lines to the Richardson plot from a projected profile.

easiest to see for a structure like a Menger sponge, consisting of an object containing pores of various sizes. As the resolution improves, ever-smaller pores become detectable. In the limit, there is a pore everywhere and the volume remaining to the structure vanishes.

There are some structures which appear to exhibit fractal characteristics, particularly a kind of scale independence, that fall short of strict self-similarity. Instead, as the resolution or measurement scale becomes smaller, the volume converges to a finite limit. Vicsek (1992) describes several such situations, such as the branching of coral colonies and bronchia in the lungs, and the structures produced by ballistic aggregation. It is interesting that this latter example may be related to the agglomerations of particles discussed by Kaye as giving rise to a multifractal or mixed fractal.

Vicsek describes the situation in terms of the total density of particles ρ within a radius r of any point in the agglomerated structure. If the particle size is negligible, then the agglomerate density falls off with cluster size r according to a power law $\rho(r) = Ar^{-\beta}$, where the value of β has been studied extensively for both real clusters and models (Stanley and Ostrowsky 1986). For instance, in the simplest case of diffusion-limited aggregation on a square grid in two dimensions, as illustrated in Chapter 6, a log–log plot of the mass as a function of radius produces a straight line, indicating that the density drops as cluster size increases. But if the individual particles are not negligible in size, the density of a large cluster does not approach zero but instead approaches a finite value. This can be written as $\rho(r) = \rho(\infty) + Ar^{-\beta}$. For 3D clusters of particles agglomerated with ballistic motion instead of simple diffusion, the values of β are in the range from 0.55 to 0.66.

Networks can be modeled in a number of ways, one of which uses L-systems as illustrated in Chapter 1. These structures are not strictly fractal in that the self-similar regression does not continue to infinitely small dimensions, but they do capture much of the shape of many natural objects including plants, river systems, blood vessels and bronchia, etc. Chapter 1 showed the evolution of several networks with a natural appearance using a simple set of iterative rules. The volume of the network vanishes as the resolution improves and the branches shrink to mathematical lines. But real trees, blood vessels, etc., possess some width, as shown in Figure 8. If the L-system is redrawn with branches whose width decreases with each iteration, then the result (Figure 9) looks natural but has a nonvanishing volume. This example justifies the use of the name "fat fractals" for these objects. The slope of the log–log plot may still be used to obtain a fractal dimension even though the plot does not continue as a straight line but asymptotically approaches a constant value at small dimensions.

Kaye's analogous example to Vicsek's fat fractal is a cluster (in 2D) formed by circles of finite size. The length of the perimeter of the cluster does not continue to increase as the stride length is reduced below the size of the circles, but becomes constant (equal to the product of the circle dimension and the number of circles, which can be thought of as representing particles). This produces a Richardson plot which rises linearly at large dimensions and then curves over to a lower slope or to a horizontal line at small stride length. The extension to a 3D cluster of spheres is not so obvious. Orbach (1986) describes behavior in which the fractal plot may level off at either large- or small-scale distances beyond which the structure acts as a homogeneous or Euclidean object, and presents several examples of percolation networks and particle aggregates. In the case of percolation networks, leveling off can occur at both large and small scales.

When this type of departure from the ideal straight-line fractal log–log plot is observed at small distances, it is most often described either based on Kaye's structural/textural explanation, or by assuming that the measurement technique simply ran out of resolution. It is also possible that during the process of generating some surfaces, such as a fracture in a material, the fracture process may itself cause some changes. Changing the applied load or strain rate on the remaining material might alter the physical process of crack propagation or

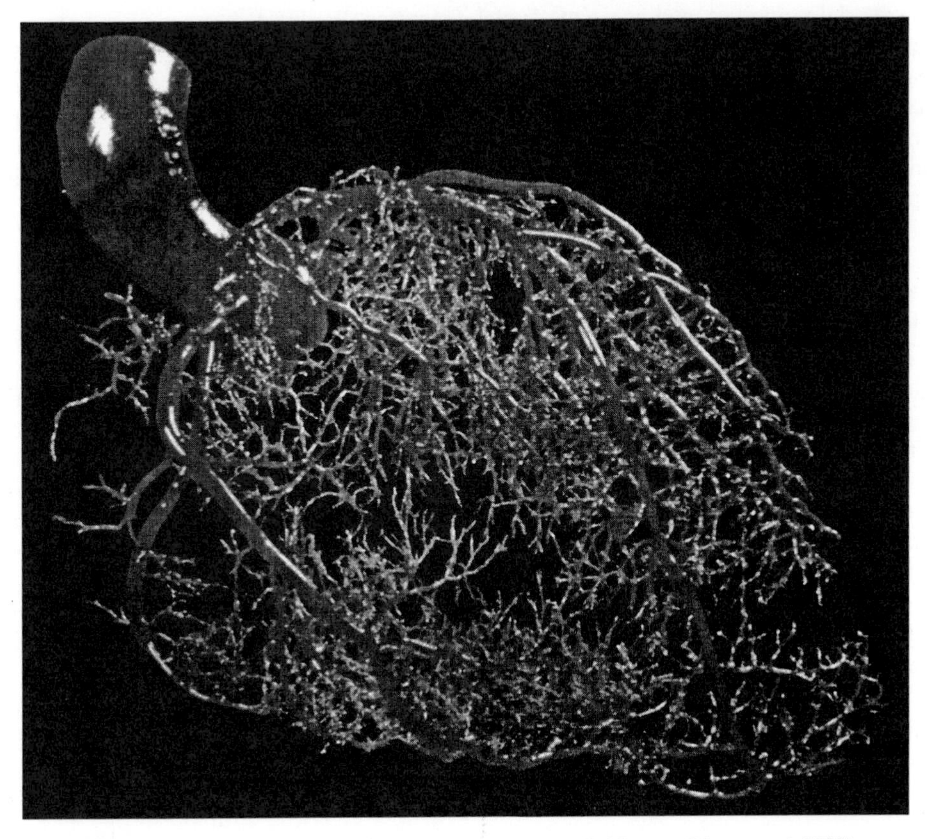

Figure 8. Cast of the blood vessels serving the heart muscle (Goldberger, Rigney et al. 1990).

Figure 9. An L-system network drawn using rules shown in Chapter 1, but with the thickness of the lines varying with distance from the origin, produces a "fat fractal."

the material properties, and hence the fractal dimension. We have observed fractal plots on machined surfaces using data measured by interference light microscopy which showed this leveling-off phenomenon. However, when higher magnification images were obtained using an atomic force microscope, the log–log plot continued as a straight line with the same slope for another two orders of magnitude in scale dimension (Russ 1991a). In effect, the resolution limit of the measurement tool gives lines a finite width, thus converting the data to the appearance of a fat fractal.

When such leveling off at small-scale dimensions is observed with Richardson plots, or the other measurement methods and fractal descriptors which are available, it is necessary to determine the cause. It may not be possible to distinguish between a true fat fractal and inadequate resolution, except by employing an additional imaging tool with higher resolution (if one is available).

There is an additional consideration for fractal surfaces. As mentioned before, any single-valued function, such as the elevation of surfaces, must be self-affine rather than self-similar. It is possible that some surfaces may actually be formed in ways that produce self-similar internal structures, but the highest elevation value at each location (as determined by most measuring instruments) will ignore the undercuts and internal voids, and the data will still be self-affine. Such surfaces must approach a Euclidean limit at very large dimensions, so that when viewed from far away the surface looks like a slightly rough but essentially planar object. This is our common experience with surfaces. The global dimension for such a structure is thus the topological limit of 2 for a surface or 1 for an elevation profile. This must control the asymptotic slope of the various kinds of measurement plots for the large-scale limit.

On the other hand, at small scales (the local dimension) the fractal nature of the surface will emerge. This suggests that there must be a crossover scale for the process. The essential fact about this crossover scale is that is not intrinsic for the surface, but is an artifact of the measurement units which happen to be selected for horizontal and vertical directions (Vicsek 1992).

Mixed Fractals

The descriptions above implicitly assume that the fractal surface or boundary being studied is uniform and isotropic. Many natural objects may not conform to these limits, and yet may contain a good deal of fractal character which it would be useful to measure. Four specific examples of mixed fractals will be described which give rise to rather different consequences. One of these, the addition of two or more different fractals, is discussed by Vicsek in addition to his presentation of the fat fractal.

Superposition of fractals can be imagined as wrapping a fractal surface with fine-scale roughness onto a larger structure with a different (but still fractal) surface roughness. Since any fractal surface has an irregularity which persists to infinitely fine scale, this word description is necessarily somewhat loose. However, surfaces or lines with different fractal dimensions are characterized by different rates at which the roughness decreases in magnitude as the scale dimension is reduced. It is possible to add together two surfaces or lines with different fractal dimensions as may result from different physical phenomena.

A related mixing of fractal dimensions will consider the case in which the exponent in the relationship $P = B \cdot \lambda^{\alpha}$ is not a constant but actually varies with λ. This may either happen abruptly at some particular scale dimension, as was originally discussed above in conjunction with Kaye's definition, or gradually as a function of scale. Perhaps this latter situation should not even be referred to as a fractal, but for the moment we will include it. It will be interesting

to see the consequences of this change in α on the measured values of fractal dimension according to several definitions and procedures.

Mixing can also occur spatially. We have studied a number of examples of fracture surfaces in materials which do not consist of a single homogeneous crystal structure. Either there are many distinct grains with essentially random orientation, or two or more distinct phases with different crystal structure and composition. It is reasonable to suppose that these differences, which affect other mechanical and physical properties, might change the fracture characteristics and consequently the local fractal dimension. If these "islands" of different dimension are present in different small regions within the measured fractal surface, what is the consequence for the measured fractal dimension of the surface as a whole? This can be explored with line profiles as well, by combining different segments of profiles with different dimensions.

Finally, many surfaces are not isotropic. Machining, wear, and many other natural phenomena create rough surfaces which have a clear directionality and yet much fractal character. Ways have been proposed to measure the dimension as a function of direction, and these have been applied with some success as shown in Chapters 4 and 6. It is still an unanswered question as to whether a single dimension measured for the surface as though it were isotropic has any simple or fixed relationship to the directional values.

In order to explore these questions, a series of simulations using line profiles and surfaces have been generated using several different methods which have been discussed in Chapter 6. Each line or surface is an ideal fractal, and the surfaces are isotropic and uniform. These have then been combined in the various ways discussed to produce new lines and surfaces for measurement using several techniques. This in turn produces fractal dimensions that have slightly different meanings.

The Mandelbrot Conjectures

It will be helpful to compare the simulation results to each other and to the ideas expressed in several of the Mandelbrot "conjectures." These have been mentioned in several previous contexts, and are summarized below:

Conjecture 1. A fractal of dimension D when projected onto a lower dimension space m produces a fractal of dimension D unless D is greater than m, in which case the dimension is m. This can be best imagined as a cluster of points in three-dimensional space, for example produced by the diffusion-limited aggregation described in Chapter 6. If the cluster has a fractal dimension of D (less than 3), then looking at a projection of the points onto a plane will produce a fractal pattern whose dimension is also D, unless D is greater than 2 (the topological dimension of the plane). If it is, then the projected pattern will also have a dimension of 2, meaning that it appears as a solid smooth-bounded region with no internal holes. In Chapter 6, clusters of particles formed by diffusion-limited aggregation in three dimensions were shown to have a mass fractal dimension greater than 2 (but less than 3). The projection of these particles has a lower dimension. In the limit of a very large cluster, the center of the cluster will be completely opaque in the projection and only the edge of the projection will be irregular. Such a solid cluster has a fractal dimension of exactly 2, since the area increases as the square of the diameter. Correction factors for finite cluster size have been proposed for diffusion-limited aggregation.

This result seems to be related to the problem discussed above for a projected outline from a rough-surfaced particle. Since the fractally rough surface of the object has a dimension greater than 2 but less than 3, projecting it onto a plane surface (dimension 2) should produce a

projection which has a dimension of 2; in other words the particle should appear as a disk with a smooth boundary which does not show any of the fractal character of the surface because all of the protrusions and intrusions should cancel out. This is surely the case in the limit where the projection is performed parallel to the nominal surface orientation and has an infinite extent. It is not clear whether any quantitative analysis of results for intermediate cases is possible.

Conjecture 2. Intersecting a fractal surface of dimension D (which exists in a three-dimensional space) with a two-dimensional plane produces an intersection boundary whose dimension is $D - 1$. This is actually a special case of the more general rule for intersections, namely that intersecting a fractal set with dimension D_A by another of dimension D_B produces a result with dimension $D_A + D_B - m$, where m is the dimension of the embedding space. For the example of a rough surface (D_A between 2 and 3) cut by a plane ($D_B = 2$) the result is $D_A + 2 - 3 = D_A - 1$ as given.

At a lower dimension, a fractal profile of dimension D (between 1 and 2) cut by a straight line (dimension 1) produces a series of points. Technically these are a Cantor dust. If the distances between successive points where the straight line crosses the fractal profile are measured and a plot of frequency vs. distance is constructed, the result is a power law relationship (a straight line on a log–log plot) whose slope is the expected $D - 1$ (between 0 and 1). This is generally a rather inefficient way to measure a fractal dimension for a line, of course. The spacing of points along the line does model the occurrence of noise in transmission lines (in which case the horizontal axis is time rather than position).

We have already seen an analogous way to measure surface dimensions. If a fractal surface ($2 < D < 3$) is cut by a plane ($D = 2$), the size distribution of the islands which appear should also have a power law relationship with a slope related to the surface fractal dimension. This can be written as $N(A > a) = (a_{max}/a)^{D/2}$, which states that the number of islands with area A greater than any value a rises as a decreases. Mandelbrot has reported finding such a power-law behavior for islands in the oceans of the world (where the ocean surface is the plane), implying a dimension D about 1.3 for the earth's surface.

This idea has also been extended into three dimensions to account for the size distribution of ore bodies (Kubik 1986; Burrough 1989). It even has an analogous relationship to Zipf's law for the frequency of appearance of words in the English language. This states that if words are ranked according to their frequency of appearance, then the frequency is given by $f(R = \text{rank}) = 1/R \cdot \log_e (1.87\ R)$. Of course, it is at the very least a stretch of the concept of fractal dimension to apply it to every instance in which power-law behavior or linear plots on log–log axes are found.

Returning to surfaces, the comparatively simple case in which the intersection is between a fractal surface and a plane is not the only one of interest. A similar analysis has been used to examine the contact areas between two fractal surfaces, corresponding to electrical contacts in a switch or other similar device (Majumdar and Bhushan 1991). They also find a power-law distribution for the area of the islands of contact.

Conjecture 3. Combining two fractal sets can be done in several ways. From a mathematical point of view, multiplying two sets of dimension D_A and D_B produces a result with dimension $D_A + D_B$. The example cited in Mandelbrot and Vicsek is the extrusion of a fractal profile of dimension D along a straight line of dimension 1, to produce a surface of dimension $D + 1$ as illustrated in Chapter 6. The second way to combine sets, which seems more relevant to most of our examples, is union. The union of two sets with dimension D_A and D_B produces a set with dimension $D = D_A$ (provided that D_A is greater than D_B). In other words, most of the ways in which we will be combining two fractal surfaces or profiles should produce

a new surface or profile which has the fractal dimension of whichever of the original surfaces or profiles was "rougher." The information from the original surface or profile with the lower dimension seems to be lost. We will see below that this is not exactly true in practice.

Measurement of Fractal Dimensions

While the Richardson technique for performing a structured walk around a boundary and constructing a log–log plot from the data is conceptually straightforward, it is not the only nor necessarily the easiest method to apply. As indicated above by the description of a plot of mass vs. radius, there are other ways to measure fractal dimensions. Many of these have been exhaustively described in previous chapters, with the caution that they do not all measure the same thing. There are a variety of definitions for fractal dimensions which are not mathematically identical, and give rise to different numeric values as well. The use of discrete data points as opposed to ideal, continuous data, introduces an additional variation in the measured values, which may vary considerably. It is useful to distinguish at least the following definitions, which have been introduced in earlier chapters:

1. The similarity or Hausdorf dimension, which is determined from a Richardson plot as discussed above. It cannot be properly applied to a self-affine profile such as a vertical section through a surface.
2. The Minkowski dimension. This can be measured for a line by passing circles of various radii along the circle and plotting the area swept out vs. radius on a log–log plot. For a surface, the analogous operation would use a sphere and plot the volume. For self-affine profiles or surfaces, the circle and sphere are replaced by a line and disk, respectively.
3. The Kolmogorov or box-counting dimension. This is determined by placing grids with different spacings over a line and plotting (on log–log axes) the number of grid squares through which the line passes as a function of grid size. Again, extension to surfaces is straightforward, but the method is inappropriate for self-affine cases.
4. The Hurst coefficient. For a line represented as $y(x)$, this is the slope of a log–log plot of the greatest difference in y value for any range Δx. It was initially applied to time-based phenomena but can also be used for elevation profiles. The extension to surfaces gives information on the fractal dimension as a function of direction.
5. Fourier analysis. For a fractal profile, a log–log plot of the magnitude vs. frequency for the Fourier transform of the function is a straight line whose slope is related to the fractal dimension. For a surface, a 2D Fourier transform can be used and the slope measured as a function of direction (Russ 1990b).

These are in addition to the mass dimension (from the cumulative plot of mass vs. radius for a cluster), plots of rms (or variance) vs. windows, the Korcak dimension (from the plot of number of islands vs. area), and dimensional analysis (the slope of a plot of log (Perimeter) vs. log (Area) for the same islands).

Not only do these definitions and measurement methods produce different values, but they also respond in different ways to the kinds of mixed fractals which have been described. In order to examine these differences, a series of line profiles and surfaces with different fractal dimensions were generated using the midpoint displacement technique. A few additional profiles and surfaces were produced using other simulation methods, such as Mandelbrot–Weierstrass functions (Feder 1988), in order to assure that the generation method had no effect on the results. In all cases, the same types of results and conclusions were obtained from these other data sets.

Addition of Fractals

Midpoint displacement is an iterative technique which first displaces the midpoint of a straight line or square up or down using a random number generator, and then does the same for the midpoint of each of the line segments as described in Chapters 1 and 6. To test the effect of adding together two fractals with different values of α, the iterated midpoint displacement technique was used to generate 1024 point line profiles with values of $\alpha = 0.1, 0.5, 0.7$, and 0.9. Table 1 shows the fractal dimension of these profiles measured by several different methods, and it is evident that the values do not agree, nor correspond to the expected $(1 - \alpha)$. This is partly due to the limitation of the generating method, which works best for values of α close to 0.5, and partly to the bias of each measurement technique. However, in all cases the relative changes in the measured D value show the same qualitative effect of adding the profiles together.

Adding the profiles in various combinations produces new data sets which were measured using the same tools. In all cases, the log–log plots used to obtain the dimensions showed straight lines which fit the 50 points used very well, with no visual hint of a break point or curvature. The results, shown in the table, are not simply explained. The Mandelbrot conjecture that the union of two independent fractals should produce a result with the greater of the two dimensions is approximately followed, but there is clearly some effect of the smaller D value as well. This may be the consequence of either a finite number of points in the profile (1024 in the examples shown) or a discrete array of points rather than a true continuous line. However, experiments with larger profiles (up to 8192 points) show the same qualitative effect.

This addition can be thought of as the superposition of two relationships, each a straight line on log–log axes corresponding to the Richardson plot for a single α value. As shown in Figure 10, because the magnitude of values decreases rapidly on a logarithmic scale, adding together two straight lines on a log plot produces a result which closely follows the lines with a narrow transition region between them. However, the location of the transition point has nothing to do with the slopes of the lines. As shown in the equation below, it depends on the multiplicative constant for each of the exponential terms:

$$P = B_1 \cdot \lambda^{D_1} + B_2 \cdot \lambda^{D_2}$$

In this equation, the total perimeter length of a profile is expressed as the sum of two terms, in which λ is the scale or stride length and D_1 and D_2 are the fractal dimensions. We

Table 1. Addition of Fractal Profiles

Profile(s) $(2 - \alpha)$	**Measured Fractal Dimension**		
	Richardson *D*	**Hurst *D***	**Minkowski *D***
1.9	1.7553	1.7113	1.5106
1.5	1.5726	1.5797	1.4033
1.3	1.2970	1.4379	1.2316
1.1	1.0686	1.1578	1.0705
(1.9 + 1.5)	1.7291	1.7001	1.4875
(1.9 + 1.3)	1.7321	1.7313	1.4983
(1.9 + 1.1)	1.7479	1.7036	1.5074
(1.5 + 1.3)	1.5085	1.4958	1.3631
(1.5 + 1.1)	1.5301	1.5575	1.3929
(1.3 + 1.1)	1.2810	1.3835	1.2238

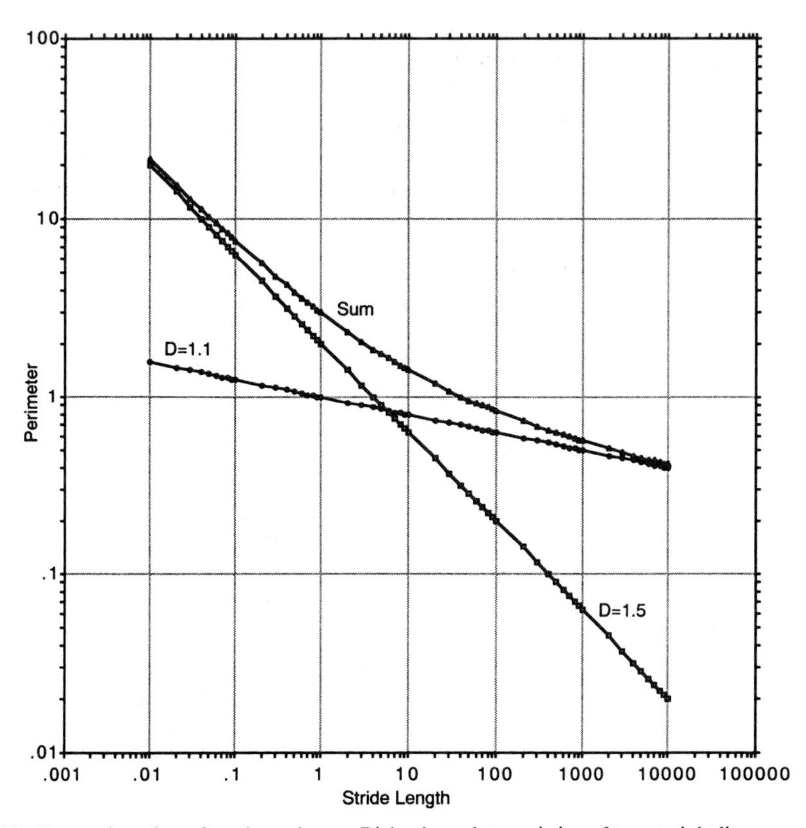

Figure 10. Summation of two fractals produces a Richardson plot consisting of two straight line segments with a curved transition region, whose location depends on the relative vertical position of the lines rather than their slopes.

have ignored the B values until now since they were not needed to define a fractal dimension. But varying the relative magnitude of these coefficients changes the relative heights of the two lines and shifts the crossover point at which their sum shows a change in slope. Since there is no information on B in the usual measurement or description of a fractal, the crossover point has no relationship to the fractal dimension. This was mentioned above in connection with Kaye's definition of multifractal behavior.

Just what is the meaning of the B values? Mandelbrot has used the term "lacunarity" to describe a property of fractal surfaces and profiles which in some cases corresponds to the B value. (This term has also been used for another property of these surfaces and profiles, related to the shape of the distribution of random values used to produce the midpoint deflection.) In other contexts, the term "topothesy" has been used for a quantity which includes B along with the fractal dimension (Thomas and Thomas 1988).

Vicsek suggests that the crossover scale where the two lines shown in Figure 10 cross and the slope appears to change is not intrinsic and depends only on the measurement units employed. He shows examples for the case of self-affine structures in which the measurement methods cause the crossover scale to be of the same order as the smallest λ value, so that the change in slope or fractal dimension cannot be observed. It should also be noted that most of the definitions of fractal behavior rely on a limit as λ becomes small, and of course this limit will not reveal the effect of different slopes at larger scales.

As shown in Figure 10, when two different slope lines on a log–log plot are added together, the slope at small λ is always greater than that at large λ. This is different from the textural/structural model proposed by Kaye and opposite to all of the published data for agglomerated particles, which show a lower slope at small λ. It is possible to imagine an agglomerate of packed spheres which would exhibit the opposite trend, but it does not seem to have been observed. That may indicate that Kaye's model of the turbulent agglomeration of small subunits produced in a thermally shifting environment is valid for his observations, but is unrelated to the problem of mixing two fractal dimensions together on a surface. If the crossover point is within the range covered by the plot, and if multifractality results from adding together two different fractals, then we must expect the change in slope to show the opposite behavior to that reported.

Variation with Scale

One of the possible kinds of multifractal of interest is the one in which the roughness actually does vary with dimension, as discussed above. This can be simulated with the midpoint displacement method by varying the magnitude of α during the iteration process. Two different cases will be considered: a sudden change in α at a particular iteration; or a continuous variation between two limiting values. In both cases, line profiles and surfaces are generated which have the rougher (smaller α, higher fractal dimension) values at larger dimensions and also at smaller dimensions. These data sets are then measured using the various fractal definitions and procedures listed above in order to determine how the results were influenced.

Figure 11 compares three simulated fractal profiles, one with constant α and two in which the value of α is changed in the final two steps of the iteration. It might be expected that a Richardson plot constructed from these profiles would show a change in slope (to greater or

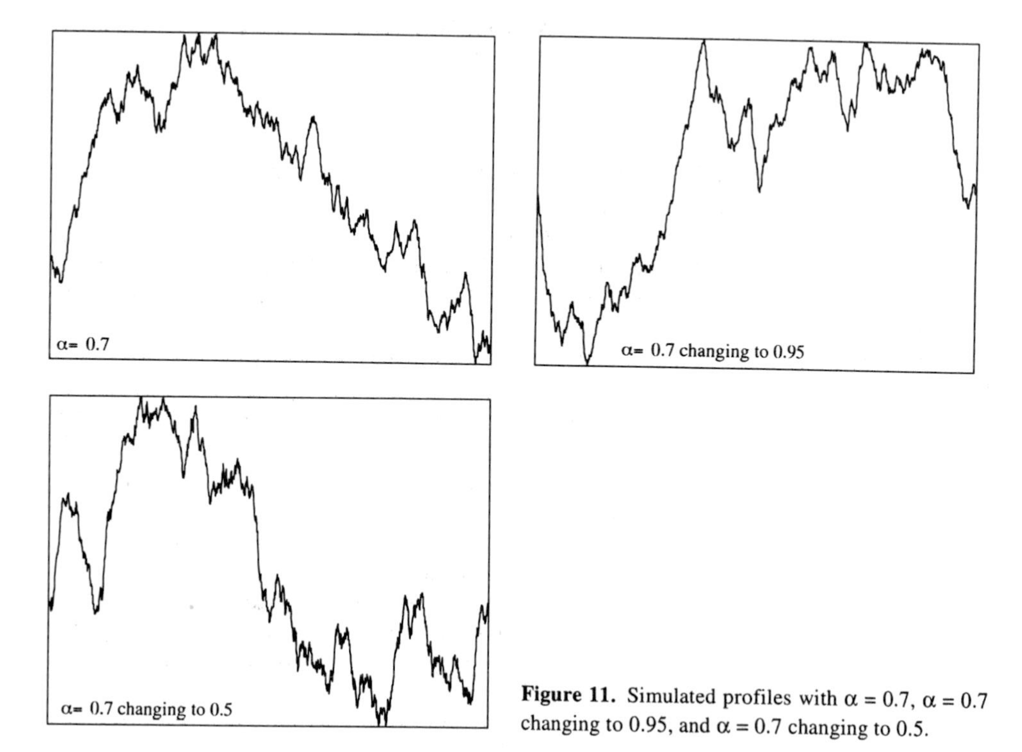

Figure 11. Simulated profiles with $\alpha = 0.7$, $\alpha = 0.7$ changing to 0.95, and $\alpha = 0.7$ changing to 0.5.

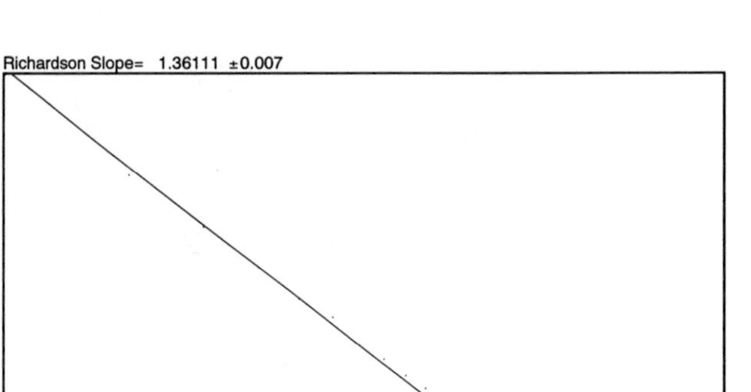

$\alpha = 0.7$

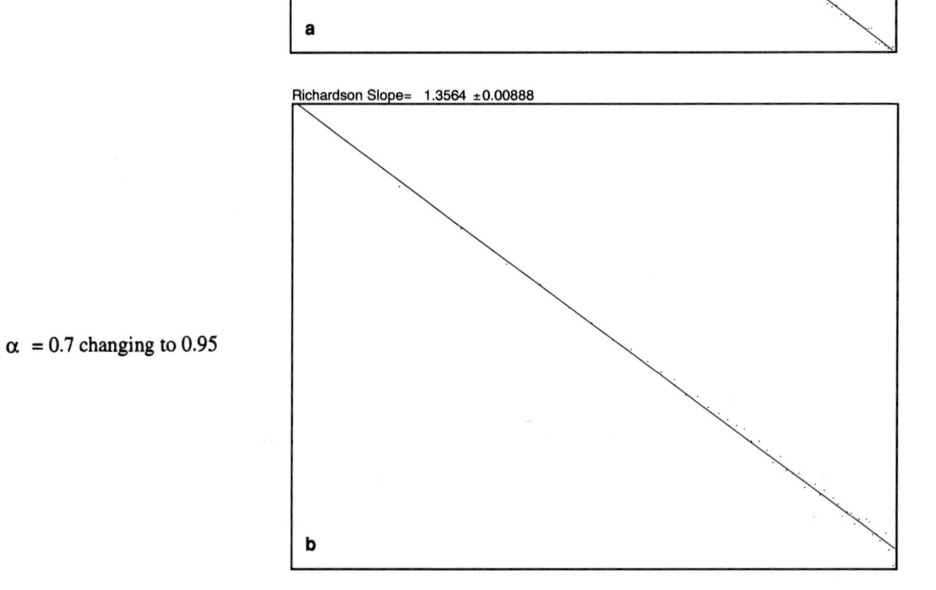

α = 0.7 changing to 0.95

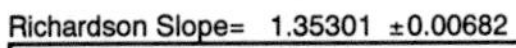

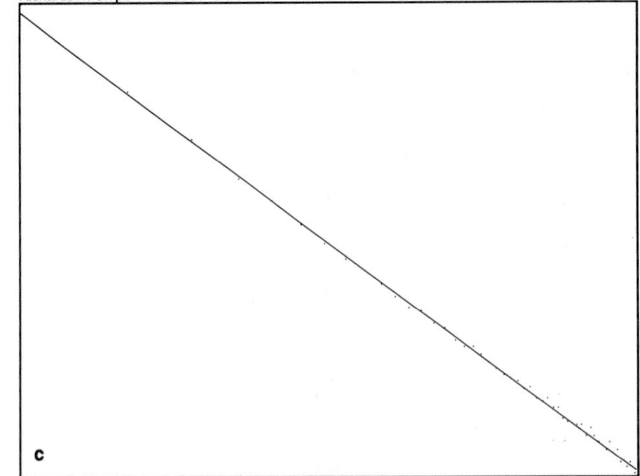

α = 0.7 changing to 0.5

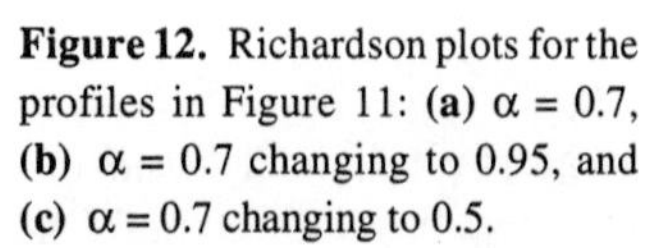

Figure 12. Richardson plots for the profiles in Figure 11: **(a)** $\alpha = 0.7$, **(b)** $\alpha = 0.7$ changing to 0.95, and **(c)** $\alpha = 0.7$ changing to 0.5.

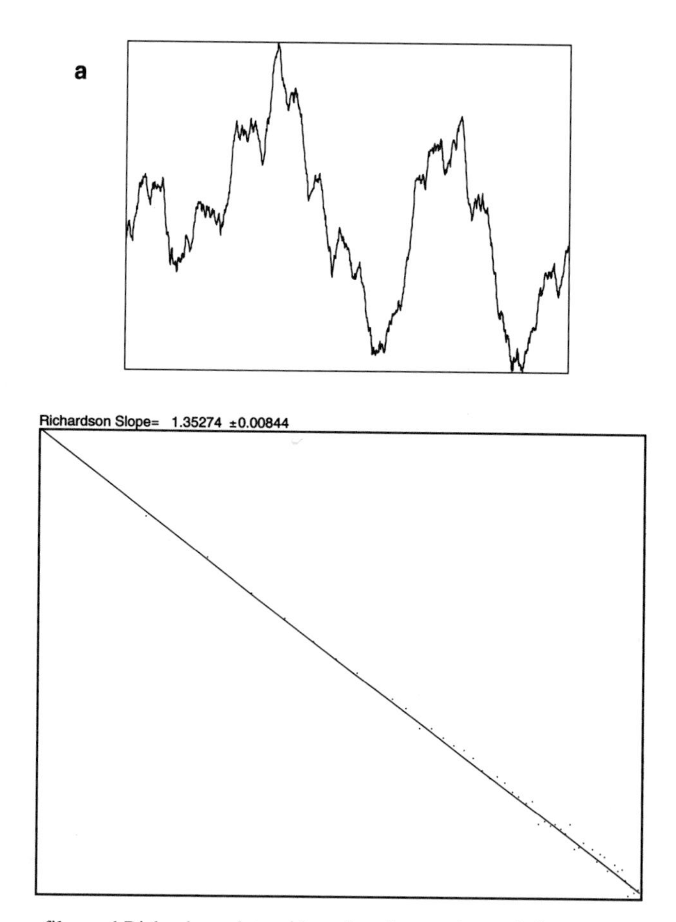

Figure 13. Line profiles and Richardson plots with α changing continuously between two values: **(a)** $\alpha = 0.7$ changing to 0.95 and **(b)** $\alpha = 0.7$ changing to 0.5.

smaller values) at small dimensions. But as shown in Figure 12 the deviations can hardly be discerned and would normally be ignored as no greater than the usual statistical fluctuation of the points around the line. When the variation is continuous, as shown in Figure 13, the overall curvature of the Richardson plot is also difficult to detect.

When the same data are measured by Minkowski, Hurst, or Fourier methods, the plots have somewhat different appearances and interpretation (Figures 14 and 15). All show essentially straight lines with no evident break or curvature. The Hurst method generally has more scatter in the plotted points, but this is usual. As noted before, the numeric values of the different coefficients do not agree. Plotting the Fourier results shows some increase in the spread of the points at high frequency, but this is hard to interpret as these plots generally have poor precision and considerable scatter.

Splicing Fractals Together

Even if a boundary line or surface is an ideal fractal with a single, well-defined dimension, there are many realistic situations in which it might be expected that regions of different fractal

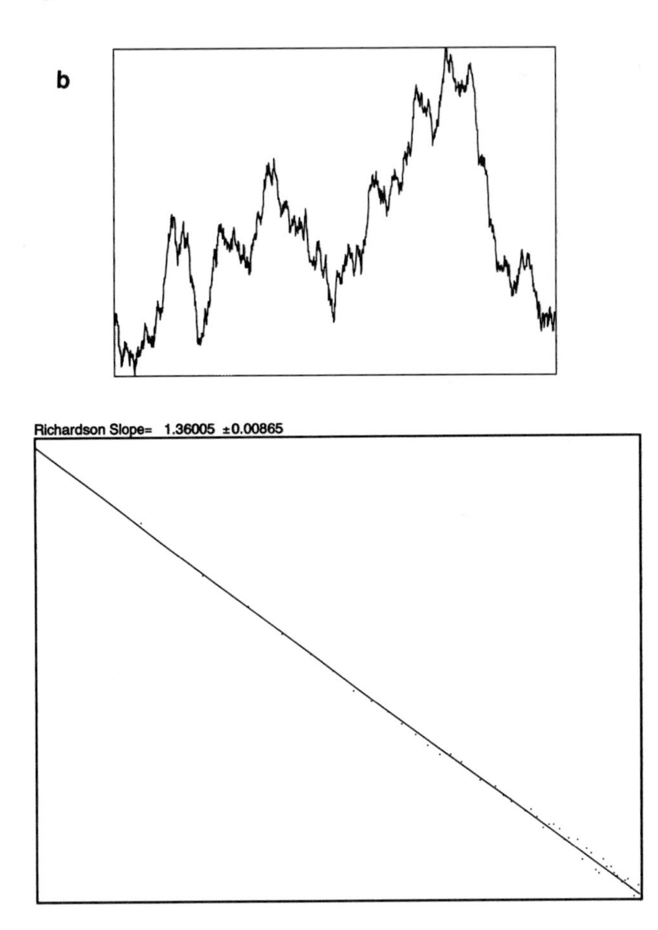

dimension might be mixed together spatially. One example is the study of fractal surfaces produced by the fracture of two-phase brittle materials (Fahmy, Russ et al. 1991). The individual single-phase materials have surfaces which are fractal, with different fractal dimensions corresponding to their different fracture toughness. For intermediate compositions which contain both phases, the portion of the surface which passes through each phase can be expected to exhibit its characteristic fractal roughness.

It is not reasonable to assume that the portion of the total fracture which passes through each phase is the same as the volume fraction of that phase in the material. It is not even clear how to describe the portion of the surface which has each fractal dimension, since for a fractal the surface area is not defined. Measurements of overall fractal dimensions for the surface show values which are intermediate between the two limits, and which correspond to the macroscopic fracture toughness data for the specimens. This is very satisfying in one sense, but leaves many unanswered questions about the proper interpretation of this type of mixed fractal.

To investigate this, a series of line profiles were generated using midpoint displacement, with $\alpha = 0.5$, 0.7, and 0.9, each consisting of 1024 points. Portions of these profiles were spliced together to construct a series of profiles with 20, 40, 60 and 80% of each fractal dimension. Notice that in this case the fraction of the total profile is specified in terms of the horizontal projected

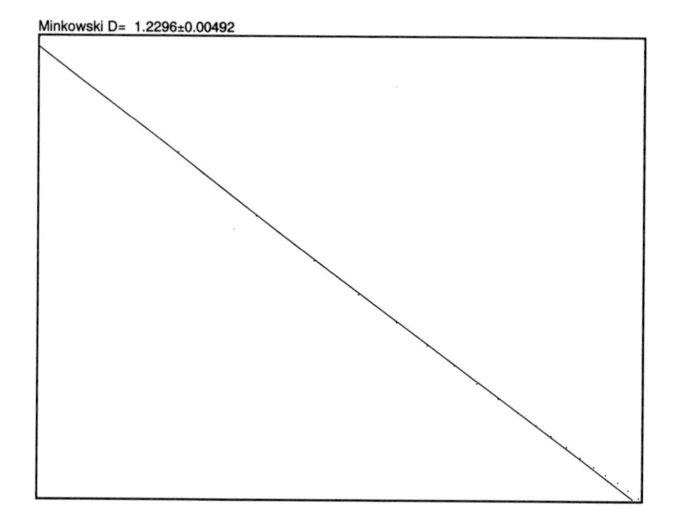

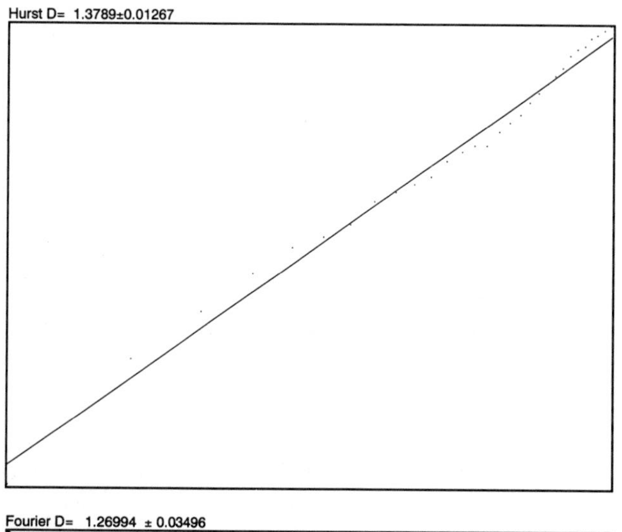

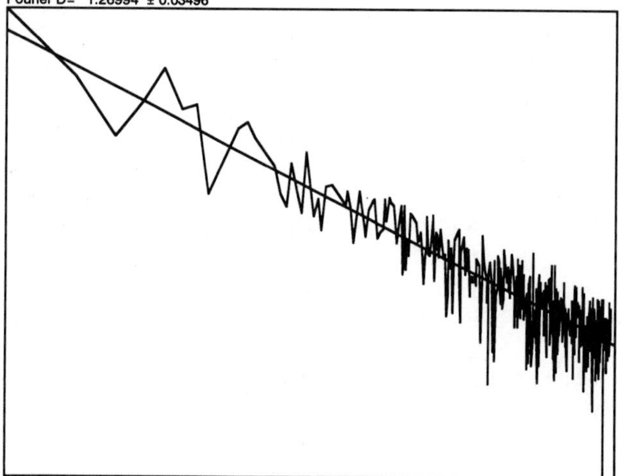

Figure 14. Minkowski, Hurst, and Fourier plots for a profile which changes from $\alpha = 0.7$ to $\alpha = 0.95$.

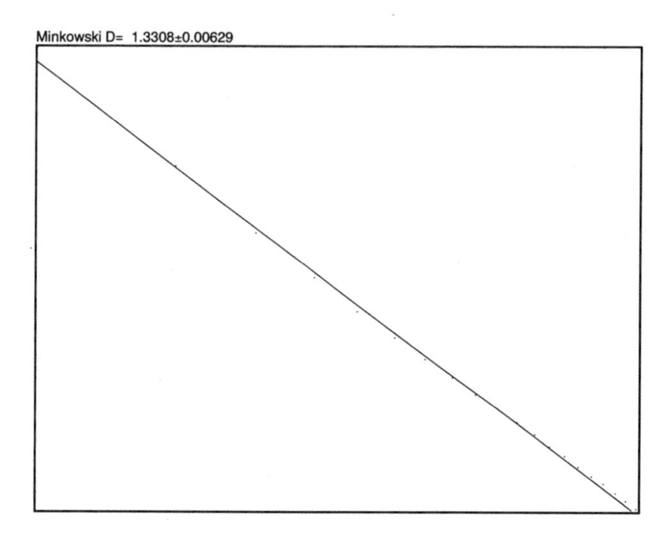

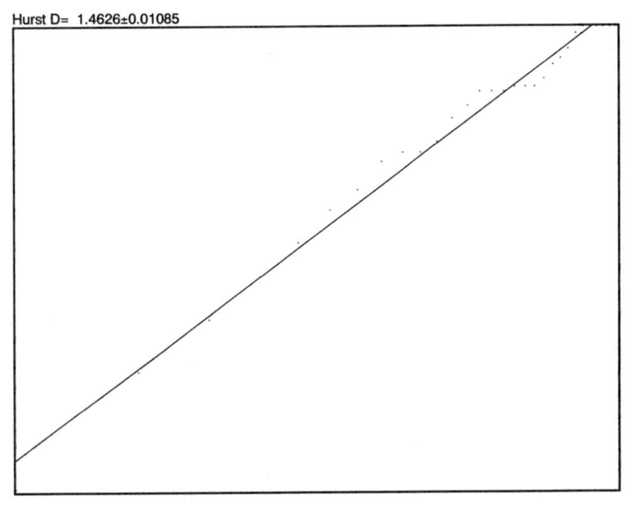

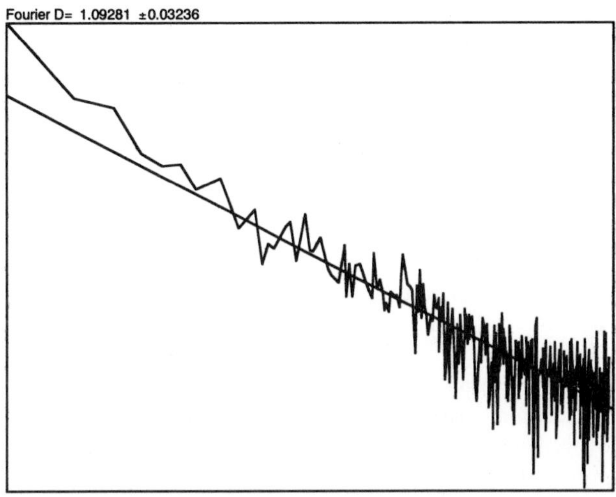

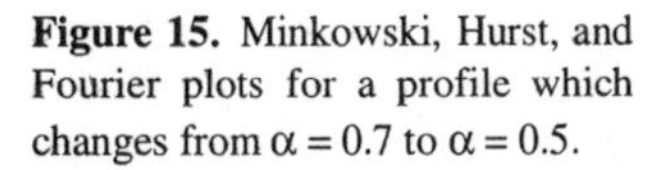
Figure 15. Minkowski, Hurst, and Fourier plots for a profile which changes from $\alpha = 0.7$ to $\alpha = 0.5$.

distance, since the actual length of the profile is undefined. These profiles were then measured using Richardson, Minkowski, and Hurst techniques. In all cases, the log–log plots showed excellent straight-line relationships with no significant curvature or sudden change in slope.

The results are shown in Figure 16. As usual, the numerical values produced by the various methods are different, but it is still reasonable to examine the variation of values for the mixed profiles. The Minkowski and Richardson data show a smooth transition between the two end values, although it is not a straight-line relationship. The Hurst results are quite different. The Hurst dimension remains essentially constant and equal to the higher value for all of the intermediate cases. This is not surprising since the Hurst method finds the maximum differences between points, and so responds to the roughest portion of the surface.

Directionality

Fractal surfaces may contain another type of fractal mixing in addition to those discussed above. Many real surfaces have fractal dimensions which vary with direction. It is not clear how the directional information which can be obtained, for instance, from a profile in one direction is related to the overall surface dimension. Directional variations in roughness and fractal dimension occur for a variety of reasons. Different material properties due to crystallography, or environmental variables such as directional deposition of particles or chemical gradients, may be present. But the most common cause of anisotropy is some directionality in the process that produces or modifies the surface. Fractures typically proceed from a starting point, and wear and machining are directional.

A typical high-precision machined surface can serve as an example of extreme anisotropy, with fractal behavior in both the transverse and machining direction. Figure 17 shows images of a surface of precision machined nickel, obtained by both an interferometric light microscope and an atomic force scanning microscope on the same specimen (but at different locations). FT analysis of vertical elevation profiles in the machining and transverse directions gives slopes that agree between the two instrumental techniques, and indicate the difference of dimension with direction, as shown in Figure 18.

The Minkowski cover dimension for these anisotropic surfaces can be calculated in the usual way, as shown in Figure 19, but while the numeric values are similar for the two different instrumental data sets, the values do not seem to have any meaning. Figure 20 shows HOT and 2DFT analyses of the two range images. The magnitude of the anisotropy is similar in all of the rose plots, and will serve as a point of reference for comparison of simulated surfaces.

A simple model which might be used to examine machining and other anisotropic surfaces utilizes extrusions. A simple extrusion can be generated by first creating a fractal profile, and then moving it at right angles to generate a surface, as discussed in Chapter 6. Figure 21 shows an example. The profile was produced by midpoint displacement using $\alpha = 0.7$, which corresponds to a fractal dimension of 1.3. Measurement of the profile using Richardson and Minkowski methods confirms this value. Measurement of the surface produces a value of 2.3, which agrees with Mandelbrot's conjecture that extrusion adds 1.0 to the fractal dimension of the profile. Measurements of the fractal dimension of a profile passed through the surface in several different directions (parallel, perpendicular, and at 45 degrees to the extrusion direction) are tabulated in Table 2. The fractal dimension of this kind of extruded surface is equal to 1.3 (the same as the profile) in every direction except exactly parallel to the extrusion direction; in that direction it equals 1.0.

Instead of extruding the fractal profile along a straight line, it is possible to create a surface by moving the first profile along a path perpendicular to its plane that follows a second fractal profile. This is equivalent to adding together two simple extruded surfaces with perpendicular

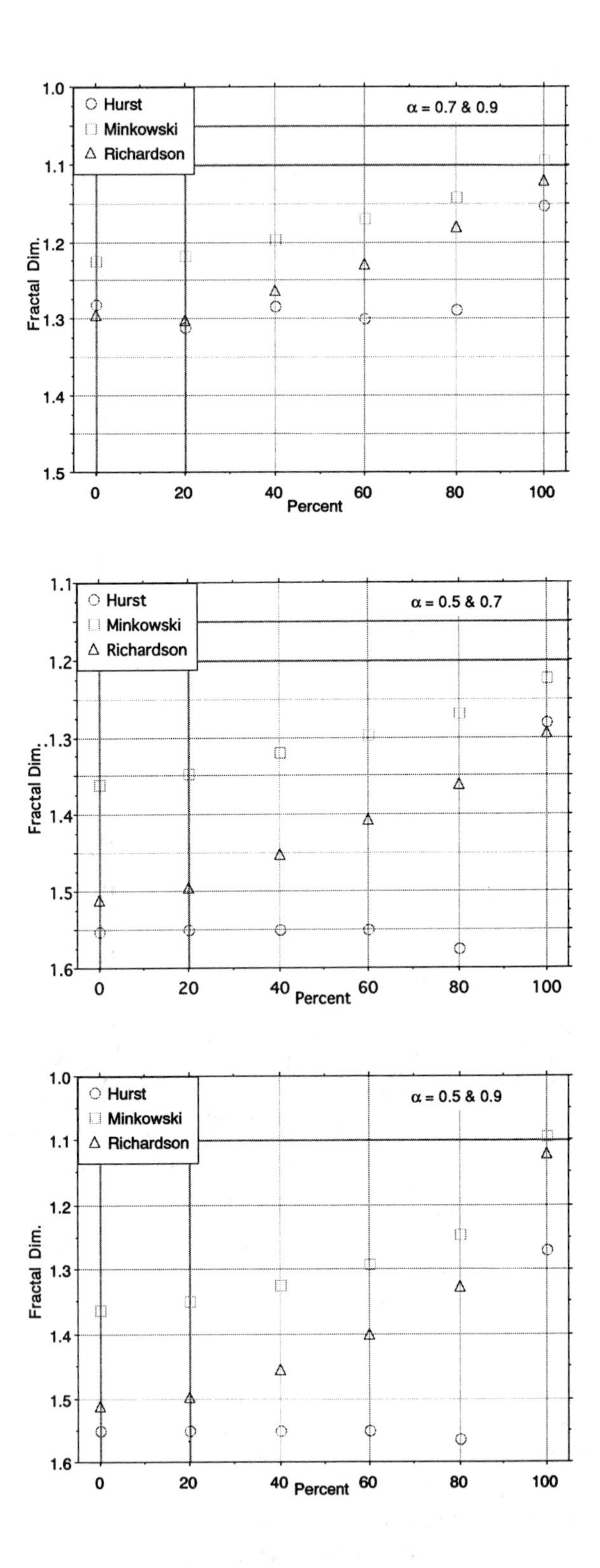

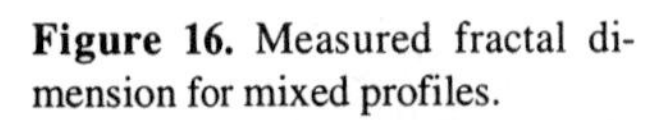
Figure 16. Measured fractal dimension for mixed profiles.

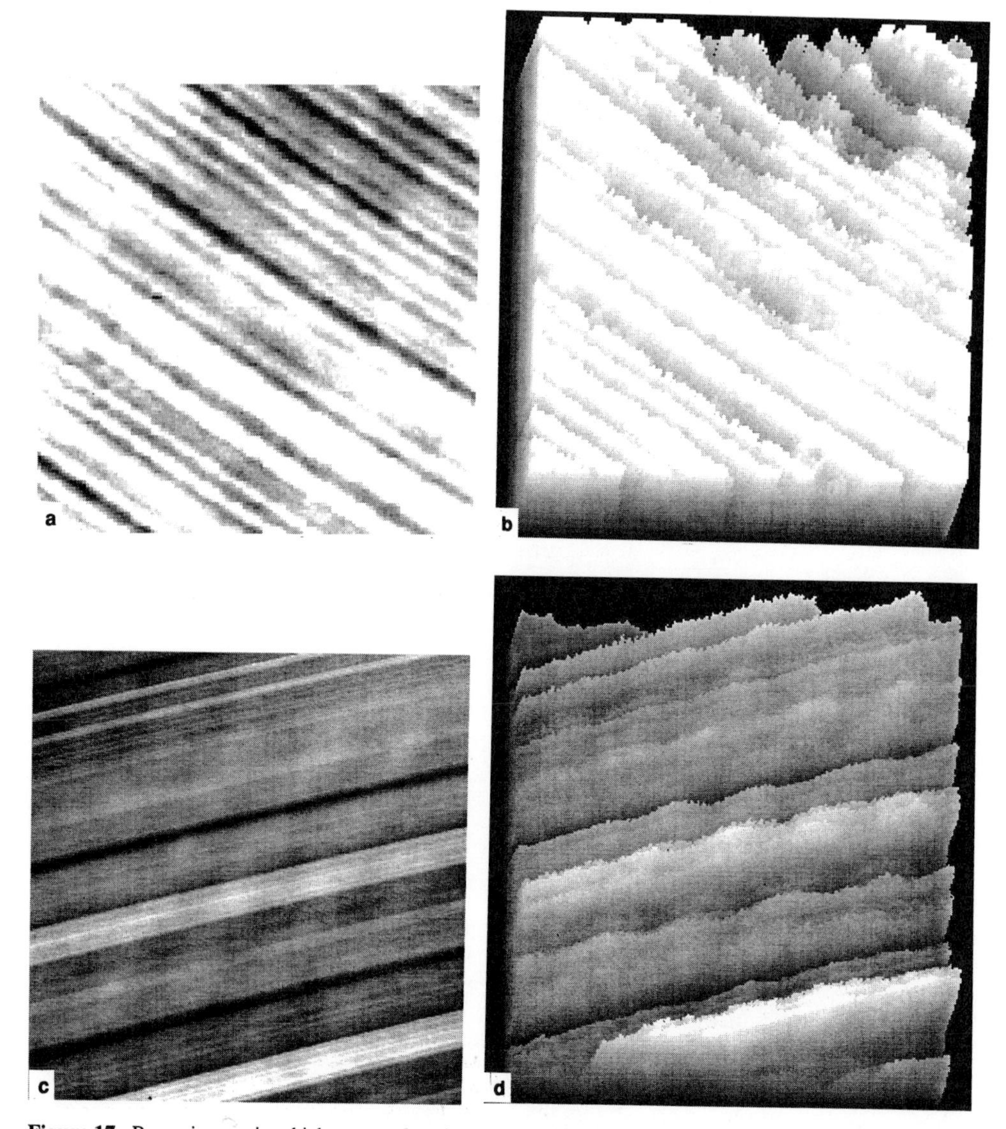

Figure 17. Range images in which grey-scale value encodes elevation, and isometric images combining vertical displacement and shading, showing a machined nickel surface using an interferometer (Zygo) and atomic force microscope (Nanoscope). The magnification of the interferometer image is 0.37 μm/pixel, and that of the AFM image is 50 nm/pixel; the image fragments shown are 256 × 256 pixels. The isometric presentations make it easier to see the roughness. **(a)** Interferometer range image; **(b)** interferometer isometric image; **(c)** AFM range image; **(d)** AFM isometric image.

extrusion directions. Figure 22 shows an example of such a surface using two profiles, each generated with $\alpha = 0.7$. The measurement data shown in Table 2 agree with Mandelbrot's conjecture that the dimension of this surface should be the sum of the two profiles. Notice that the profile measurements in various directions are nearly the same, although the surface is certainly not an isotropic one. Also, since the surface is not uniform and isotropic, the profile dimensions are not related to the surface dimension by the $D + 1.0$ relationship discussed above.

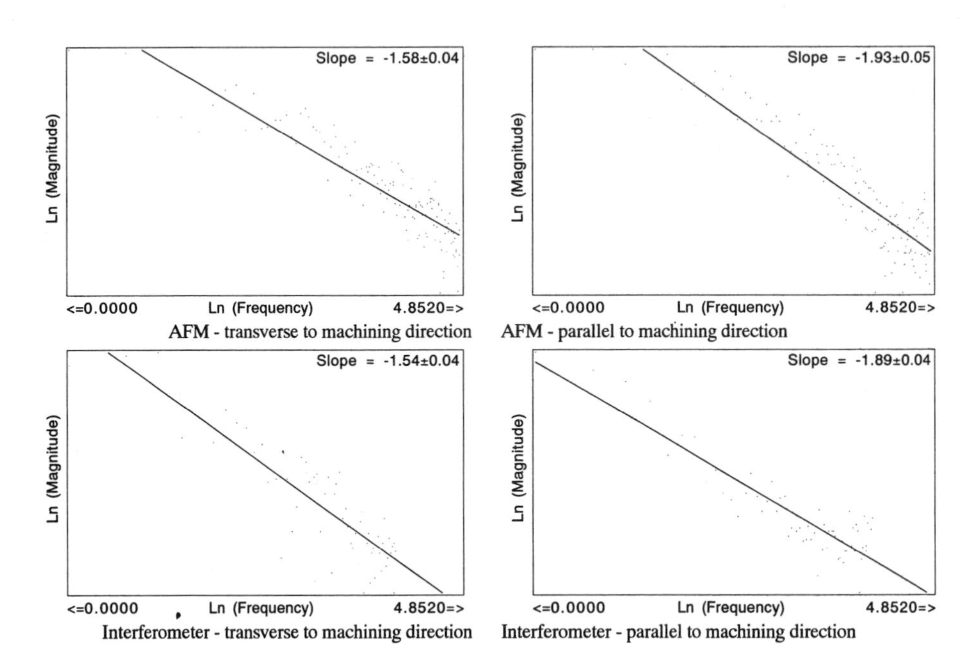

Figure 18. Fourier power spectrum plots from 1D elevation profiles, in directions parallel and transverse to the machining direction, for the AFM and interferometer images in Figure 17.

The situation becomes much more complicated when the two cross-extruded profiles have different dimensions. Figure 23 shows the result of using $\alpha = 0.5$ and 0.9. The dimensions of the profiles should be 1.5 and 1.1, producing a surface of dimension 2.6. The Minkowski measurement of the surface is in approximate agreement, and the directional profiles show the larger value in every direction except the one corresponding to the smaller dimension. Thus, the profile dimension follows the rule for union while the surface dimension corresponds to the rule for multiplication.

Figure 24 shows the result of using $\alpha = 0.5$ and 0.1. The two profile dimensions should be 1.5 and 1.9, with a sum of 3.4. Since this exceeds 3, the topological dimension of the space in which the surface resides, the Mandelbrot conjecture predicts that the surface dimension

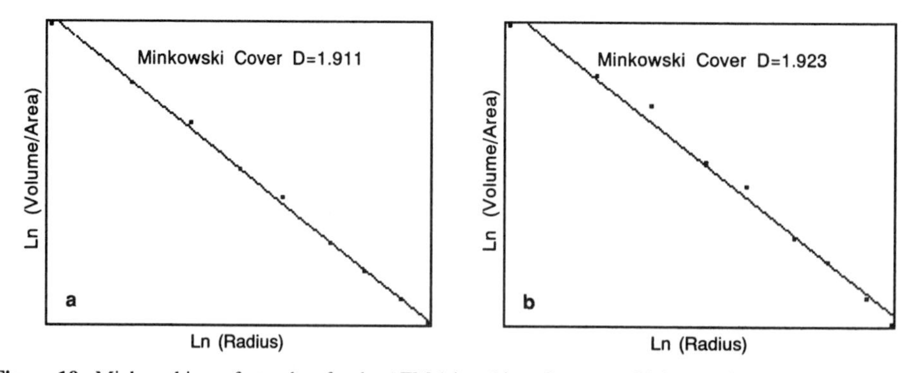

Figure 19. Minkowski comforter data for the AFM (**a**) and interferometer (**b**) images in Figure 17. The values for the two images are very similar but do not seem to represent the actual surface roughness.

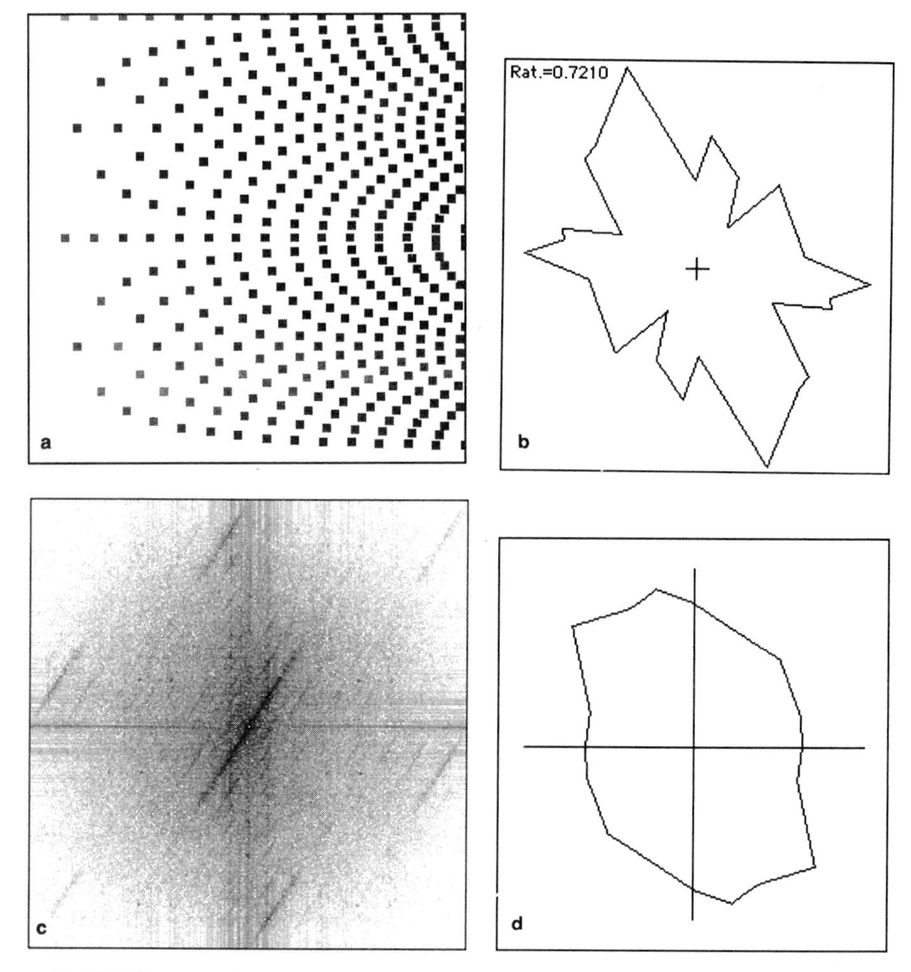

Figure 20. 2DFT images and rose plots, and HOT images and rose plots for the interferometer and AFM images in Figure 17. **(a)** Interferometer HOT image; **(b)** interferometer HOT rose plot; **(c)** interferometer 2DFT image; **(d)** interferometer 2DFT rose plot; **(e)** AFM HOT image; **(f)** AFM HOT rose plot; **(g)** AFM 2DFT image; **(h)** AFM 2DFT rose plot.

Table 2. Fractal Dimensions Measured on Extruded Surfaces

Surface	Minkowski (Covering) Dimension	Profile Dimensions: Direction (Minkowski Method)				Direction (Richardson Method)			
		45°	90°	135°	180°	45°	90°	135°	180°
$\alpha = 0.7$	2.319	1.294	1.348	1.003	1.347	1.332	1.358	1.001	1.361
$\alpha = 0.7\&0.7$	2.621	1.290	1.306	1.248	1.280	1.332	1.380	1.331	1.312
$\alpha = 0.5\&0.9$	2.757	1.413	1.440	1.143	1.440	1.539	1.560	1.189	1.564
$\alpha = 0.5\&0.1$	2.492	1.447	1.489	1.470	1.484	1.556	1.696	1.722	1.683
not perpendicular	2.576	1.555	1.157	1.556	1.546	1.446	1.136	1.446	1.398
4-way	2.935	1.287	1.304	1.317	1.318	1.357	1.369	1.436	1.432

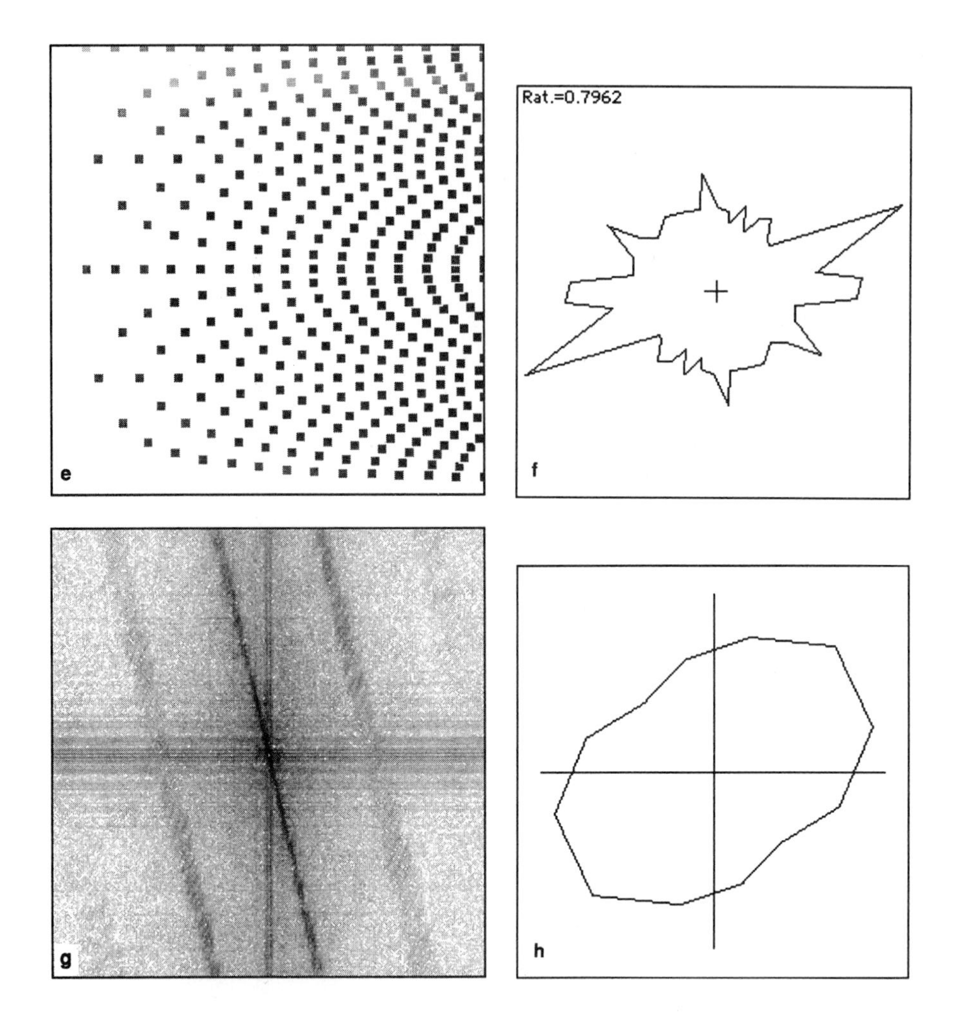

should equal 3. Instead, the measured value is actually less than that for the case shown above. In addition, the directional variation is more complicated.

When two profiles with $\alpha = 0.5$ and 0.9 are extruded, but not at right angles to each other (Figure 25), the result is the same as when the projections were perpendicular. However, even more complex situations arise when more extrusion directions are added. Figure 26 shows the result of combining four extrusions, using profiles with $\alpha = 0.1, 0.2, 0.3,$ and 0.4. Measurement of the surface fractal with a Minkowski method reports a value of 2.935, and directional profiles have dimensions that are all close to 1.35 and show little variation with direction. In the limit, if extrusion in many directions with the same profile dimension is performed, the surface becomes isotropic and the surface dimension is 1.0 greater than the profile dimension in accordance with Mandelbrot's second conjecture.

From these few examples, it appears that the relationship between profile and surface dimensions can be more complex than predicted by the Mandelbrot conjectures. This at least occurs when the sum of profile dimensions exceeds that of the embedding space, or the anisotropy of the surface is more complex than that produced by one or two extrusions.

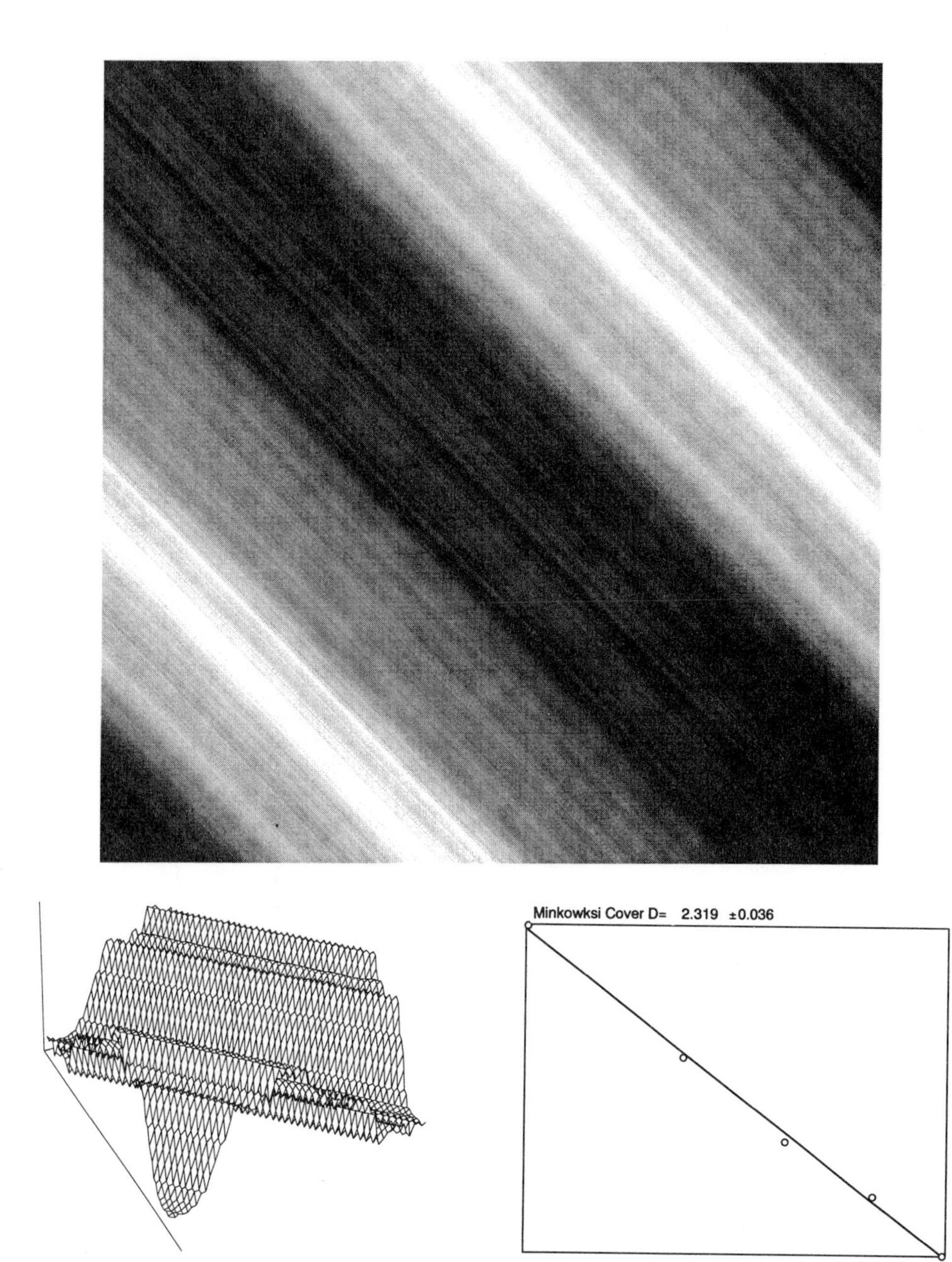

Figure 21. Extrusion of 2D surface from a profile with $\alpha = 0.7$, shown as a grey scale and perspective image, and the log–log Minkowski plot.

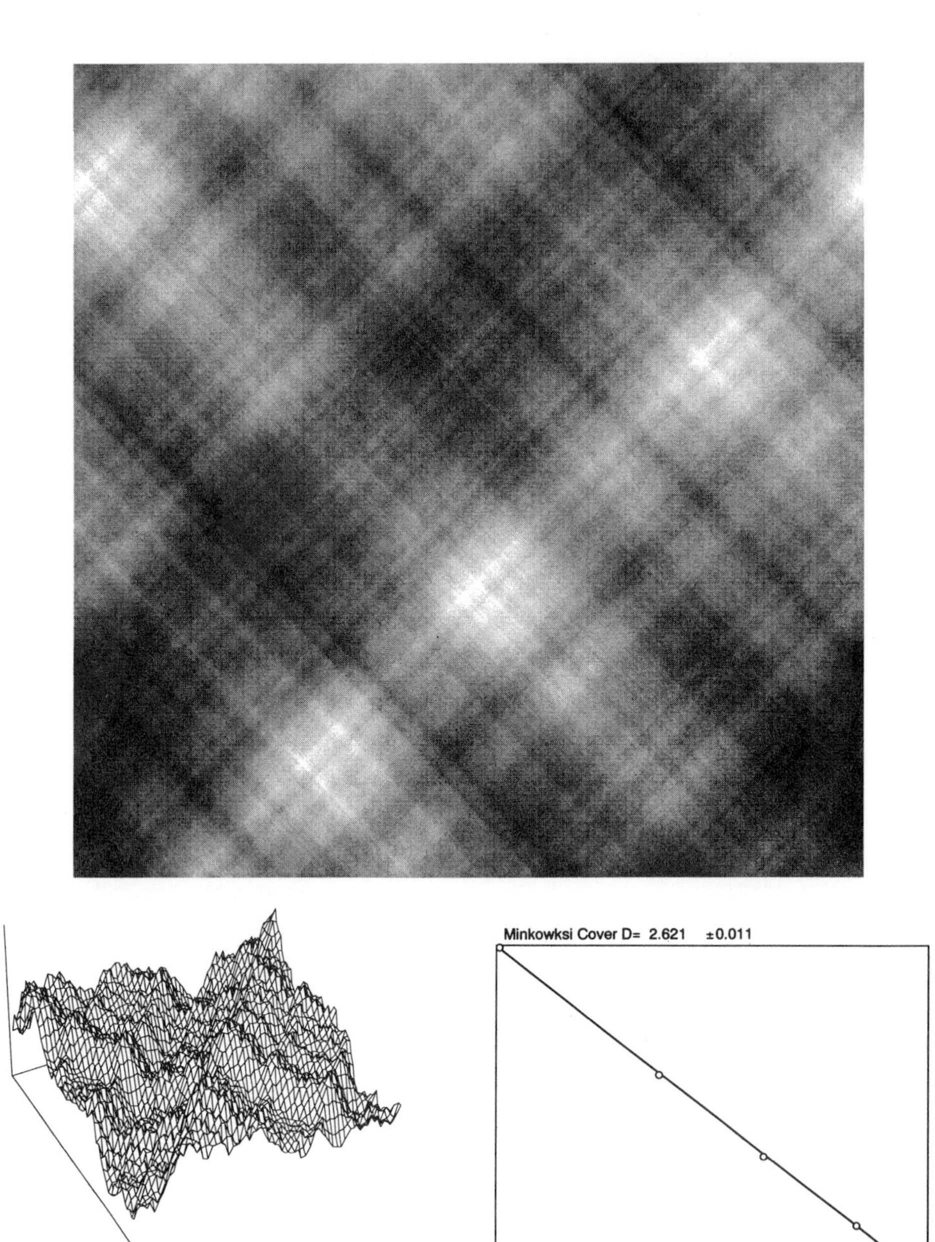

Figure 22. Cross-extrusion of 2D surface with $\alpha = 0.7$ in each direction, shown as a grey scale and perspective image, and the log–log Minkowski plot.

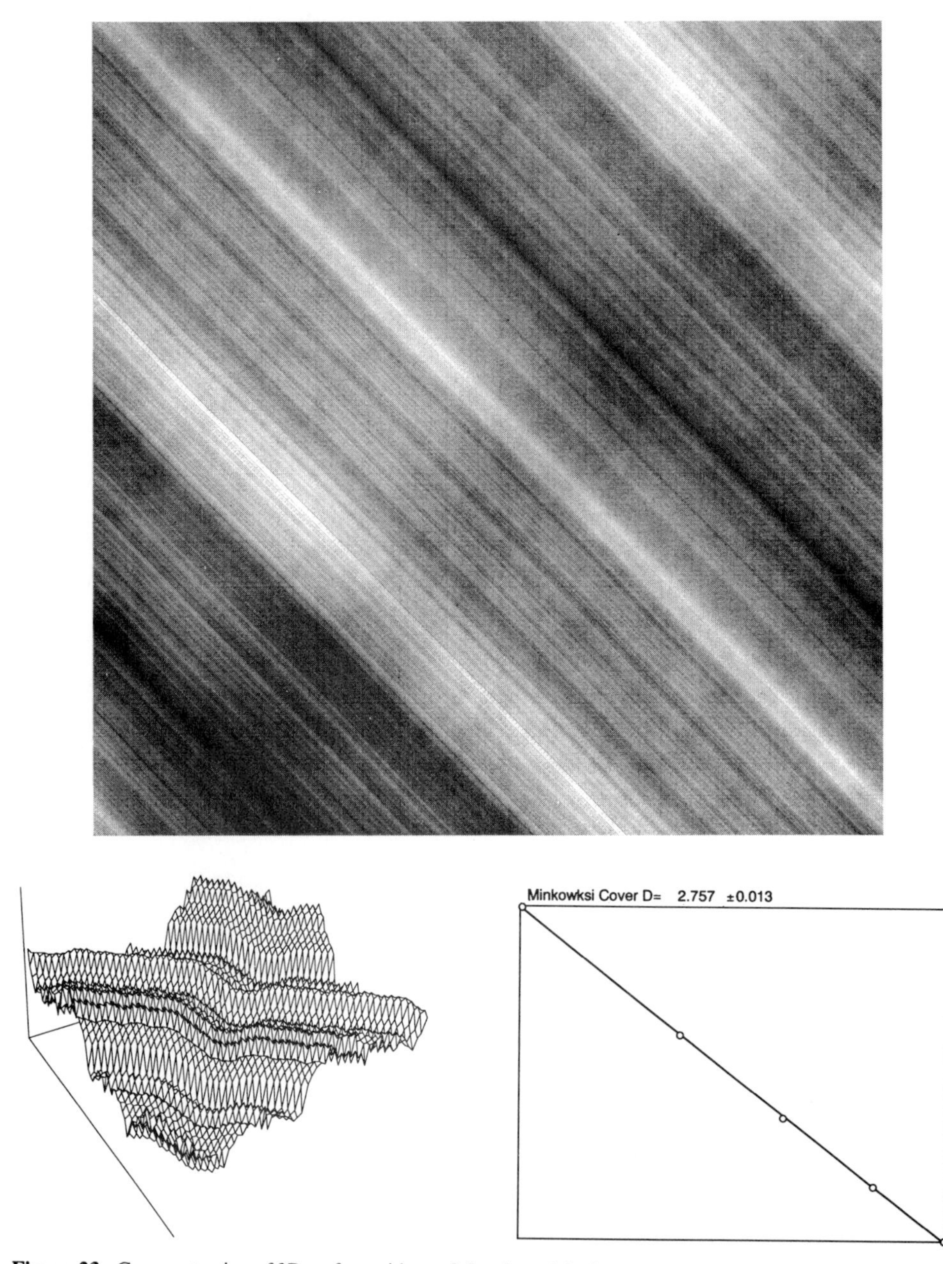

Figure 23. Cross-extrusion of 2D surface with $\alpha = 0.5$ and $\alpha = 0.9$, shown as a grey scale and perspective image, and the log–log Minkowski plot.

Figure 24. Cross-extrusion of 2D surface with $\alpha = 0.5$ and $\alpha = 0.1$, shown as a grey scale and perspective image, and the log–log Minkowski plot.

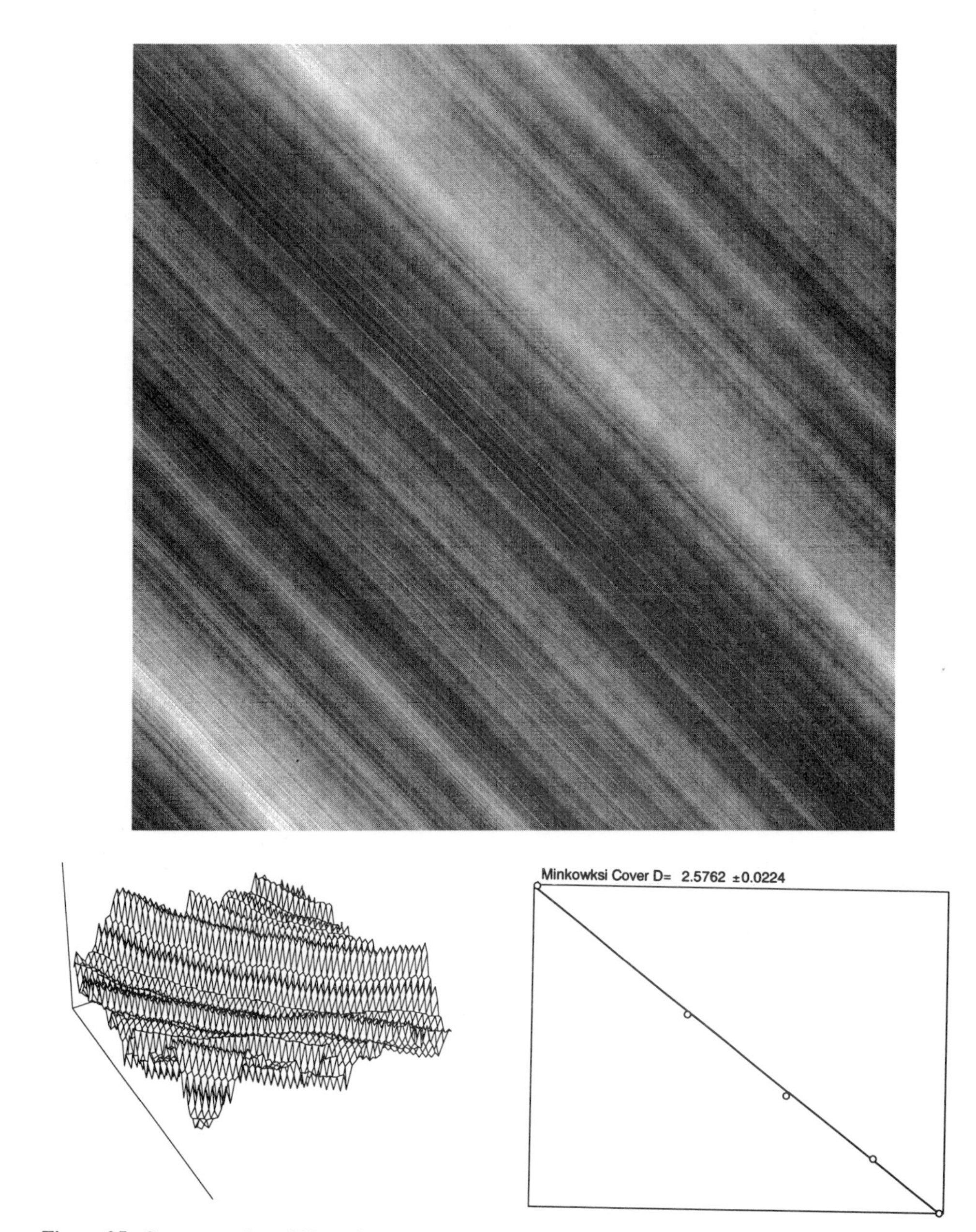

Figure 25. Cross-extrusion of 2D surface using profiles with $\alpha = 0.5$ and $\alpha = 0.9$ which are at 45 degrees rather than orthogonal.

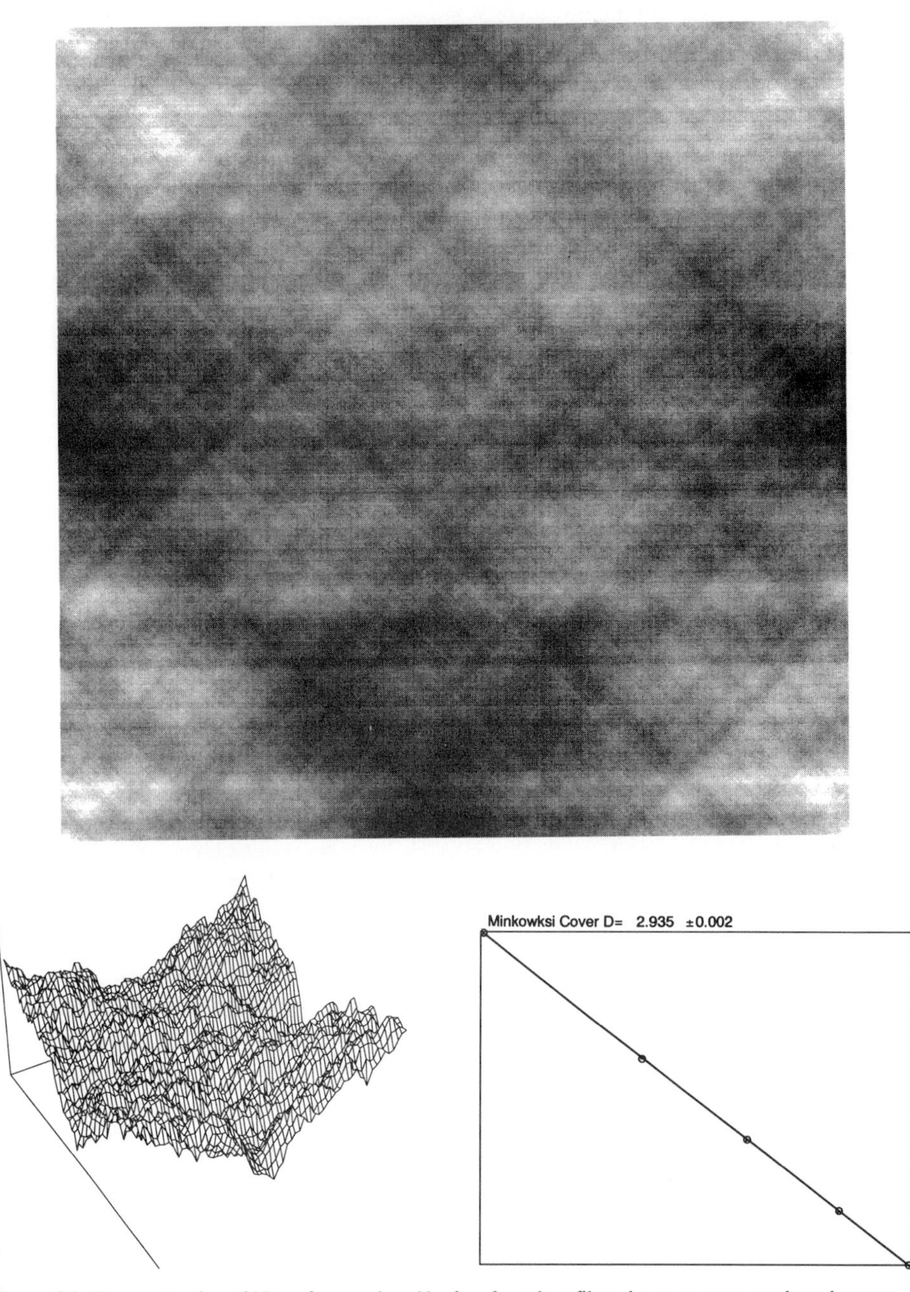

Figure 26. Cross-extrusion of 2D surface produced by four fractal profiles, shown as a grey scale and perspective image, and Minkowski plot.

Tentative Conclusions

The different possible interpretations of mixed fractal behavior illustrated here give rise to very different measurement results, using the various methods commonly used to evaluate fractals. It does not seem possible to draw any comprehensive conclusion about the interpretation of such values. It may not be possible from these measurements to distinguish the various ways that fractals can be combined, nor in some cases to distinguish them from single or ideal fractals. Finally, it is not really clear which if any of the various interpretations of multifractality or mixed fractality may correspond to physical surfaces and boundaries.

Nor are these problems confined to the case of surface dimensions and measurements using tools such as Minkowski or Fourier methods. They also apply to particulates, and to direct surface measures such as gas or dye adsorption. Pfeifer (1984) shows a mixed fractal behavior for adsorption of dye on a soil as a function of particle diameter. The soil is a mixture of coarse particles (feldspar, quartz, limonite) and fine (kaolinite). A plot of log (milligrams of adsorbed dye per gram of soil) vs. log (particle diameter) shows a plot with two definite slopes. Although the slope of the curve for small particles is lower, so that the plot looks superficially like a typical Kaye "structural/textural" example, in fact the corresponding fractal dimensions vary the other way for this measurement method as shown on the graph in Figure 27. The fine particles produce a structure with so much internal porosity that can be reached by the dye that the surface area increases almost in proportion to volume (dimension 3.0). The fractal dimension of 2.92 for small particle sizes is much greater than that (2.46) for the coarser particles.

Stanley (1991) discussed multifractals in terms of clusters. The dimension of a random walk on a cluster can be either the range or the mass (number of sites visited) as a function of number of steps. The two dimensions are different. Also different is the scaling dimension of the minimum path. The author distinguishes between additive and multiplicative processes: the former should produce a single dimension while the latter produces a multifractal. Mandelbrot and Evertsz (1991) further distinguish between these multiplicatively generated or cascaded cases (which are renormalizable) and a newer class of multifractals in which there is a partition function that has a scaling property. This is where the Hölder exponent comes in and the function $f(\alpha) \geq 0$ is needed to characterize the fractal. This seems not to be related to surfaces, but rather to various types of clusters on fractal supports.

Because of these uncertainties, it is probably wise to avoid the use of the term "multifractal" unless a clear description of the context, measurement, and physical meaning is included. The most widespread use of the term to date is Kaye's, but this is somewhat suspect because it is usually applied to data from projected boundaries, and/or to a possibly inappropriate use of the Richardson plot for self-affine data. Fractal dimensions from such projections do not accurately characterize the surfaces, and at least in some cases this geometry may produce the appearance of two different slopes as an artifact. In addition, another completely different definition of multifractal exists in terms of the Lipshitz–Hölder coefficient, which has precedence and a firmer mathematical interpretation.

The physical meaning of surfaces and boundaries which may correspond to the various types of mixed fractal behavior must be investigated and modeled. The existence of a power-law plot (straight line on log–log axes) does not, of course, "explain" the fractal. The observation that multiplying together a large number of independent random values can produce a power-law plot offers no physical explanation, either.

Publishing surveys of actual surfaces and profiles which exhibit simple or complex fractal behavior should include enough information to enable the interpretation of the values. Anecdotal evidence that a few specimens of a given type can be measured by one or a few of the available fractal dimensions to yield a number is not adequate. It is essential to show the

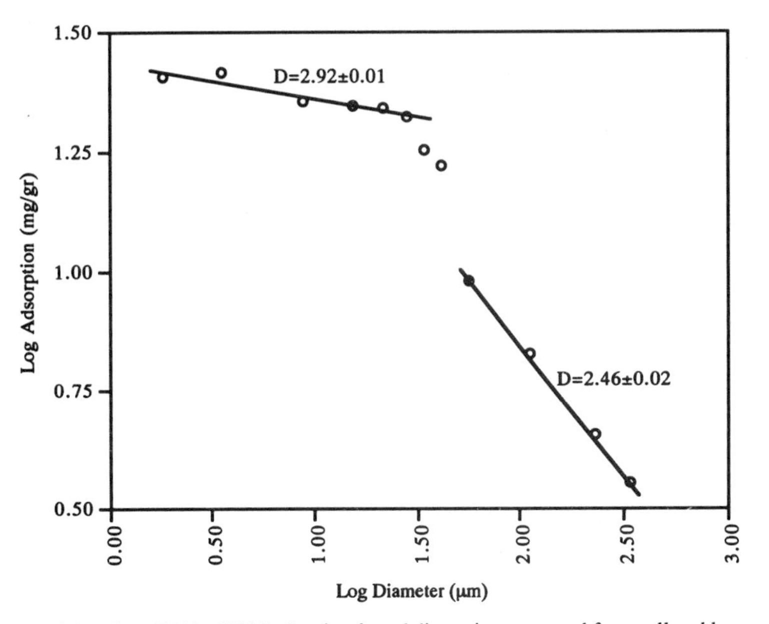

Figure 27. Plot of data from Pfeifer (1984) showing fractal dimension measured for small and large particles in a mixture.

linearity and quality of fit of the log–log or power-law plot, with actual ranges of distance scales. Care in sampling procedures, and full discussion of this important step, is also needed. The comparison of D values with each other should not use simple parametric tests (e.g., Anova or t-test) since the presumption of normality is not met. Correlation of D values with various parameters describing surface properties or history should likewise include raw data and not make parametric assumptions (e.g., linear regression).

The fact that fractal geometry describes many real situations has been demonstrated in a large number of situations. It is clear that this approach provides a more meaningful way to look at natural objects than is possible with classical Euclidean geometry. The tools of fractal geometry, the various measurement methods, and their response to different kinds of surfaces (especially mixed fractals) need to be investigated. This will require both modeling and mathematical analysis, in addition to continued application to real examples.

8

Examples of Fractal Surfaces

Discussion of some of the situations in which real surface fractal measurements have been made can be organized in two different ways. The preceding chapters have discussed the various measurement tools, including the kinds of microscopes (light, electron, scanned probe) used to acquire surface range or elevation images and the mathematical procedures applied to obtain the various fractal dimensions. Some mention of the findings from these studies and the use of these tools and simulations for real-world samples has been essential to explain the various approaches, but the emphasis has been on measurement techniques.

In this final chapter it is intended to emphasize a few of the variety of applications in which fractal characterization has been utilized. This includes the types of surfaces studied (various materials, the effects of the surface history, surface properties, etc.). The material is organized primarily by specific applications such as fracture or wear surfaces, human skin, etc. Of course, it will still be necessary to identify and compare some of the measurement methods (kind of microscope, calculation procedure, etc.) since, as we have seen, these can influence the results. Many of the surface images from these applications have been presented in earlier chapters and will be referred to here but not duplicated. Consequently, there will be more references but fewer illustrations in this chapter.

It cannot be stated too strongly that there is no attempt here to "explain" any of these occurrences of fractal surface behavior. In most cases, the physical reasons why surfaces should arise having this kind of geometry are quite unknown (Maddox 1986; Mandelbrot 1990). Many of the published results have simply (and frustratingly) shown that certain kinds of surfaces, often naturally occurring ones, *are* fractal. In some cases, attempts have been made to go beyond that to show that the fractal dimension can be correlated with some other piece or pieces of information, such as the material properties, the variables controlling the formation of the surface, or the resulting properties of the surface. Such correlations may suggest some causal linkage, but as in all such cases they do not prove it.

In a few instances, efforts have been made to use physical principles, models, or simulations in attempts to relate the surface geometry and fractal dimension to parameters of history, properties, or behavior. It is not the purpose of this book to investigate the models too closely, but brief descriptions and references will be offered as appropriate. Some of the models are still quite crude and seem greatly to oversimplify the real-world phenomena at play, while others are numerically precise, offer accurate predictions, and may capture most of the important aspects of the physics. Formation of clusters of particles by aggregation is an example of the latter, but as it is only tenuously connected to surfaces, which are the primary topic of this book, and because there are comprehensive published treatments of the subject (see for instance Meakin 1983; Sander 1986a; Kaye 1989a; Vicsek 1992), it is not dealt with

extensively in this chapter. Some examples of surfaces generated by these methods are shown in Chapter 6.

The physical mechanisms that seem to naturally give rise to fractals, and can be exploited in physically based models to generate them, make one of three basic assumptions. First are those situations in which Laplacian fields ($\nabla^2 = 0$) are present, such as electrical fields, diffusion, etc. In such a field, any advance of a diffusion front, crack, etc., becomes unstable. This leads to viscous fingering of fluids, branching of electrical discharges, dendritic solidification of crystals, and so on. Observed phenomena such as the branching of retinal blood vessels (which are believed to follow chemical concentration gradients), percolation of fluids, formation and growth of neural networks, and other apparently diverse applications can be related to this situation. However, few of these are related to the central interest of this book, which is fractal surfaces.

A second assumption that can be used to generate fractal response is to assume a superposition of a random set of events. The classic example of this is Brownian motion, which has been used in earlier chapters as a generating method for fractal profiles, invoked as a source of noise, and also employed to generate aggregates by the random walking of many particles or clusters of particles. It is a basic statistical principle, known as the central limit theorem, that if a great many independent sources of variation are present, the superposition of all will produce a Gaussian probability distribution function (pdf). Consequently, it is most common to use such a Gaussian pdf when generating Brownian fractals. However, that is not an essential feature of the process, and any distribution function may be used.

A third way to generate fractals is to begin by assuming a power-law distribution of some kind. For instance, cratering by particle (or meteorite) impact will produce a fractal surface if the size distribution of the particles follows a power-law. The Sierpinski gasket and Menger sponge, or their randomized natural brethren, can be treated as the removal of material following a power-law distribution. The existence of power-law distributions in nature is quite common, but this does not "explain" the fractal; it simply moves the arbitrary assumption back one step. A partial justification may lie in the fact that the product of many independent events can be used to produce a power-law distribution, just as the sum of many independent processes produces a Gaussian distribution.

Fractal behavior is suspected in many complex situations such as turbulent fluid flow, based on an argument much like the preceding, that there is a superposition of activity at all scales, that the various processes are independent, and therefore "should" give rise to power laws and fractal behavior. This approach has not yet been reduced to a formal physical model with quantitative predictive power, however. The fact that nature frequently produces self-

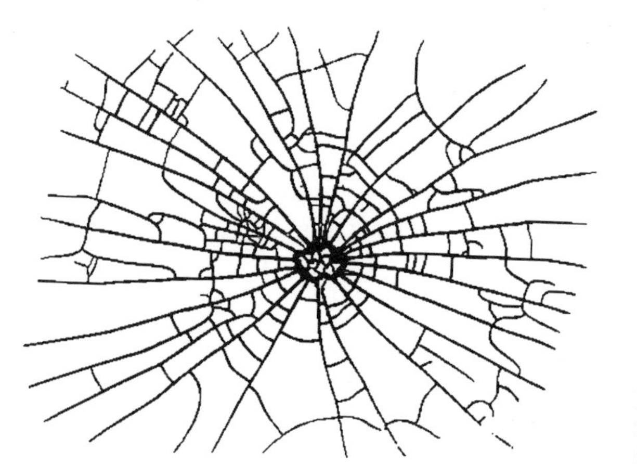

Figure 1. A network fractal produced by impact on a sheet of polymethyl acrylate (Vicsek 1992).

similar coastlines or self-affine time records of tree ring growth which can be efficiently and compactly characterized by fractal geometry is incompletely understood, but does justify the use of these methods to study the phenomena.

Brittle Fracture

The fracture of materials is an important topic of study in many branches of engineering, and the art of fractography (the examination and study of fractures) is well established; for a review, see Chermant and Coster (1983) or Russ (1990d). There are few quantitative rules, however, that relate the morphology of the fracture surface to the properties of the material (particularly its resistance to fracture), or to the circumstances of the fracture (sudden and catastrophic, gradual by fatigue, etc.). The primary interest has been directed to fracture surfaces that represent a crack that propagates through material, more or less transverse to an applied force, and results in separation of the parts and the formation of a new surface (or pair of surfaces). Other kinds of fracture may also produce fractals, such as the network of cracks shown in Figure 1 created by impact on the surface of a sheet of polymer. This may be related to the formation of networks of cracks in drying mud, and other geological features, but this is a different kind of fractal.

Recognizing that fracture surfaces are usually quite rough, and that fractal geometry often applies to rough surfaces, quite a few researchers have tried to apply fractal concepts to the study of fracture surfaces, beginning with Mandelbrot himself (Mandelbrot, Passoja et al. 1984). His original work used two methods to determine a fractal surface dimension. The first is Fourier transform (FT) analysis of elevation profiles across the fracture surface, and the second is "slit island" analysis, in which a flat polished section parallel to the nominal surface creates islands and lakes.

As discussed in previous chapters, these islands can be used in several ways to obtain a surface dimension, including measuring the profile dimension (by any method, including a Richardson plot, since the zeroset created by the sectioning makes the boundaries self-similar even if the surface itself is self-affine), performing dimensional analysis in which a plot of the log (Area) vs. log (Perimeter) data for many islands is used, or the Korcak method in which the number of islands whose area exceeds A is plotted against the value of A. Mandelbrot used the second of these methods, a plot of log P vs. log A (dimensional analysis), commenting that lakes within islands should be included and islands within lakes be skipped in performing slit island analysis to plot log P vs. log A. He showed for the fracture surfaces examined a close agreement between the two values (e.g., 2.28 by dimensional analysis, 2.256 by FT).

Unfortunately, many of the papers (especially in the "early years") misapplied the measurement procedures, for instance by using Richardson plots on elevation profiles, although they are inappropriate for surfaces that are self-affine. For instance, Clarke (Clarke 1986; Clarke and Schweizer 1991) calculated D using a Richardson technique for vertical elevation profiles in x and y directions along each row and column of a complete array of elevations, and then averaged the values and added 1.0 to obtain a dimension for the surface. The paper claims the result is "robust," but admits it does not agree with other methods. There is in fact no justification for: a) using the Richardson method for vertical profiles; b) averaging the dimension from several lines, especially ones in perpendicular directions; or c) adding 1.0 to estimate the surface dimension, since it may not be self-similar or isotropic.

Denley (1990a) tried to apply the Richardson method to determine the variation of measured area to measuring scale directly for fracture surfaces of steel and epoxy resin. The elevation data were determined by STM. Triangles fit to the regular grid of x, y points were used to estimate the area. Although the different materials exhibited visually different rough-

nesses and quite different mechanical properties, the measured surface fractal dimensions could not distinguish them. The plots also show considerable nonlinearity; this is almost certainly due to the fact that the points were uniformly spaced in x, y and not along the surface, so that the triangles varied considerably in size. Also, the surface elevation data as measured by the STM are single-valued, hence self-affine rather than self-similar regardless of whether the surface itself is actually self-similar.

Many authors worked with vertical elevation profiles through fracture surfaces (Chermant, Chermant et al. 1987), although Long, Suqin et al. (1991) have criticized the use of vertical sections to examine anisotropic fractal surfaces. Alexander (1990) misapplied the Richardson method to vertical elevation profiles and obtained log–log plots that are curved, taking the slope of the central portion as an estimate of the dimension. While this is not a correct measurement of the surface fractal dimension, the values were still found to be positively correlated with the fracture toughness. Rather than use physical sectioning, Antolovich, Gokhale et al. (1990) used computed tomography to image sections through cracks within materials. However, the resolution was inadequate to permit quantitative analysis. Lung (1985) and Lung and Mu (1988) have used horizontal sections (slit island) with dimensional analysis to obtain D values to correlate with fracture toughness values.

Working with profiles produced by vertical sectioning, Banerji and Underwood (1984) reported that fracture surfaces in a high-strength steel were fractal, with the lowest dimension corresponding to the heat treatment that produced embrittlement of the material. However, Underwood and Banerji (1986) later reported that fracture surfaces were not true fractals. Instead, they plotted the ratio of profile length to projected length (often called the "roughness" R) vs. the length of the measuring scale and found a "sigmoidal" curve instead of the linear (log–log) plot expected for a fractal. But Ling (1989) has shown that for a variety of surfaces R varies with the measurement unit as $R = \exp(2.3\varepsilon/\varepsilon_c)^{-\alpha}$, where ε_c is surface-specific and α is typically 2 for a variety of surfaces. This suggests a possible fractal behavior for the surface, but it is not easy to determine a fractal dimension from such a result (Majumdar and Bhushan 1991) because the surfaces are self-affine rather than self-similar. It may be expected that the flattening of the curve at small dimensions is an artifact of limited imaging resolution, and that the misapplication of a Richardson approach to a self-affine vertical profile is a factor in Underwood's report. It is also not clear that the fracture reported was actually brittle.

Wang, Dong et al. (1990) also employed a series of heat treatments on steels to vary the microstructure, measuring the fractal dimension by a slit-island dimensional analysis technique. A change was reported in the surface fractal dimension with recovery and recrystallization in the crystalline grain structure after cold working.

In addition, many of the papers have been content to report that a fracture surface in some particular material was observed to be fractal, or at least that it gave a more-or-less linear plot on log–log axes. There was little effort devoted to comparing the fractal dimensions of materials with different mechanical properties, microstructures, etc. Furthermore, the studies of this type that were made used a wide variety of different parameters to describe the fracture resistance of the materials, ranging from reasonably well understood quantities such as the fracture toughness K_{IC} to simple tests that produce a complex result, such as the impact energy in a Charpy test. It is noteworthy that these reports are not consistent in how they measure fracture properties *or* fractal surfaces, and that some of the reports show increasing fractal dimension with increasing fracture toughness, while others are opposite.

Ray (Ray, Mandal et al. 1990; Ray and Mandal 1992) reported a positive correlation between D (measured by the slit island and dimensional analysis method) and impact energy for steels. However, measurements on vertical profiles (using a Richardson method inappropriate for self-affine profiles) did not give a straight-line log–log plot, and the fracture mechanism may be different from that investigated by Mandelbrot. As discussed above,

Mandelbrot found a negative correlation between fracture toughness and fractal dimension using a slit island method in which lakes within islands are measured and islands within lakes are not, while Huang, Tian et al. (1990) reported a negative correlation when measuring lakes, but a larger dimension and a positive trend of increasing dimension with toughness when measuring islands. Mu et al. (1993) have criticized the log *P* vs. log *A* (dimensional analysis) method as flawed because the perimeter (*P*) values are determined for both large and small features with a fixed ruler or scale length, which biases the data. Further, Mu blames this for his own report (in Mu and Lung 1988) that *D* and K_{IC} are inversely related, and reports that when this is corrected, the correlation becomes positive. Przerada and Bochenek (1990) report straight-line Richardson plots over more than 2 decades ($\lambda = 10$ to 1000 μm) and correlation of the *D* values with the K_{IC} fracture toughness and microstructure (produced by varying the cooling rate in heat treatment).

Imre, Pajkossy et al. (1992) used a novel method to measure surface dimension. They measured the time dependence of the flux of molecular species with sizes from 1 to 100 μm from surfaces of Charpy impact fractures of carbon steels. For *D* values in the range 2.0 to 2.4, this gave a precision of ±0.02 and showed a slight decrease in dimension with increasing impact energy produced by varying tempering temperatures.

Not all researchers agree about the existence of any correlation, positive or negative, between fractal dimension and fracture properties. Baran, Roques-Carmes et al. (1992) examined the fractal dimension *D* of fractures and the fracture toughness K_{IC} of glass and porcelain. K_{IC} was measured by an indentation method, *D* by the Minkowski variational method applied to profiles. They found no universal relationship between the two values and therefore suggested that the relationship must vary for different materials. Bouchard, Lapasset et al. (1990) measured the same dimension value for all fracture surfaces evaluated and concluded that materials properties had no effect. Davidson (1989) found no correlation of the fracture surface fractal dimension with any material property, and Richards and Dempsey (1988) found no correlation between fractal dimension of the fracture surface and tensile strength, ductility, or microstructural features in titanium alloys.

Pande, Richards et al. (1987a; 1987b) compared several measurement methods on fractures in titanium alloys, including slit island dimensional analysis, Richardson plots on vertical elevation profiles (incorrect for self-affine surfaces), and a Richardson plot applied to the secondary electron brightness profile for a line scan across a fracture surface in the SEM. This latter technique is expected (as discussed in Chapter 5) to produce fractal self-affine profiles for a fractal surface, but there is no reason to expect the numerical dimension to be uniquely related to the geometric surface roughness. Yet the authors claim all of the values are "reasonably consistent." No comparison to mechanical properties is given.

The situation is still somewhat confused, but some generalizations are beginning to emerge. It is generally recognized that as a material's resistance to fracture increases, the energy absorbed in the fracture process increases and so does the visually observed "roughness" of the fracture surface. It is not obvious that the visual observation of increased magnitude of roughness necessarily corresponds to an increase (or in fact, to any predictable variation) in the fractal dimension of the surface. Many materials, such as most ceramics and glasses, and some polymers and metals, exhibit brittle fracture in which there is a minimal plastic distortion of the surface. However, even in brittle fracture, the two sides of the broken part can still not be reassembled to form the original.

Many metals undergo ductile fracture in which there is considerable plastic deformation of the surface. Instead of a single crack propagating through the matrix, ductile fracture is generally described as the coalescence of voids which form at many locations within the material. This often produces a surface consisting of many "dimples," concavities correspond-

ing to the original voids. Ductile fracture is generally not considered to produce a fractal surface, due to the large amounts of local deformation which modify the surface geometry.

However, Ishikawa (1990) has reported that a dimple rupture (ductile) fracture exhibits a power-law relationship between the number of dimples and dimple size, such that the number of dimples of diameter d is proportional to $d^{-1.5}$. Figure 2 shows a typical dimple rupture fracture, with a variety of dimple sizes. While this is in some sense a fractal relationship, it is not suggested that the surface of the fracture is a fractal, nor that the exponent is related to material properties. Williford (1988) has attempted to extend the dimensional arguments of Mecholsky discussed below to apply to ductile fractures. He shows characteristic dimensions in the Mecholsky relationship that range from a few nm to 0.1 mm for various steels, which may be compared to the < 1 Å up to about 50 Å values reported by Mecholsky for brittle materials, but again there is no correlation of these values with any microstructural feature.

Harvey and Jolles (1990) have reported a correlation between the total area under the ductile stress–strain curve for a material and the fractal dimension of the surface. Also for ductile slip in materials, Gobel (1991) has reported a correlation between fractal dimension and temperature.

Most of the interest in fractal analysis of fracture surfaces has centered on brittle fracture modes (Kagan 1991; Cahn 1989), which for many materials may be either transgranular (transecting the grain structure of the metal or ceramic) or intergranular (following the more or less polyhedral grain boundaries). It is the former which have been primarily analyzed. For noncrystalline materials such as glass, or single crystal materials such as silicon, there are no grain boundaries and the fracture can be considered as transgranular. Long, Suqin et al. (1991) reported an increase in surface fractal dimension with increasing transgranular fracture and increasing energy required to form a crack. They also observe that in slow crack growth, the fractal dimension D increases until it reaches a critical value at which crack propagation becomes unstable.

A number of models for fracture have been proposed that might produce fractal behavior. For instance, models that either try to describe fracture as a solution to a Laplace equation, or as a problem in network percolation, can produce fractal crack lines in two dimensions but are difficult to extend to crack surfaces in three-dimensional solids. There is a superficial similarity between the shape of cracks in brittle materials and the branching shapes of dielectric breakdown in insulators (Takayasu 1985; Takayasu 1990). Both phenomena have a fractal character and dimension (Mandelbrot and Given 1984). The latter process is governed by

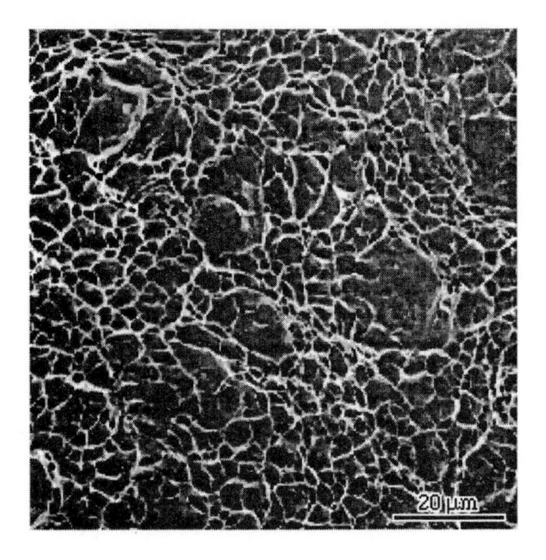

Figure 2. SEM image of a dimple rupture (ductile) fracture in metal.

Laplacian equations for the electric field (Weismann and Pietronero 1986), which has also been shown to govern another fractal phenomenon, viscous fingering of liquids (Van Damme 1989; Vicsek 1992). However, there seems to be no reason to suspect that a Laplacian field is relevant to fracturing.

The fracture process can be modeled and shown to generate a fractal boundary in a 2D plane using either a triangular or square lattice, by growing the brittle fracture in a way similar to bond percolation modeling (Hurd, Weitz et al. 1987; Huang 1989; Balankin and Bugrimov 1991; Panagiotopoulos 1992). Figure 3 shows an example of such a modeled crack growth. Note in particular the large number of multiple cracks, which are not usually an important feature in transverse fractures, although they may be in some other geometries (Yuhong, Berry et al. 1989). Also, the dimension of the lattice unit does not correspond to any physical or structural unit in real metals or ceramics, and there seems to be little if any predictive power to relate the fractal dimension to materials properties, except perhaps in the case of the fracture of granular materials (Arbabi and Sahmi 1990). For granular compacts, Liu, Shih et al. (1990) have reported a relationship between the fractal structure and the modulus of elasticity, but this may not be related to fracture processes.

In addition, it is not easy to see how such fracture models can be generalized to three dimensions (Herrmann and Roux 1990; Herrmann 1991). These kinds of models seem more appropriate for the formation of radial or network cracks from a point of impact on a surface, as shown in Figure 1, or to cracks formed due to shrinkage in drying, rather than to the kind of fracture that separates a specimen into two pieces. This network modeling approach is also used, with some success and greater plausibility, to dielectric breakdown. Most of the models that use analogy to percolation or dielectric breakdown predict the formation of a great many cracks (Herrmann 1988), only one of which becomes the final fracture surface. Examination of most real specimens does not agree with this behavior.

A similar set of stochastic stepwise fracture models uses a network of microscopic (but much larger than atomic scale) "bonds" which have a probability of breaking. This has been shown to be capable of generating a fractal boundary or network (Louis, Guinea et al. 1985; Solla 1985; Guinea, Pla et al. 1986; Louis, Guinea et al. 1986; Termonia and Meakin 1986; Guinea, Pla et al. 1987; Rosenfeld 1987; Brown 1990; Guinea and Louis 1991) in much the same way that percolation fractals form in lattices based on the probability of site occupation by pores. For instance, the model of Louis predicts a fractal surface with dimension $D \approx$ 2.62–2.64 based on elastic strain and no ductility or void coalescence. There is no evidence that this is a realistic model for brittle crack formation, although it may conceivably be related to ductile failure by void coalescence, which does involve the nucleation of many local cracks

Figure 3. Model of crack growth by bond-breaking on a triangular lattice.

but does not usually produce fractal surfaces as mentioned above. In the absence of a physical model, this may be simply another case in which many different processes can give rise to fractal behavior for different reasons.

Percolation modeling (either bond or pore) is discussed below. It is perhaps more relevant to corrosion processes, with some additions to reflect the mobility of ions on the surface (Daccord 1989). This has been shown to predict fractal corrosion surfaces as well (Corderman and Sieradzki 1986; Herrmann 1989). A special case of corrosion is stress corrosion cracking in which microcracks are advanced by chemical attack operating preferentially at the highly stressed material at the crack tips. Cross sections through such cracks typically have a branching, tree-like appearance. Network percolation models have also been shown to describe some aspects of ferromagnetism (Birgenau, Cowley et al. 1987; Malozemoff 1987), and phase separation in thin films (Deutscher, Lereah et al. 1987).

There is no apparent connection between the parameters used in the models and any characteristic of the material or its microstructure. In other words, these models do not seem to describe reality nor to offer much predictive power for real materials. Li (1988) has proposed a theoretical relationship between toughness and fracture energy proportional to the surface fractal dimension D, but without supporting data.

Passoja (1988) reported measurements of fractures in several ceramics and metals using Fourier analysis of surface elevation profiles, and determined a characteristic dimension a_0 that relates the crack energy per unit volume expended at the crack tip as

$$\gamma_c = \frac{1}{2}E(D^* \cdot a_0)$$

where E is the modulus of elasticity and D^* is the fractal increment (the fractional part of fractal dimension, expressed in Chapter 1 as 1.d for a profile or 2.d for a surface). The characteristic values of a_0 vary from < 1 Å to more than 50 Å. For the ceramics, the larger spacings are associated with glass-ceramics, the intermediate spacings with the polycrystalline aluminas, and the smallest spacings with brittle intermediate compounds such as calcium fluorite and spinel, which have a complex unit cell and little opportunity for deformation due to slip. It is not clear what specific meaning can be assigned to these dimensions, which are not the size of the atom, unit cell, crystalline grain, etc.

Mecholsky et al. (Mecholsky and Passoja 1986; Mackin, Passoja et al. 1987; Mecholsky, Mackin et al. 1987; Mecholsky and Mackin 1988; Mecholsky, Mackin et al. 1988; Mecholsky, Passoja et al. 1989; Mackin, Mecholsky et al. 1990; Mecholsky and Freiman 1991; Tsai and Mecholsky 1991; Mecholsky 1992) obtained extensive measurement data, primarily on glasses and ceramics, and developed the model further, relating K_{IC} to D^* as

$$K_{IC} = E \cdot a_0^{1/2} \cdot D^{*1/2}$$

where again E is the modulus of elasticity and a_0 is a parameter having the units of length. It is plausible to argue that the characteristic length is associated with some step length for the crack, such as the distance over which voids must join (a sort of percolation fractal model for fracture), but it is not clear that this has any such simple geometric meaning. The model does imply a linear relationship between the energy to initiate a crack and microbranching of a crack. The model has been shown to produce consistent results for glasses, ceramics, intermetallics, and silicon.

It is very interesting that this model appears to offer a unification between fracture surface geometry and fracture resistance that is independent of the details of microstructure or atomic bonding. Some of these materials are polycrystalline, some single crystalline, and some amorphous, and the atomic bonding is metallic, ionic, and covalent. This implies that a successful continuum mechanics model for fracture propagation should be able to predict the

generation of fractal surfaces whose dimension correlates with material properties. The process of crack formation, a highly dissipative dynamic system, produces surface configurations which are at least partially independent of the microstructural details. This is a promising direction for future work. Modeling of fracture processes remains an elusive goal, and the introduction of fractal concepts provides a constraint but not a guideline.

Additional results which agree both qualitatively and quantitatively with these were obtained for metals by Fahmy, Russ et al. (1991). They used a modified slit-island approach by embedding fragments from each fractured metal specimen in a metallographic mount, sectioning, and measuring the boundaries of the islands with a Minkowski (sausage) method. Figure 4 shows representative cross-sections of particles with different boundary dimensions. The surfaces were technically mixed fractals (as discussed in Chapter 7) since they pass through two different metal phases expected to have different fracture properties. The two extreme material compositions tested were single-phase, and showed quite different surface roughness for the fractures (Figure 5). All of the other specimens tested consisted of mixtures of the two phases in various proportions, but it is not reasonable to expect the area fraction of the fracture passing through each phase to be the same as the volume fraction. Still, the correlation between fracture toughness (K_{IC}) and fractal dimension for these metal alloys was consistent with other materials measured using other testing procedures by other researchers (Figure 6).

Mixed phase fractures were also reported by Wang and Chen (Wang, Chen et al. 1988; Chen, Wang et al. 1989) for a varying volume fraction of martensite (a hard, brittle phase) in steels. They found that fractal dimension (fracture roughness) increased with martensite content up to about 35% and then decreased. The paper qualitatively explains this in terms of the fracture passing through or around the martensite. Regardless of the mechanism, the authors find that the fractal dimension correlates well and linearly with the threshold stress for fatigue fracture. Of course, fatigue is a different fracture mode than the brittle fractures discussed above. Surface pitting under cyclic loading has also been reported by Zhou, Chen et al. (1989) to exhibit fractal characteristics.

Fatigue fracture (propagation of a crack due to cyclical application of a stress lower than that required to cause simple failure) was also suggested to have a fractal character by Altus (1991) based on the cumulative breaking of "microscale elements" (which are not defined in terms of any specific microstructural feature) that are assumed to have a Weibull strength distribution (Anderson 1989; Barton 1989; Wight and Laughner 1989). Breaking the weakest of these under stress causes a "cross-reaction" to weaken neighboring links. The model is able to predict fatigue stress-life relationships, but it does not produce a fractal surface with a meaningful D value. Also, there is no spatial correlation of the breaks into macroscopic cracks, and no confirming data are provided. In fact, Fourier analysis of profiles across fatigue fracture surfaces is sometimes used to determine a mean spacing for the striations in order to determine the approximate time scale for the fracture (Passoja and Psioda 1981), but the power spectrum does not exhibit the $k^{-\beta}$ form characteristic of a fractal. Analysis of fracture surfaces using the power spectrum has been reported by Lanzerotti, Pinto et al. (1989; 1990).

It is also important to extend the understanding of fracture and fractal surfaces to other scales and materials. The fracture of rocks is described by Trottier and Beswick (1987), who related the observation of fractal surface geometry to the power-law distribution of particle sizes, and to the mutual fracture of impacting particles, and by Kumar and Bodvarsson (1990), who tried to model the surfaces with Mandelbrot–Weierstrass functions. Baker, Giancola et al. (1992) have also derived a basis for the common observation of power-law size distributions in fragments. They compared log normal and other distributions for particle sizes produced by fracture and interpret the log normal distribution as fractal in nature, the product of many independent processes with arbitrary distributions.

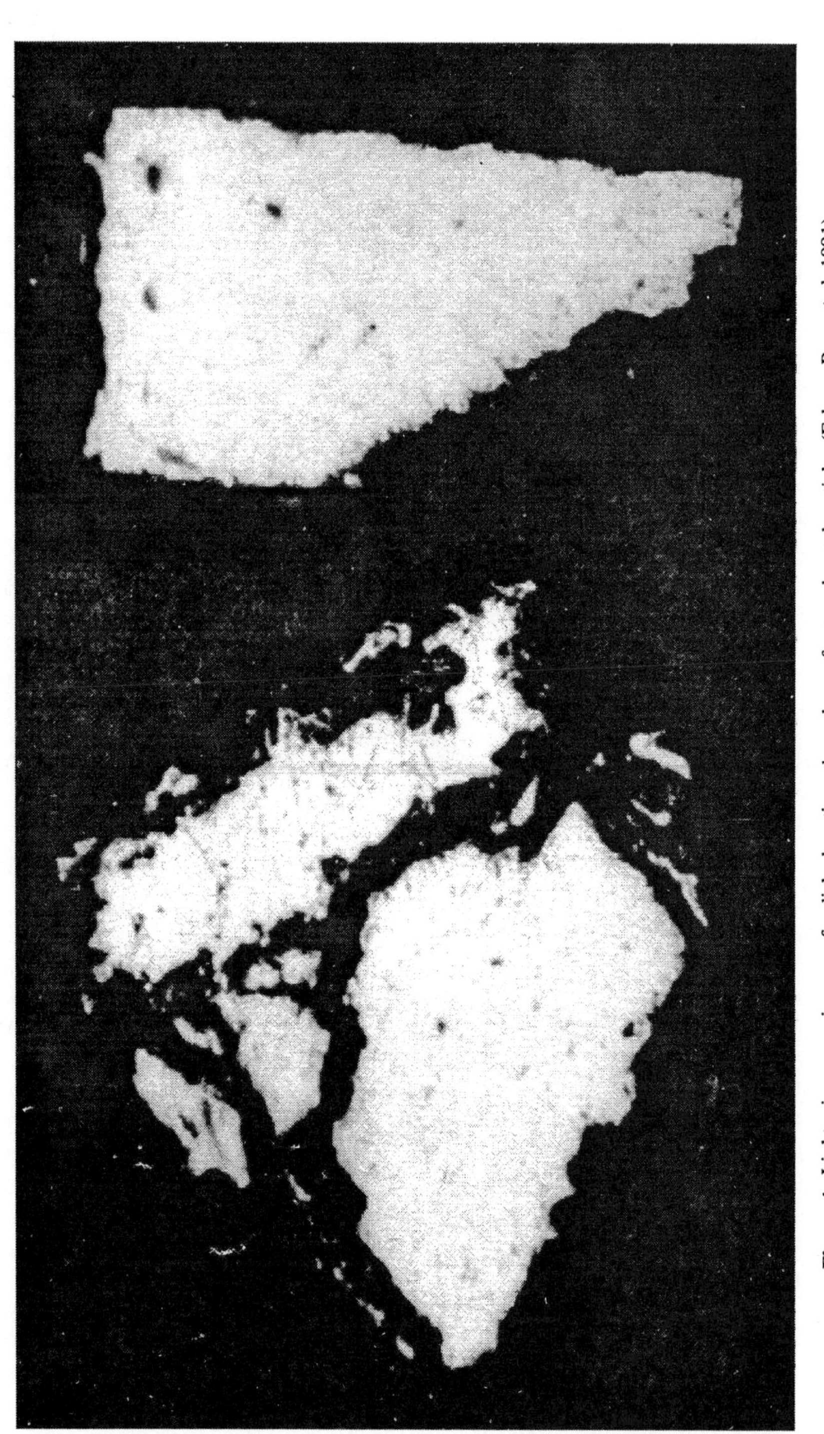

Figure 4. Light microscope images of polished sections through two fractured metal particles (Fahmy, Russ et al. 1991).

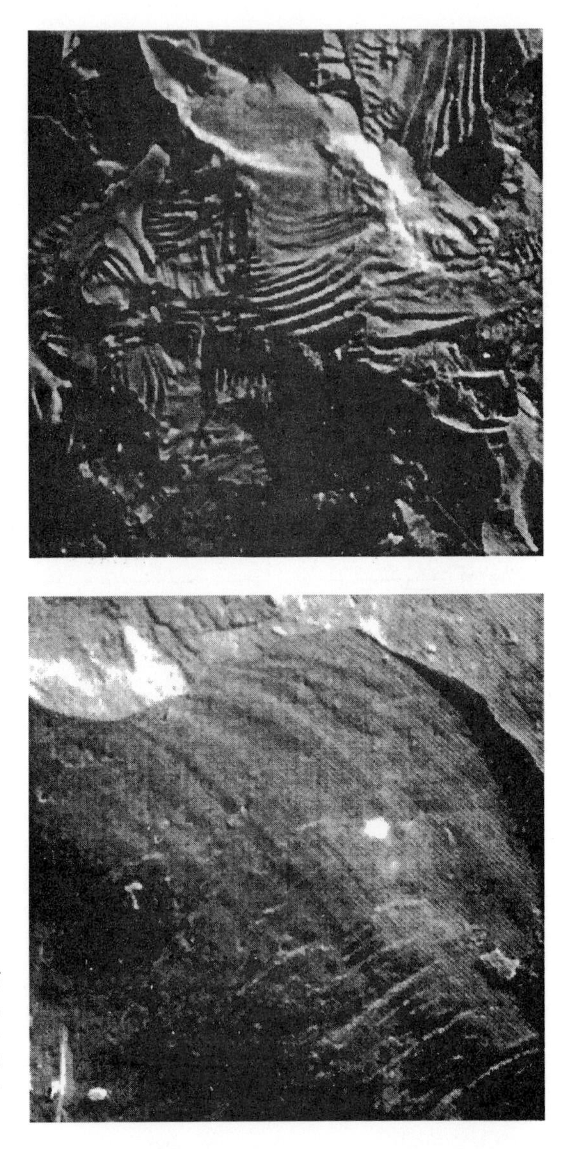

Figure 5. SEM images of fracture surfaces of particles from single-phase materials (Fahmy, Russ et al. 1991). These have values of fracture toughness and fractal dimension which bracket all of the intermediate samples consisting of a mixture of the phases.

At a very different scale, Passoja (1988) compared images of cracks in metals to a satellite image of the Gulf of Elath, which is a rift fracture in the earth's surface. The Fourier transform dimension of the rift is similar to that for some of the metal fractures. Effects of weathering are considered to have modified only the small-scale features along the rift so that the power spectrum still gives a meaningful measure of the roughness.

Interpretation of fractal surface and fracture of materials must inevitably be related to microstructure. Kaye (Kaye 1983; Kaye, Beswick et al. 1986) uses fractal geometry to describe the structure of concrete, as well as powder metallurgy specimens. Pekala, Hrubesh et al. (1990) have related the mechanical properties of gels to fractal dimensions. Dauskart, Haubensak et al. (1990) relate fracture roughness (fractal dimension) to microstructural variables. Hornbogen (1989) and Su, Zhang et al. (1991) use fractal dimensions to describe many aspects of metal microstructures, including dislocations, grain boundaries, distributions of precipitate particles,

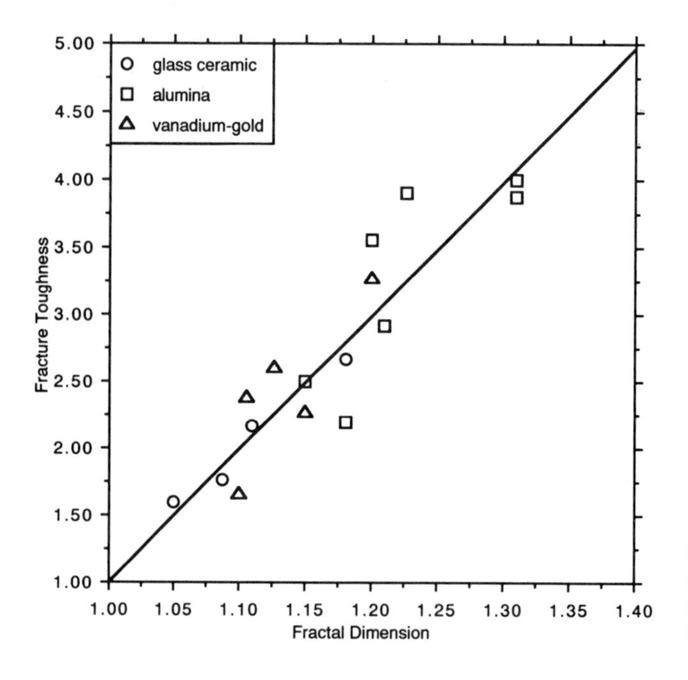

Figure 6. Observed correlation between fracture toughness and profile fractal dimension for different types of materials (Fahmy, Russ et al. 1991).

dendrite spacings, martensitic microstructures, and spacing of slip steps. The paper shows several possible demonstrations of ways that fractal geometry might be related to the formation of these microstructures, or at least may provide descriptive tools to characterize them, but no quantitative data are included to indicate that actual microstructures are really fractal dimensions. Even if these structures are compactly described by fractal geometry, the structures would be networks, sponges, or dusts. The possible relationships between these microstructural features and the fracture surfaces that form remains, at best, speculative.

Machining and Wear

No other single problem of material surfaces has been studied as widely as fracture. However, there are quite a few other ways in which surfaces of materials may be created or modified, and some of these are also believed to be fractal in nature. It might seem that machining of a surface is one example of a Euclidean geometry, and certainly it is usually the purpose of a machining operation to produce a surface whose geometry conforms to some design that consists of Euclidean planes, cylinders, etc. Such surfaces would have a fractal dimension the same as the topological dimension, and elevation profiles would show a fractal dimension of 1.0. But machining is a complex process in which the tool creates the new surface by a combination of mechanisms, and at high magnification such surfaces are certainly not ideal planes. Berry and Hannay (1978) examined machined surfaces to characterize the roughness as a fractal dimension. The appearance of fractal surface geometry is also true of other operations such as polishing, wear, or erosion that remove material from surfaces. In fact, we will see below in the discussion of characterizing very smooth surfaces, that it is extraordinarily difficult to produce a "flat" surface that is ideally planar to use as a calibration tool. Examples of a precision machined surface imaged by AFM and interferometry were shown in Chapter 7.

There are many different modes of material removal, some of which may generate fractals and some of which apparently do not. Most are quite complex, involving the deformation, fracture, and removal of surface material. For instance, grinding is typically defined as an operation in which hard particles are firmly attached to one surface and remove material from another, softer surface. Machining typically uses a single tool (or at least a small number of edges) to remove material. Lapping, on the other hand, places a large number of hard particles between two surfaces which move relative to each other. A combination of rolling, gouging, indenting, and other processes result in the removal of material from one or both surfaces, but at a much lower rate than grinding. Polishing also uses loose particles but with ductile deformation of surface layers. The use of particles harder and/or softer than the surface can drastically alter the mode of surface modification.

Scott (1989; 1990; 1991) has pointed out that machining is a damped, driven system in which a great deal of energy is dissipated at the contact of the tool and workpiece. This is the condition in which chaotic behavior often arises (Becker and Dörfler 1989; Devaney 1989; Devaney and Keen 1989; Baker and Gollub 1990), and that in turn frequently gives rise to fractal geometry (Brown and Savary 1988; Wehbi, Roques-Carmes et al. 1992). This seems to be the case for machining operations. For the most part, these oscillations are extremely local to the contact point between the tool and the surface. Such chaotic systems often give rise to power-law behavior (Bak and Chen 1989; Ben-Jacob, Garik et al. 1990; Tang, Bak et al. 1990; Bak and Chen 1991; Bak and Creutz 1991; Diab and Abboud 1991; Schroeder 1991), and it should not surprise us that the result is a surface that has oscillations of many frequencies superimposed. The addition of many sinusoidal terms with magnitudes which decrease as a power of the frequency gives rise to fractal profiles. Indeed, this method is used for generating fractals mathematically as discussed in Chapter 6. Scott has suggested an analogy between a cross section in a turned surface and a Poincaré section in phase space, which is a standard tool for analysis of chaotic time series phenomena (Broomhead and Jones 1990).

An additional complexity for most man-made surfaces such as by machining or polishing is that they are not isotropic. Machined surfaces are usually fundamentally anisotropic. The direction along the path of the moving tool has a much different elevation profile and typically a very different fractal dimension than the direction perpendicular to the machining direction. This is true whether the tool is a simple cutting edge or a series of particles of varying shape and orientation as in grinding. There may also be different properties for the material in different directions, as for instance in cutting natural materials such as wood. To a lesser extent such orientation sensitivity is also present in grossly isotropic materials such as rock or metal, since the local grains have crystalline axes with different mechanical properties of strength, ductility, cleavage, etc. Most of the measurement and modeling methods discussed in previous chapters assume isotropy and do not deal correctly with anisotropy. Among the first to recognize this limitation were Thomas and Thomas (1986), who also commented on the inadequacy of simply measuring many profiles in different directions, and the lack of a relationship between the surface fractal dimension and the profiles. Russ (1991a) has proposed several methods for measuring the anisotropy of such surfaces, as described in earlier chapters.

The fractal nature of surfaces produced by machining, polishing, or other operations designed to produce a specified (Euclidean) shape has important consequences for the behavior of the surfaces in their intended application. Several examples will be considered below, such as electrical contacts, reflection of light from mirror surfaces, and friction. Relating the fractal dimension of the surface back to the parameters controlling the creation of the surfaces is more difficult, and little general progress has been made.

Bush, Gibson et al. (1978) have modeled elastic contact of rough surfaces as a series of protrusions much like a Takagi model (Chapter 6), but shaped like parabolas of revolution with an elliptical cross section. Deformation of such surfaces during contact depends on material

properties (primarily modulus of elasticity, but mistakenly described as hardness), and on the distribution of the sizes of the protrusions. Majumdar and Bhushan (1991) extend this approach to deal with fractal surfaces. The electrical contact between such fractal surfaces produces a power-law distribution of contact island areas. Majumdar and Bhushan (1990) relate the surface fractal dimension to the distribution of contact areas as a function of applied load (producing both elastic and plastic deformation). The model also agrees with the observed effects of surface contact on friction. Majumdar and Tien (1991) use a similar model for thermal conductance between contacting surfaces.

Contact between fractal surfaces is equivalent to the contact between one rough surface and a plane (although predicting the dimension of the equivalent rough surface requires some care; rules for combining fractals were discussed in Chapter 7). This produces a series of islands that obey the Korcak relationship $N(A > a) \sim a^{-D/2}$, where the relationship between the boundary fractal D_P and the surface dimension is assumed to be $D_P = D_S - 1$. The total area of contact is $a_L \cdot D_P/(2 - D_P)$, where a_L is the largest island. The infinite number of infinitesimal contacts makes a negligible contribution to the total. However, locating the largest contact in a finite search area presents a difficulty.

Existing models for contact between rough surfaces, the effect of pressure, etc., include terms such as "rms" or "peak-to-peak," "height," "slope," "curvature," and other classical measures of rough surfaces that are functions of the surface magnification used for measurement. It is clear that the rms values, which ignore the spatial organization of the elevation data and depend on the area sampled for measurement, do not capture the important information about the surface. The use of such measures in the analysis of elastic and plastic contact between rough surfaces is flawed and new models will need to be developed.

This is made more complicated by the fact that in real contact, the large spots are elastic and the small ones are plastic. This is because small contact spots are asperities with a small radius of curvature; as the load is increased, these small spots merge to form larger elastic contact spots. It is worth commenting that this model of superposition of asperities of ever-smaller sizes is similar to a Takagi model for surface generation, rather than the Mandelbrot–Weierstrass functions used by several authors. The classic Greenwood–Williamson contact model assumes that all contact asperities are of the same radius, and hence predicts that the large ones undergo plastic deformation and the small ones elastic deformation. This is the opposite behavior to that predicted by a superposition of protrusions for which the radius of curvature varies.

Majumdar and Bhushan (1991) present the beginnings of a new theory based on fractals, whose predictions are in reasonable agreement with experiments for low loads (it ignores work hardening during deformation and frictional forces). Experimental data for two Pyrex glass surfaces pressed against each other show an exponent of 1.3 in the relationship between area fraction in contact A_a vs. normalized load $P^* = (P/EA_a)$. This corresponds in the authors' model to a surface fractal dimension of 2.4. The model reduces to P^* proportional to $A_a^{(3-D_P)/2}$, where D_P is the profile dimension, nominally equal to $D_S - 1$.

Majumdar and Tien (1990) suggest that machining a surface flattens the power spectrum (reduces the low-frequency terms) and thus raises the fractal dimension of the surface, but does not modify the small-scale features which remain unprocessed. This could be related to the observation noted below that several measurement tools (interferometry and AFM) produce plots that indicate surfaces are fractal, but at the highest magnification the dimension approaches 2.5, corresponding to a $1/f$ noise or Brownian motion. These plots show for stainless steel surfaces (produced by polishing) and for a thin-film magnetic disk (produced by deposition) that at the highest frequencies, the dimension approaches $D_P = 1.5$ or $D_S = 2.5$ (Brownian). This may indicate that the smallest features on such surfaces are naturally

produced with this dimension and remain unmodified by subsequent macroscopic processing, or they may be artifacts of the imaging system.

On the other hand, Kaye (1989a) shows a series of profilometer traces on polished copper which progressively become smoother with polishing. The numeric values are suspect because a Richardson method is inappropriately applied to self-affine profiles, but still the magnitude of the roughness and the fractal dimension both decrease with progressive polishing. This may indicate that the polishing of copper causes plastic deformation of the surface, which is not the case for some other materials and surface preparation procedures.

Kaye (1989a) also shows that sandblasting of surfaces produces a progressive increase in the surface roughness. The numeric values of the fractal dimension reported cannot be used directly because they were determined using a Minkowski method with a circular structuring element, rather than a horizontal line on the self-affine elevation profiles. However, remeasurement of the published profiles using a Fourier method indicates that the limiting roughness has a fractal dimension close to 1.5, corresponding again to the idea that the limiting roughness may be Brownian.

Most surfaces are produced either by fracture, deposition, or solidification. Each of these processes has been shown to produce, at least under some circumstances, a fractal structure. Wear (rubbing of one fractal surface by another) is not observed to produce a fractal surface at microscales, although some fractal characterization has been reported (Gagnepain and Roques-Carmes 1986). This may be related to the presence of ductile deformation in many cases of wear. It was mentioned above in connection with fracture surfaces that ductile deformation may produce surfaces that are not fractal. Some other surface processing methods, such as aligned particles on magnetic tape or disks, may also not be fractal.

Precision machining and polishing are used to produce mirror surfaces. Since the scattering of light (or other radiation) can be used to measure the fractal dimension of a surface, it is not surprising that the performance of the mirror surface is influenced directly by the value of the dimension. Hogrefe and Kunz (1987) and Kjems (1991) discuss scattering experiments with light, neutrons, and X-rays to measure the fractal dimension of a surface. Starting with a collimated, monochromatic beam, the scattering is measured as differential cross sections. The different particles sample different dimensions, and scattering arises from different physical phenomena (variations in refractive index, electron density, etc.), but the same scattering formalism and interpretation of results apply (Lin, Lindsay et al. 1990; Montagna, Pilla et al. 1990; Rarity, Seabrook et al. 1990). Scattering experiments measure the Fourier transform of the pair correlation function of some property (elevation, slope, reflectivity, etc.) (Charalampopoulos and Chang 1991). The result is a plot of log (Intensity) vs. log (Frequency) which is a straight line, whose slope gives the fractal dimension.

Diffraction of light from a Koch (structured) fractal was computed by Uozumi, Kimura et al. (1991), extending the work of Allain and Cloitre (1986), who showed that scattering from a random fractal follows a power law k^{-D}. In this case, peaks appear in the pattern whose height does not follow the power law, but the averaged value or the integrated sum can be used. Additional modeling of scattering has been done by Markel, Murativ et al. (1990) and Pouligny, Gabriel et al. (1991). Berry (1979) describes waves that have been scattered from fractal surfaces as "diffractals." As usual, structures on scales much smaller than the wavelength do not influence the wave.

If there is a self-similar structure on a range of scales that includes the wavelength, then fractal modeling is appropriate. This would apply to sound or radar diffracted by trees, radio waves scattered by ionospheric turbulence, etc. Even for the case of a monochromatic and coherent incident wave, a Brownian ($1/f$) fractal, and examination of the diffractal far from the scatterer, the structure of the diffractal is rich. Introduction of additional variables adds additional complexity. Jakeman (1982) has solved the diffraction pattern from an infinite

corrugated surface with fractal slope variations. The paper uses "corrugation" to describe extruded surfaces as discussed in Chapter 6; of course, this is a one-dimensional analysis.

These models seem to be particularly applicable to scattering of light or other radiation passing through turbulent media rather than scattering from surfaces. Jakeman compares surfaces with fractal height distribution (type I), fractal slope distribution (type III), and Gaussian distribution (type II). Type I surfaces produce geometric optical effects. Type III produces only diffraction effects. Type II is intermediate but can be solved analytically. This approach has been applied to the effect of sea waves on microwave radar echo. It assumes all important scattering facets are larger than wavelength of radiation. Ignoring the smaller dimensions creates a practical limit to using fractals as an effective description for surfaces. However, the model predicts that the scattered brightness distribution is also fractal. Such brightness patterns were used in Chapter 5 to characterize surfaces.

Analysis of optical surfaces in frequency space is discussed by Church et al. (Church, Jenkinson et al. 1979; Church 1980; Church 1982; Church and Berry 1982; Church, Howells et al. 1982; Schmidt 1982; Church 1983; Church, Vorburger et al. 1985; Schmidt 1989) in terms of light scattering. A portion of the 2D Fourier image at frequencies from 1/MD to 10/MD (where MD is the mirror diameter) is controlled by "figure" of mirror. The portion from 10/MD to 1/wavelength depends on the "finish" of mirror. This is statistical in nature. It is difficult to precisely measure the "tail" of the distribution accurately enough to distinguish various statistical models of surface irregularities.

Church (1986) adds a discussion of correlation length. Surfaces have two different attributes—vertical "roughness" and transverse "length." Roughness is conventionally described by the rms surface height σ, while λ is the correlation length, generally little considered and hard to measure by traditional methods. This is the distance between successive crossings of the profile through some set level (i.e., the zeroset of the function), which turns out upon examination to be an artifact of the measurement procedure rather than a characteristic of the surface (of course, this is also true for σ).

Church uses the example of the distance between successive maxima calculated in terms of elevation, slope, and curvature for a Gaussian random surface. Finite bandwidth in the measurement process distorts the measured values. Indeed, for a fractal profile the zeroset is also fractal with dimension $D_S - 1$. Church shows results for limiting cases of λ. A small λ produces a white power spectrum (if smaller than minimum wavelength, the rms roughness tends to zero). A large λ produces a featureless $1/f^2$ form. This was discussed in Chapter 5.

The relationship between the fractal dimension of a surface and the distribution of sizes of "lakes" on the surface has been used as a measurement tool (the Korcak dimension) for wear and contact surfaces as mentioned above. These cavities on the surface are also important for lubrication and wear. Hence, the fractal dimension may be expected to correlate with performance in some way. Ling (1989) mentions the extreme case of fluid-filled composites such as human joints (with cartilage) in which the hydrostatic fluid pressure would also influence load bearing. Kirk and Stachowiak (Kirk and Stachowiak 1990; Kirk, Stachowiak et al. 1991; Kirk and Stachowiak 1991a; 1991b; 1991c) have shown that particles taken from synovial fluid in joints are indeed fractal, and that there may be a correlation between the particle dimension and the arthritic behavior of the joint.

A fractal surface model may also serve in studying friction through the relationship between applied force and contact area (Rehr, Gefen et al. 1987). Ling (1990) models wear surfaces with randomized Koch functions, measures real profiles (showing linear Richardson plots over nearly four orders of magnitude), and discusses areas of contact between surfaces. This is, of course, of great interest in electrical systems in terms of the contact problem described above, but it may also be applicable to sliding contact and the analysis of friction

and wear. It has even been proposed (Bush, Gibson et al. 1978; Cates and Witten 1986) that lubrication effects may be better understood in terms of fractal geometry.

Other modes of surface shaping or wear, studied under the general topic of tribology, also involve fractal geometry (Wehbi, Roques-Carmes et al. 1987; Stupak and Donovan 1990). Erosion by particle impact produces steady-state surfaces shown by Srinivasan, Russ et al. (1990) to be fractal and to have a dimension that varies only slightly with large variations in erosion rate produced by using particles with different relative hardnesses (SiC harder than the sapphire substrate, Al_2O_3 softer than it). However, the surface texture did vary significantly when the erosion rate was varied by changing the impact velocity of the particles. This was interpreted in terms of the effect of particle size and velocity on the size of the subsurface plastic zone, and a threshold dimension for crack nucleation. Figure 7 shows an example of such a surface.

Spark-eroded surfaces (Thomas and Thomas 1988) are superpositions of essentially circular, shallow craters of varying size. The fractal nature of cratering should not be surprising. The familiar surface of the moon (and other planetary bodies including earth, although subsequent weathering has produced alterations) is due to impact erosion. A simple random superpositioning of craters whose sizes vary according to a power law (equivalent to assuming a power-law distribution of sizes for the impacting bodies, in agreement with observation) will produce a fractal surface with fractal elevation profiles. This provided one set of the tools used in Chapter 6 for the modeling of fractal surfaces.

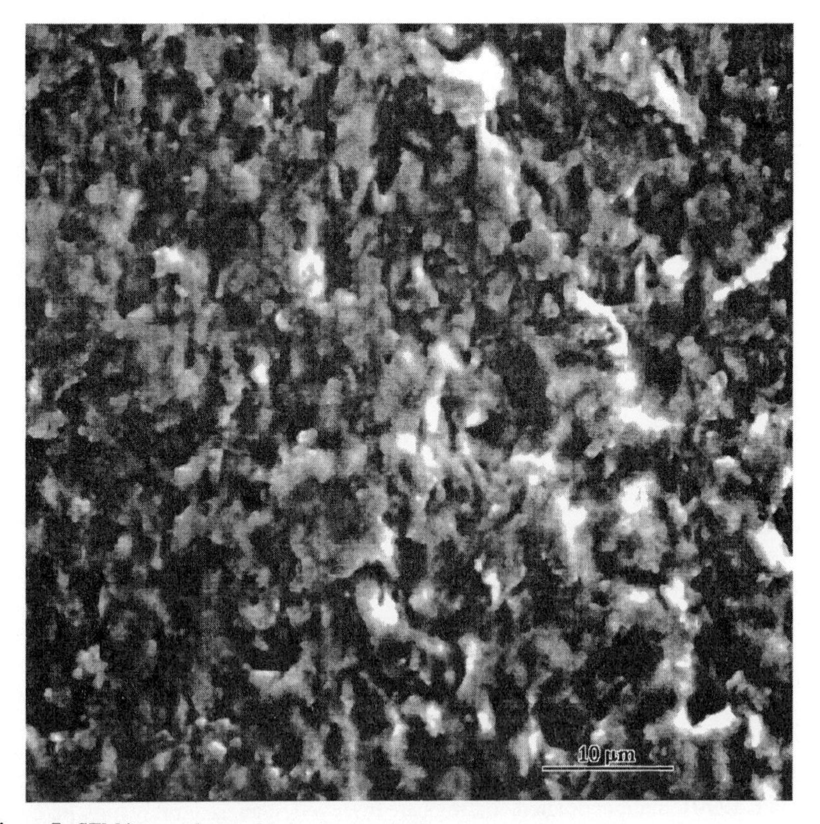

Figure 7. SEM image of a sapphire surface eroded by particle impact (Srinivasan, Russ et al. 1990).

The evolution of wear surfaces and corresponding changes in the fractal dimension have been studied for the case of stone tools (Bueller 1992; Rovner 1993; Russ 1993a). The tools were made and used as part of studies of wear patterns in archaeological tool use. Tools made from two different materials (flint and chert) were worn by repeated use in various specific tasks, such as cutting hide or scraping wood. Periodically, photographs were taken of the wear surface. These were then digitized and the brightness fractal dimension measured as discussed in Chapter 5. Wear of the tools causes the fractal dimension to change progressively, with different rates for each material and use. For up to 1600 cutting strokes on each tool, there was no indication that any end point had been reached. Not surprisingly, the most marked change occurred when cutting hard wood, and the least when cutting leather. However, whereas it might be expected that the wear would reduce the fractal dimension of the surface, in fact the change is to increase it. We reason that the cutting process removes the larger asperities on the surfaces but does not reach the smaller ones, so that the amplitude of the roughness decreases while the fractal dimension increases toward the Brownian noise value of 2.5. Beauchamp and Purdy (1986) showed that heating chert before toolmaking coarsens the grain structure and changes the fracture toughness and surface morphology.

In another example of progressive wear with broader application, the surface fractal dimension of precision-machined metal surfaces has been measured as a function of the cumulative wear on a machine tool cutting tip (Tyner 1993). The machined surface is anisotropic, as was shown in Chapter 7, Figure 17, due to the direction of tool motion. However, a mean fractal dimension and intercept can be determined using the 2DFT approach. This value depends on many parameters, including the hardness of the metal being machined. With a new diamond tool, the fractal dimension for machined copper is 2.24, and the intercept which describes the magnitude of the roughness is about 13 nm. For soft aluminum, the fractal dimension is 2.20 and the intercept is about 20 nm. After the tool has been subjected to 1 kilometer of total wear, the fractal dimension of the surface is reduced (to 2.21 for the copper and 2.17 for the aluminum). However, the intercept value decreases for the copper (to 11 nm) while it increases for the aluminum (to 22 nm). Clearly, the notion of a smooth surface is a complex one that cannot be described by a simple "rms roughness" parameter, and varies in interesting ways with the variables of the machining process.

Stupak, Kang et al. (1991) have studied wear of rubber, applying a variational (Minkowski) measurement method to profilometer traces. He finds $D \approx 1.5$ for all surfaces, but the intercept value for the log–log plots depends on magnification and varies with wear rate, material, velocity, frictional work input, lubrication, etc. This raises again the subject of the intercept or topothesy value discussed in Chapter 4, which is often overlooked in fractal measurements.

Deposited Surfaces

In many instances, surfaces are produced by deposition of material. This can be done chemically or electrochemically from solution, or by evaporating the coating material in a vacuum so that it collects on the intended surface, or by direct bombardment of the surface. The deposited species may be laid down atomically, or as molecules of varying sizes, or as discrete particles, and there may be a single deposited species or a mixture of species, which may or may not react with the substrate or with each other. Solidification of a phase or phase mixture from the liquid can be considered as falling within this category of processes (Schaefer, Bunker et al. 1990). Such deposited coatings are involved in an extremely high percentage of all modern engineered materials. Surface deposition is widely used to produce tribological,

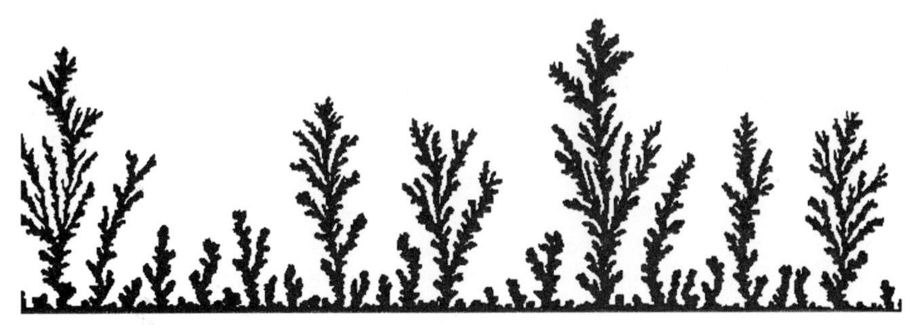

Figure 8. Cross section of zinc metal grown on a carbon electrode (Matsushita 1989).

magnetic, optical, electronic and other devices. The roughness of the substrate as well as the deposition parameters all influence the structure.

For instance, West, Marchese-Rugona et al. (1992) measured the fractal dimension of clay coatings applied to paper. The deposited particles in this case are comparatively large and have definite shapes of their own, yet the overall surface has a fractal roughness. The authors used a slit-island approach with dimensional analysis, which was flawed by the anisotropy of the surface which results from the underlying structure of the paper. No quantitative interpretation of the dimension or correlation with processing conditions was reported.

It was pointed out in Chapter 6 that diffusion-limited aggregation (DLA) is a classic example of a mass fractal, producing a highly dendritic and branched structure. This is usually modeled in two or three dimensions as a growing cluster around some initial seed particle. However, DLA deposition on a surface has the same fractal dimension as DLA deposition of a cluster around a seed (Meakin 1983; Meakin 1984a; Meakin 1984b; Meakin 1986; Meakin 1989b; Meakin and Tolman 1990). The change in the boundary condition does not change the dimension, which is 1.7 in a plane and 2.5 in 3-space. A mass fractal for either kind of DLA

Figure 9. SEM image of zinc metal leaves grown on a carbon electode (Hibbert and Melrose 1990).

obeys the rule Mass $\propto$ Distance^D. For a cluster this is the radial distance, and for a surface it is the vertical or normal distance.

Some forms of chemical deposition on a surface apparently correspond to this rather simple model (Hibbert and Melrose 1990; Sapoval 1991). As shown in Figures 8 and 9, the morphology of the resulting structures of zinc metal leaves grown on a carbon electrode, at the interface between a 2 molar aqueous solution of zinc sulfate and *n*-butyl acetate (Matsushita 1989), are similar to those produced by the model (in this case a 2D simulation of diffusion-limited deposition). Chapter 6 has numerous examples of 2D and 3D simulation of diffusion-limited and ballistic deposition. Other examples of electrodeposition of copper were shown in Chapter 5.

In addition to the mass fractal relationship for the diffusion-limited deposition, there is also a power law for the number of separate trees in the deposited coating as a function of size. Vicsek (1992) shows that the number of trees N of mass M varies as $N[M] \propto M^{-\tau}$, where $\tau = 1 + (S - 1)/D$, S is the space dimension (equal to 2 for a planar model and 3 for a space-filling one), and D is the mass fractal dimension. Since the classic diffusion-limited deposition dimensions are 1.7 and 2.5 as mentioned above, this gives $\tau = 1.59$ in $S = 2$ and $\tau = 1.80$ in $S = 3$, which is confirmed by simulation and consistent with experiment.

Of course, it is clear that surfaces produced by such a procedure are not examples of the kind of surface fractal we have been concerned with here. Instead, they are the second of the three types of fractals as discussed in Chapter 1. In the ideal case of a mass fractal, the surface area is proportional to the volume occupied by the deposited material, which would produce a fractal surface dimension $D = 3.0$ according to our definition. Indeed, measured dimensions very close to 3 have been experimentally determined by BET methods for materials such as silica gels, as discussed in Chapter 3. The mass fractal dimension $M(r) \sim r^D$ can also be determined by nuclear magnetic resonance (NMR) of paramagnetic impurities present in silica gels. The time dependence of magnetization recovery follows a power law whose exponent is proportional to D (Devreux, Boilot et al. 1990).

However, modifications to the rules for deposition can produce results that can be described by the surface fractal geometry we have been using. This was illustrated in Chapter 6. The changes made to conventional DLA models include varying the sticking probability for the particles, introducing directional preferences to mimic crystalline structures, and allowing the particles to form separate clusters which then may aggregate together (Meakin 1984a; Meakin, Ramanlal et al. 1987; Sevick and Ball 1990). For instance, varying the sticking rule in diffusion-limited deposition can produce compact structures that are not mass fractals. Instead, they have mass that increases in direct proportion to the occupied volume at short ranges, but eventually at large dimensions the fractal nature becomes evident. Meakin says it "postpones" the fractal dimensionality (Meakin 1986).

An Eden model growth pattern is also quite different from diffusion-limited aggregation. In this model, each successively deposited particle fills in an empty space next to a randomly chosen occupied site. The site may be chosen in several ways, for instance giving all sites with any occupied neighbor an equal probability, or weighting the probability according to the number of occupied neighbor sites. All of these methods produce a compact structure which is not fractal but dense. However, the surface of the cluster is not trivial and may be a self-affine fractal (Vicsek 1992). An example of measuring such a cluster is shown in Chapter 6. It has also been observed that the scattering pattern from a fractal changes as it is compressed to a compact structure (Sinha and Ball 1988).

Some results suggest that ballistic deposition is equivalent to this, at least in the early stages of the process. Ballistic deposition is radically different from diffusion-limited deposition (Voss 1984; Messier and Yehoda 1985; Yehoda and Messier 1985; Messier and Yehoda 1986; Sander 1986a; Sander 1986b; Meakin, Ramanlal et al. 1987; Ball, Blunt et al. 1990).

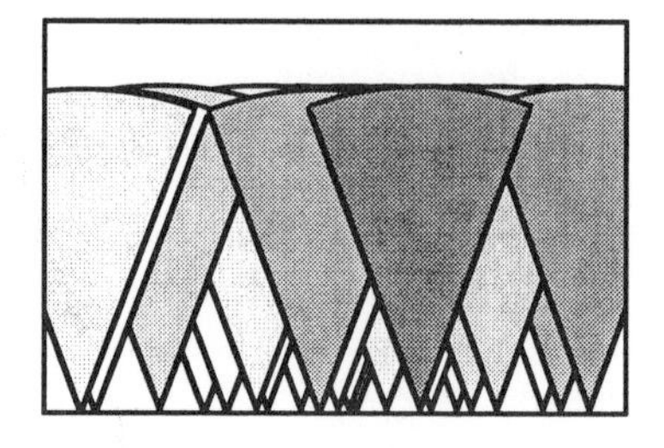

Figure 10. Schematic diagram of the growth of conical structures by ballistic aggregation, with larger ones dominating smaller ones.

The particles arrive with a fixed direction of motion (or narrow range of directions). They may either stick on contact or have associated probabilities of scattering laterally when they encounter a particle. This produces compact structures which grow upwards as inverted cones. As some cones close off the space above others, the number of cones is reduced and the surface structure coarsens. This produces a characteristically columnar structure and/or a superposition of (point-down) conical features.

The surface is not a self-similar fractal, but instead is self-affine. Vicsek (1992) reports that it is actually a fat fractal. It is often described as a "cauliflower" surface because of the morphology. Modeling of this process, with either normal or inclined incidence angles for the particles, produces structures that are quite similar in appearance to physical coatings. Figure 13 in Chapter 6 shows a cone "grown" by computer simulation (Vicsek 1992) . Figure 10

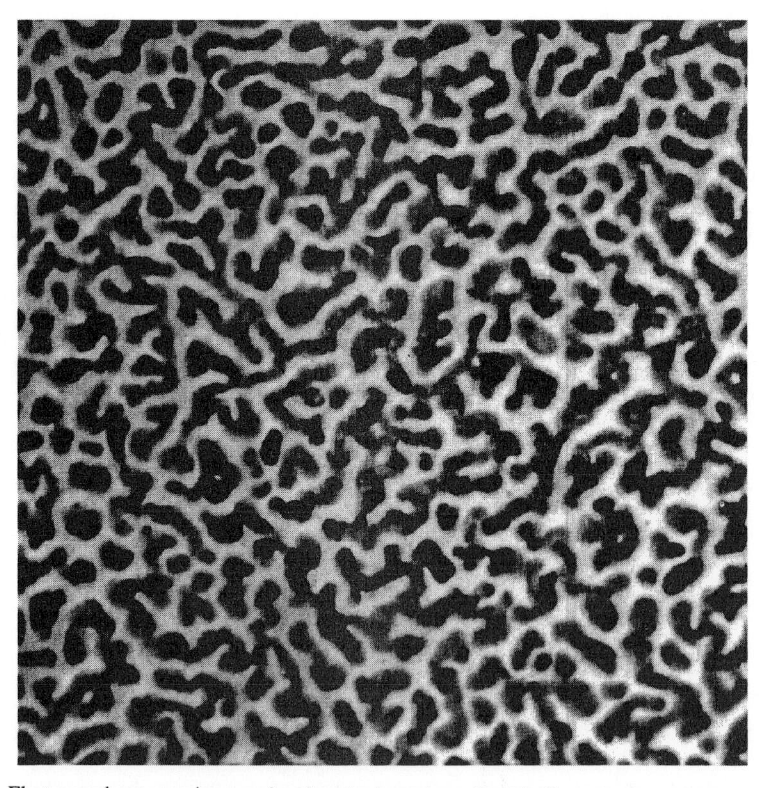

Figure 11. Electron microscope image of a deposited coating of gold (Voss, Laibowitz et al. 1982). The interpenetrating fingers can be modelled as a percolation fractal, forming an infinite cluster size at a critical value of coverage.

shows schematically how a cross section through such a deposited coating would appear, as larger cones dominate smaller ones. There is certainly a fractal character to this process, but as pointed out in Chapter 6, the surface is not fractal in the sense used in this book.

Another consideration for deposited surfaces is their ability to cover the substrate. We have already seen that classic diffusion-limited deposition produces very open structures that expose much of the substrate. Even a dense coating may not cover everywhere, however. Voss, Laibowitz et al. (1982) applied gold films to Si_3B_4 windows in thicknesses from 6 to 10 nm. Rather than uniformly coating the surface, the gold collected into interpenetrating ramified structures whose perimeter was directly proportional to the area on scales greater than the thickness. Figure 11 shows an example. A "percolation threshold" was observed at about 74% coverage at which cluster size became infinite. Below that, the size distribution of clusters followed a power law, and the fractal dimension was found to be close to that expected for two-dimensional percolation ($D = 1.89$, $\tau = 2.05$, where τ is the exponent of the power-law distribution of cluster sizes). Similar results have been reported for other metals, thicknesses and substrates (Kapitulnik and Deutscher 1982; Voss, Laibowitz et al. 1982; Laibowitz 1987; Varnier, Llebaria et al. 1991; Noques, Costa et al. 1992).

Percolation studies indicate that the infinite cluster should have a dimension of about $D = 1.89$ for $\tau = 2$ and about 2.5 for $\tau = 3$. Other dimensionalities may be associated with the outer convex hull of each island, the backbone length, the length of the minimum path between sites, etc., as these are also fractal. It is not known whether the values for these fractals are deterministically related.

Pore Networks

The third type of fractals classified in Chapter 1 was networks, including porous structures. Solids containing pore networks have surfaces with a distribution of pore openings. A "dimension" for the surface (Tricot 1989b; Tricot 1989c; Jaroniec, Lu et al. 1991) can be calculated from the slope of a cumulative logarithmic plot of the number of pore openings vs. their diameter. However, this dimension is not directly related to the usual surface dimension that has been of interest here (Pfeifer 1984; Salli 1991). Most open pore structures will have a surface fractal dimension that approaches 3.0, since the internal surface area of the pores which is accessible from the surface increases in direct proportion to the volume of the object. Of course, such networks can be fractal only over a finite size range, limited at the top by the size of the object and at the bottom by the dimension of the measurement tools (Bittner and Sernetz 1990; Bittner 1991).

Fractal pore networks are very important in many applications. Kaye and Clark (1987) use fractal geometry to describe filters, for example. One of the most intriguing areas of application includes physiological structures (Goldberger and West 1987; West and Goldberger 1987; Wlczek, Bittner et al. 1989; Goldberger, Rigney et al. 1990). The large amount of surface area required for exchange of gases with the blood leads to a branching structure for the bronchi (airways in the lung) that is fractal. Likewise, the need to bring the blood to every cell in the body, but have a circulatory system that itself occupies the minimum volume, leads to a branched capillary network (actually two networks, the arterial and venous) that is fractal. A similar sort of evolutionary selection process works to make our telephone and road networks have a fractal branching network that brings service to every home with local streets and wires that merge into ever larger arteries (Bittner 1990).

Figure 8 in Chapter 7 showed a cast of the blood vessels serving the heart muscle, as an example. Similar branching fractals are found in many other biological structures, including the kidney, the villi and microvilli of the intestines, trabecular bone, the branching networks

of neurons (Chapter 1, Figure 22), and the growth patterns of microbes (Wlczek, Bittner et al. 1989; Obert, Pfeifer et al. 1990; Wlczek, Odgaard et al. 1991). Little interpretation of the dimension of these networks, and how it may change with disease, aging, etc., has been reported. Lynch, Hawkes et al. (1991) and Jacquet, Ohley et al. (1990) apply fractal measurement to profiles of density across X-ray radiographs of leg bones. The trabecular bone varies in density to produce a pattern across the bone that gives a fractal signature with a "dimension" that is different for healthy and arthritic bone, or bone that is undergoing osteoporosis. Bale and Schmidt (1984) have discussed the use of X-ray scattering to study porous fractals. Other medical and dental radiography applications of fractal geometry have been demonstrated as well (Lundahl, Ohley et al. 1985; Lundahl, Ohley et al. 1986; Dellepiane, Serpico et al. 1987; Chen, Daponte et al. 1989; Kuklinksi, Chandra et al. 1989; Caldwell, Stapleton et al. 1990; Chuang, Valentino et al. 1991; Rigaut 1991; Fortin, Kumaresan et al. 1992). Xiao, Chu et al. (1992) have used measurement of the fractal dimensions of brightness patterns for recognition in mammograms, and Recht (1993) is doing the same for pap smears. These applications were mentioned in Chapter 5 in connection with the interpretation of fractal brightness patterns and their use in image segmentation.

Although the chief interest here is with fractal surfaces, it is worth mentioning that several aspects of fractal networks have been studied in detail. The greatest attention has been given to the formation of percolation networks. In a 2D or 3D lattice, a series of links placed at random will form clusters of connected points. The number of clusters or islands, plotted against a measure of size such as the maximum linear distance between any two points will show the usual power-law behavior expected for fractals.

When the fraction of links that is occupied reaches a critical value, equal to 59.28% for a 2D square lattice, a sudden and qualitative change in the network takes place. In addition to the isolated islands, there will be somewhere an "infinite" network that reaches entirely across the lattice, no matter what its size. Actually, there are three different types of percolation networks that are considered. The description above is a "bond" matrix where the links in the lattice are occupied (open) or not (closed), and of course the geometry of the lattice can be varied (square, hexagonal, etc.) in either two or three dimensions. It is also possible to model a percolation lattice by the occupancy of pore sites, in which it is the nodes that are occupied or not, or by using an invasion model, in which a cluster can only grow from previously occupied neighboring sites (Adler 1989). The details of the different models have been summarized by Mandelbrot and Given (1984), Stauffer (1985), Stanley (1987), Kaye (1989a), Bunde and Havlin (1991), and Stauffer and Aharony (1991).

Obviously, these percolation networks have direct importance to the recovery of oil from reservoir rocks. They predict the permeability of the rock, and the amount of oil that is trapped in unconnected islands and cannot be recovered. Percolation theory can also be used to model the spread of disease in a population or fire in a forest, and other natural phenomena. It may also provide a model for the crosslinking of polymers and gels (Sernetz, Bittner et al. 1991). The pore structure of gels used in chromatography has been modeled and measured as a fractal network (Sernetz, Bittner et al. 1989). Polymerization by an addition method in which mers are added one at a time to molecule chains or networks has also been modeled using lattice site occupancy (Vicsek 1992).

However, the branching of networks of bronchi, arteries, and other vessels is not well modeled by a percolation fractal on a lattice. Methods based on the L-systems (Prusinkiewicz and Lindenmayer 1990) introduced in Chapter 1 are more successful, and have been used to generate models which correspond to many physiological structures (Wlczek, Bittner et al. 1989; Bittner 1990; Wlczek, Odgaard et al. 1991). The flow through such networks has been analyzed to model reaction rates and metabolism in organisms (Sernetz, Gelléri et al. 1985; Sernetz, Willems et al. 1989; Sernetz, Willems et al. 1990). The vibrational dynamics of such

fractal networks have also been investigated (Orbach 1986), as have those of agglomerates (Courtens and Vacher 1990).

Other Surface Applications

Surface roughness obviously has an influence on many properties. One is the strength of adhesive bonding. Russ (1990a) has measured the roughness of surfaces of a metal alloy used in dental restorations, prepared by various chemical and electrochemical etching procedures. The fractal dimension of the surfaces was obtained both by cross-sectioning and profile measurement, and by measuring the brightness fractal dimension of surface images obtained using secondary electrons in the SEM. These values correlate well with the failure stress of adhesive bonds formed with the etched surfaces, as shown in Chapter 5. The conclusion is that regardless of how the surface is prepared, the "rougher" it is (i.e., the higher the fractal dimension), the greater the strength of the adhesive bond.

Little work has been reported on the general problem of chemical etching or corrosion of surfaces. The observation that such surfaces are often fractal (Sapoval 1991), at least over a range of dimensions smaller than the grain structure of the material, falls short of providing any comparison of the roughness with various corrosion parameters, time, etc. Gouyet, Rosso et al. (1991) suggest that modeling corrosion as an invasion percolation or diffusion front produces realistic surface morphologies. Kaye (1989a) has reported that the fractal dimension of the projected boundary of metal particles exposed to acid first drops as protrusions are smoothed, and then increases as the corrosion eats into the surface.

In a very different application area, there has been a study of the wrinkling of skin. Skin wrinkling has been of interest to the cosmetics industry for some time. The wrinkles on facial skin are dependent on superficial facial muscles, which produce facial expressions and modify the skin texture. In addition to producing a highly anisotropic texture in the wrinkles, this makes it difficult to isolate the effects of aging, environment (sun, wind, etc.), and the ameliorative effects of cosmetics. Studies of skin on the forearm have been more useful in studying the effects of aging. A close relationship between fractal skin surface patterns and the architecture of the dermis reveals a progressive change in the anisotropy of furrow patterns with age (Corcuff 1983; Corcuff, de Rigal et al. 1983; Corcuff, Chatenay et al. 1984).

In sedimentology, there has long been an interest in classifying particles based on shape. Frisch, Evans et al. (1987) applied a Richardson analysis to projected profiles of sand grains. The purpose was to classify the grains in a sedimentation study as having been deposited by one of several river systems in a bay. The data were observed to produce good Richardson plots with linear slope covering more than a decade in dimension. The slopes for each population of sand grains were shown to be distinct and to offer assistance in classification. Romeu, Gomez et al. (1986) have used the fractal dimension of small metal particles to correlate with their processing history. His measurement method used thickness fringes from the transmission electron microscope to obtain surface contours used in Richardson plots, instead of projected profiles. This method avoids the projection problems discussed in Chapter 7, but the precision with which the profiles can be determined is not known.

The usual approach to using these projected profiles is to "unroll" them as a plot of radius vs. angle, perform harmonic (Fourier) analysis, and then apply statistical classification techniques to determine the particular frequencies that can distinguish particulate classes in any given instance (Ehrlich and Weinberg 1970; Beddow, Philip et al. 1977; Flook 1982; Kaye, Leblanc et al. 1983; Barth and Sun 1985; Kaye 1985b; Russ 1989; Klasa 1991). This was illustrated in Chapter 4, Figure 21. While the method is often successful, it is both time-consuming and raises troubling questions about the meaning of the selection of particular

harmonic terms. The fractal approach is more efficient and also seems to more succinctly summarize the particle shape in a single numeric value. Of course, it still begs the question of what the dimension "means" in terms of the erosion of the particle, as well as raising concern over the use of the projected dimension rather than a section. The projection should introduce some dependence in the roughness value on particle size, but since for sand grains there is a rather narrow range of sizes involved, this may be negligible.

Some other "geological" applications of fractal geometry have also been reported. Sidescan sonar imagery of sea-bed gives range and reflectivity images which are mathematically fractal and can be segmented by automatic classification techniques based on measurement of a local fractal dimension (Linnett, Clarke et al. 1991). Similar interpretation of Landsat images has been performed by Lam (1990).

Bishop and Chellis (1989) measured the profiles of the keels (bottom surfaces) of ice in Arctic pack ice. Pressure ridges that form in such packs have a sail (the above-surface portion) and a keel (below the surface). The keel profile is measured by an acoustic profiler. They conclude that the profiles of the keels are fractal, whereas other profiles of the bottom of the ice pack are not. Bruno, Taylor et al. (1992) show that lava flows are fractal in branching and surface morphologies. Discontinuities between rock formations have been shown to be fractal by Lee, Carr et al. (1990).

And, of course, several examples have been presented in earlier chapters using the boundaries of lakes and islands, the size distribution of islands, and mountain profiles. Matshushita (1985) shows that a Korcak plot provides an excellent fit for the size of drift icebergs over more than an order of magnitude in diameter. Kaye (1989a) has constructed Richardson plots (which are not really appropriate for vertical elevation values nor to the projection of the range against the horizon) to several mountain ranges, obtaining values which vary from 1.03 to 1.08. The observations that these structures are fractal unfortunately include little quantitative interpretation of the local differences that may be present due to differences in geology (rock types, etc.) or environment (principal types of weathering, prevailing wind direction, etc.). Measurement of contours and elevation profiles of earth topography have been shown to produce fractal plots by several authors (Burrough 1981; Mark and Aronson 1984; Klinkenberg and Clarke 1990; Carr and Benzer 1991; Taylor and Taylor 1991). In one of the few quantitative uses of fractals, Goodchild (1980) showed that the fractal dimension of lake boundaries in Newfoundland increased with elevation.

Very Flat Surfaces I: Interferometry

The idea of "very flat surfaces" is a completely arbitrary category. Many different processes produce surfaces that are "nearly" flat in a macroscopic sense. These include the cleaving of crystals, polishing of silicon wafers (of great industrial interest, of course), precision machining, chemical deposition, and so forth. The unifying principle in this section and the next is the need for high resolution, both vertical and lateral, to measure the surface irregularities.

Most of the measurement tools available for very flat surfaces provide a single-valued elevation reading at each point in an x, y raster or grid. This will be blind to any undercuts that may be present, and so the surface data may be self-affine but will not be self-similar. As we have seen in earlier chapters, there are several suitable analysis tools for dealing with this kind of data. There is no agreement on a single best method, although analysis of the frequency transform offers the most flexible interpretation, especially for anisotropic surfaces.

Just as radar and sonar have wavelengths in the range of centimeters to meters, and thus are useful for measurements of large objects such as geologic landforms, so a much shorter

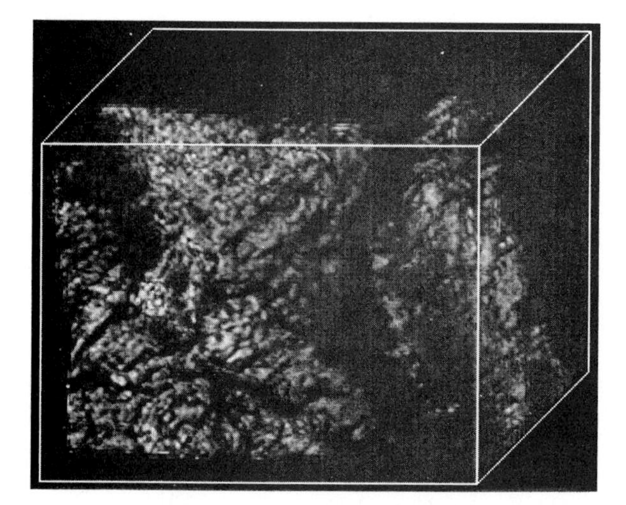

Figure 12. Reconstructed 3D image of a brittle fracture surface in a ceramic, imaged with a confocal scanning light microscope.

measuring scale is needed for high precision, or very flat surfaces. Attempts to use scanning electron microscopy (SEM), or conventional or confocal scanning light microscopy (CSLM), either on the original surfaces or on vertical sections cut through them, have been only partially satisfactory. The lateral resolution of the SEM is very good, but the depth resolution is not. Stereo pair measurements are both difficult to perform and time-consuming to convert to an elevation map or range image of the surface, and the resulting depth resolution is still much poorer than the lateral resolution (Lee and Russ 1989). Special adaptations to the SEM to obtain topographic information involve tradeoffs between lateral and vertical resolution (Thong and Breton 1992) that do not produce enough measured points in the image for fractal analysis, although limited data on fracture surfaces has been reported (Friel and Pande 1993). Interpreting the brightness patterns of reflected light to reveal the fractal character of the surface is useful only for direct comparisons of similar materials, and is sensitive to changes in instrument parameters and to surface composition or contamination.

Conventional light microscopy has a lateral resolution of better than one micrometer, but the depth of field is neither great enough to view an entire rough surface nor shallow enough to isolate points along one iso-elevation contour line. The confocal scanning light microscope improves the lateral resolution slightly, and greatly reduces the depth of field while at the same time rejecting scattered light from out-of-focus locations. The result is an instrument that can image an entire rough surface by moving the sample vertically and keeping only the brightest light value at each location, or can produce a range image by keeping track of the sample's vertical motion when the brightest reflected light value is obtained for each point in the image. It is the latter mode which is most interesting for surface measurement purposes. The resolution is better than one micrometer in all directions.

Figure 12 shows a reconstructed view of the surface of a fracture in a brittle ceramic. It is formed from 26 planes, separated by 1 μm in the z direction, each of which records a pixel only if that location is brighter than any other plane. The perspective view can be rotated to present a very realistic image of the surface. However, plotting the same information in the form of a range image in which each pixel brightness corresponds to the plane in which the brightest reflection was recorded is more useful for measurement. Figure 13 shows this presentation of the same surface, along with an elevation profile along an arbitrary line across the surface.

This is interesting for many macroscopically rough samples, such as fractures and some deposited coatings, but it is not adequate for the really flat surfaces which are currently being

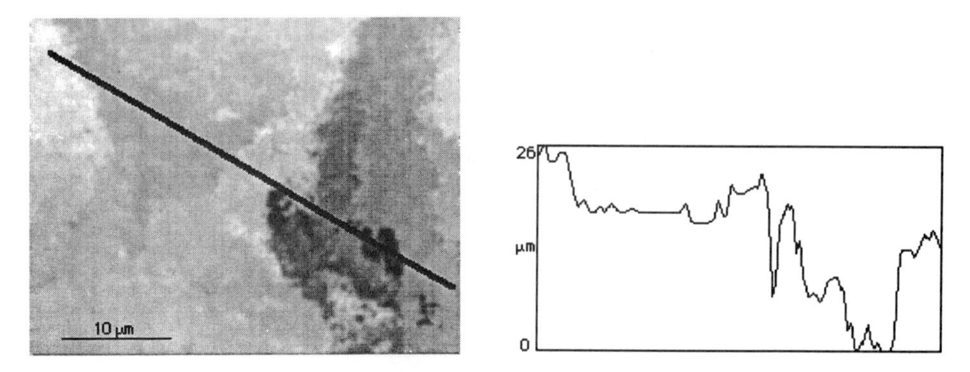

Figure 13. Range image produced from the data in Figure 12, with an elevation profile along an arbitrary line.

produced in many applications. The surface irregularities on a typical polished silicon wafer, or precision-machined mirror surface, are typically of the order of nanometers.

Three principal methods have been applied to such surfaces. Historically, the profilometer provided a tool which would accurately measure vertical elevation with a resolution approaching a nanometer. Although it has been widely used, the profilometer has two serious disadvantages for fractal surface applications. The first is that it determines elevations only along a single profile. While the analysis of such elevation profiles is straightforward, their relevance to complex surfaces which may have anisotropic properties is questionable. The second limitation is the large tip size which makes it impossible to follow steep slopes or steps accurately.

This leaves the interferometric light microscope and the atomic force microscope as the methods of choice for studying very flat surfaces. Both are somewhat novel instruments, one a modern implementation of the principles of light interference discovered a century ago and the other a technology invented and rapidly commercialized only within the past decade. Neither tool is perfect, and for examining fractal rough surfaces there are some unanswered questions which remain to be considered.

The interferometric light microscope (see the review by Robinson, Perry et al. 1991) reflects light from the sample surface as one leg in a classic interferometer, which is then combined with light from a reference leg. Figure 14 shows a schematic diagram. The usual principles of phase-sensitive interference occur so that changes in path length (due to different elevations of points on the surface) produce changes in the intensity of the light. This image is then digitized using an appropriate CCD detector array and analog-to-digital conversion, so

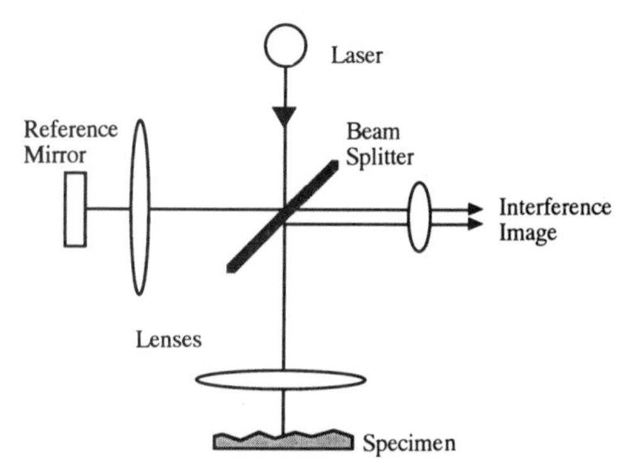

Figure 14. Schematic diagram of an interference microscope.

that a brightness value, and from it a derived elevation value, is recorded for each point on the surface. Although the wavelength of light used is typically about 630 nm, phase differences between the two legs of the interferometer of one-thousandth of the wavelength produce a change in intensity so that the vertical resolution is a few angstroms. The lateral resolution, however, is still of the order of one micrometer, limited by the wavelength of the light used and the design of the optics.

The interferometric light microscope also suffers if the surface has very high slopes or a highly specular finish, since no light will be reflected back to the detector. Such points become dropouts in the final image. For visual purposes, it is satisfactory to fill in such missing points with a median or smoothing filter, but of course this would bias subsequent fractal measurements. Fortunately, for very flat surfaces there are typically few (if any) dropout points.

The vertical resolution of the interferometric microscope is very high, approaching atomic dimensions, and although the lateral resolution is much lower it should be quite suitable for measuring fractal dimensions. This is because the basic underlying principle of fractal geometry is that deviations will scale in magnitude with lateral dimensions over many decades of dimension. Measuring at lateral dimensions of micrometers (or even tens or hundreds of micrometers) should be able to determine the slope of one of the log–log plots (depending on the analysis method selected) which should match the slope at lateral dimensions of nanometers.

However, the surfaces being examined are real, not mathematical fractals. Hence there is no reason to expect that the fractal plot must follow a straight line over this range of dimensions. It is possible that some other physical processes may operate at different scales, so that the slope (and with it, the fractal dimension) really does change as atomic dimensions are approached. To study this, it is necessary to obtain data across several decades of lateral dimension, always with enough vertical resolution to detect deviations of the appropriate scale. We will see shortly what data such experiments yield.

The response of any microscope or other imaging device can be described by means of a modulation transfer function (MTF). Two factors control the shape of this function: the difference in height (or, for a conventional image, the difference in brightness) between two points, and the distance between them. For a conventional image, the MTF commonly shows the best performance (ability to distinguish two slightly different grey values) when the points are spaced apart by several times the absolute resolution limit of the microscope. For points lying closer together, a greater difference in brightness must be present. When points are very widely spaced, it is also necessary to have a greater difference between brightnesses to distinguish them. Figure 15 shows an example. The sinusoidal variation of brightness varies in spacing from side to side, and in magnitude from top to bottom. In the middle of the chart, the lines can be distinguished higher on the chart (i.e., to a lower contrast) than at either side.

At least qualitatively, the same response is expected from any imaging or measuring system. However, it is not clear what the exact response of the interferometric light microscope looks like. At very large spacing, the absolute height difference between points is not measured precisely because the overall surface alignment and the shape or "figure" of the part is not known. It is common to deal with the problems of alignment and shape by fitting a function to the data. Determining a best-fit plane, or other low-order polynomial function, by least-squares fitting to all of the elevation data and then subtracting it is called "detrending" the data. It is then possible to display the magnitude of deviations of points from this surface. However, the absolute difference between points that are widely separated is affected by the detrending plane, and the ability to distinguish small differences between points that are far apart is reduced.

For fractal analysis, the loss of sensitivity at low frequencies (large spacings) is a minor inconvenience. In a plot of log (Magnitude^2) vs. log (Frequency) from a Fourier transform,

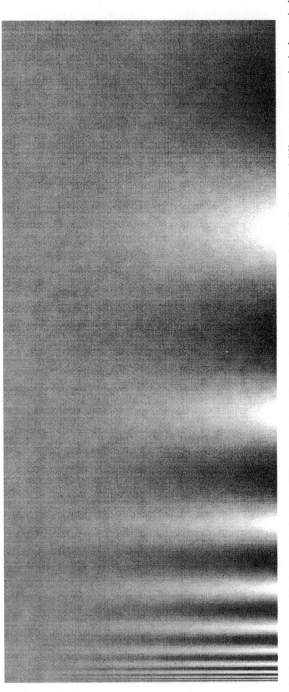

Figure 15. Example of a modulation transfer function. Different spacings or frequencies are plotted in the horizontal direction and different contrast ratios in the vertical direction. The ability to resolve the spacings depends on both factors.

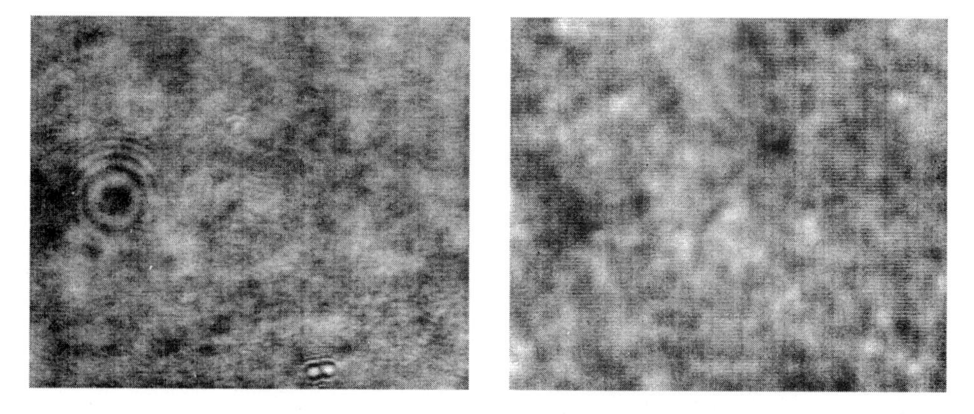

Figure 16. Comparison of interferometric range images on a flat polished silicon wafer using Fizeau **(left)** and Mirau **(right)** optics.

this would affect only the first few terms, which would also be sensitive to the choice of an arbitrary zero-elevation reference, and to the overall shape or figure of the object. At high frequencies there are many more points, and these should be sufficient to determine the slope provided the low-frequency points can be excluded from the fit, which unfortunately requires a somewhat arbitrary cutoff.

However, there are limitations here as well. The MTF of the microscope should show a falloff in sensitivity at high frequencies (small lateral spacings). In practice, this seems to take the form of a change in slope of the log–log plot at high frequencies, toward a slope of −1.0. This is the so-called $1/f$ or Brownian noise limit imposed by the inevitable noise in the light source, sensor, digitizing circuit, and all other optical and electronic components. It corresponds to an apparent surface fractal dimension of 2.5, and may mask the real behavior of the surface at high frequencies. When the magnitude of the noise is very small compared to the signal from the surface roughness, the high-frequency behavior is instead a dropoff in magnitude in the power spectrum and MTF when the frequencies exceed the resolution of the instrument.

To illustrate these effects, a very flat surface produced by polishing a silicon wafer was imaged in a commercial interferometric light microscope (Zygo). Both Fizeau and Mirau sets of optics were available, and the comparison of their performance is shown in Figure 16. The field of view is the same in both images, and the total vertical range of elevations is only about 2 nm. The two bright white spots toward the bottom of the image are probably due to dirt somewhere in the optics. These are much less pronounced with the Mirau optics. In addition, the ringing (oscillation or ripple pattern) around features that can be seen in the Fizeau image is not present with the Mirau optics. These characteristics are usually interpreted as indicating that the Mirau optics are superior for the measurement of very flat surfaces. On the other hand, the Mirau image seems to have less lateral resolution (to be "smoothed"), and a pattern of horizontal lines can be discerned that may come from the alignment of the diffuser plate with the raster scan pattern of the camera.

Figure 17 shows the 2DFT analysis from these images. The Mirau data produce a more isotropic circle for the fractal dimension results, and the numerical values for the slope and intercept of the log (Magnitude^2) vs. log (Frequency) values are quite similar. The reported fractal dimension is about 2.3, and the intercept is about 2 nm. It is not known, of course, whether the anisotropy in the Fizeau plot is an artifact, or a real characteristic of the surface that has been hidden in the Mirau data.

The appearance of the power spectrum plots for these two images is particularly revealing. The Fizeau plot shows the expected linear decrease of magnitude with frequency, on log–log axes. In the Mirau plot, the high-frequency terms have been suppressed. This is apparently what accounts for the smoother appearance of the range image. It clearly represents a change in the system MTF, and a loss of data. Whether that loss removes useful information from the sample or characteristics of the optics themselves is not obvious.

It is interesting to compare these rather high dimension values to those obtained by other methods. Figure 18 shows the dimensional analysis (log *P* vs. log *A*) plots for these range images, and Table 1 summarizes the results from several measurement methods on both

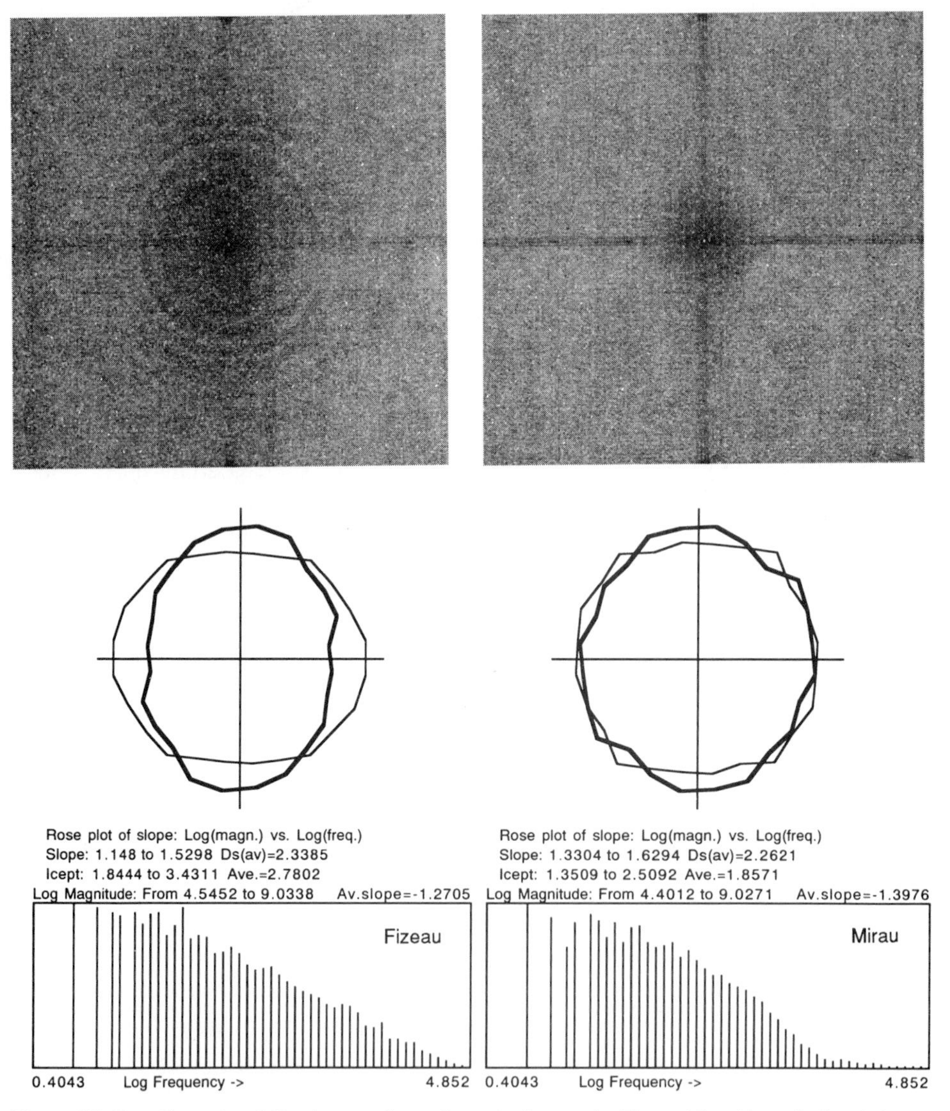

Figure 17. Two-dimensional Fourier transforms from the images in Figure 16, with analysis of slope and direction.

Table 1. Measured Fractal Dimensions for Fizeau and Mirau Range Images (Figure 16)

	Fizeau	Mirau
2D Fourier Transform	2.338	2.262
Dimensional Analysis	2.572	2.561
Slit Island/Kolmogorov	2.561	2.557
Minkowski	2.330	2.321
RMS vs. Area	2.347	2.316

images. In general, the agreement between the techniques is similar to that expected from comparisons reported in Chapter 6 for generated surface. For instance, the Minkowski values are lower than those by methods that use a zeroset approach (slit-island and dimensional analysis). The measured values seem to have been reduced only slightly by the Mirau optics and their effect on the system MTF.

If the actual MTF of the microscope were known, it might be possible to unfold it from the data or at least select the portion of the data which could be safely used. Determining the actual MTF of an optical system is usually accomplished by imaging a target containing many different line spacings and contrasts, something like the image shown above in Figure 15. Obviously, constructing such a surface with dimensions approaching the atomic is impossible. Even using an arbitrary surface such as a "very flat" silicon wafer, or a cleaved mica sheet, is hampered by the inability to determine what the "real" surface geometry is. There are no perfectly flat surfaces available, as everything contains some irregularities and for the surfaces examined so far, these seem to tend toward a fractal arrangement whose dimension is also close to 2.5, perhaps reflecting a Brownian displacement of atoms on a very fine scale. Separating the fine-scale surface roughness from the high-frequency instrument response presents a major challenge to both the instrumentation and the analysis methods. This problem has been studied for both the interferometric light microscope and the AFM, with similarly inconclusive results.

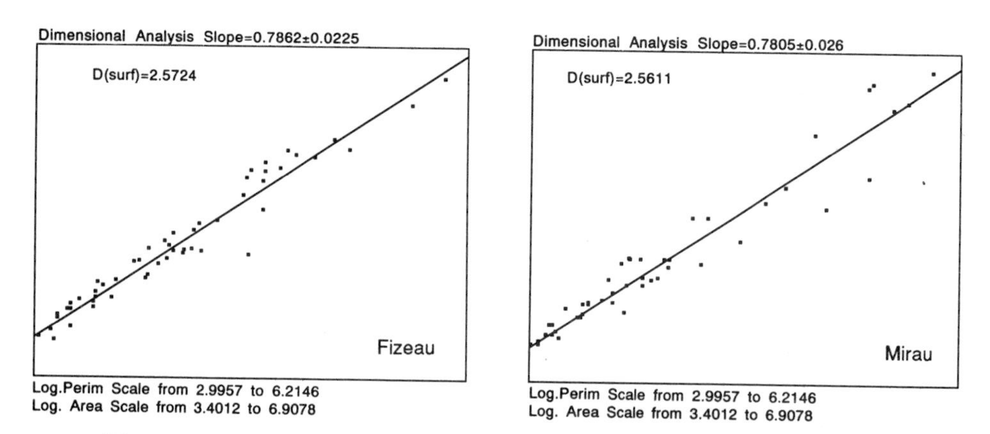

Figure 18. Dimensional analysis plot for the Fizeau and Mirau range images in Figure 16.

Very Flat Surfaces II: Atomic Force Microscopy

The AFM is in essence a profilometer which scans a complete raster over the sample surface, but with a very small tip (see the review by Wickramasinghe 1989). The standard profilometer has a tip that cannot follow very small or very steep-sided irregularities on the surface because of the tip dimensions. The AFM tip can be much smaller and sharper, although it is still not usually fine enough to handle the abrupt steps (or even undercuts) present in microelectronic circuits and some other surfaces. The tip can be operated in either an attractive or repulsive mode of interaction between the electrons around the atom(s) in the tip and those in the surface. Usually, repulsive mode in which the tip is pressed against the surface does a somewhat better job of following small irregularities, but it may also cause deformation of the surface and the displacement of atoms.

There are other modalities of interaction for these scanned-tip microscopes. The STM (scanning tunneling microscope) was the original, but it can only be used for conductive specimens and is strongly sensitive to surface electronic states, surface contamination, and oxidation. Lateral force and other modes of operation developed within the last few years offer the ability to characterize many aspects of surface composition and properties, but the straightforward AFM mode is most often used to determine surface geometry, and presents quite enough complexities for understanding.

As indicated in the schematic diagram of Figure 19, the usual mode of operation for the AFM is to move the sample in x, y, and z. The x, y scan covers the region of interest while the z motion brings the tip back to the same null position, as judged by the reflection of a laser beam. The necessary motion is recorded in a computer to produce the resulting image. Modifications which move the tip instead of the sample, or use a linear array light sensor to measure the deflection of the tip rather than wait for the specimen to be moved in z, do not change the basic principle. The motion is usually accomplished with piezoelectric elements whose dimensions can be sensitively adjusted by varying the applied voltage. However, there is a time lag or creep associated with these motions which appears in the MTF of the microscope as a loss of sensitivity at the largest or smallest dimensions (lowest and highest frequencies). Locating the x, y coordinates of the tip interferometrically instead of based on the piezo driver voltages can be used to overcome some of these problems.

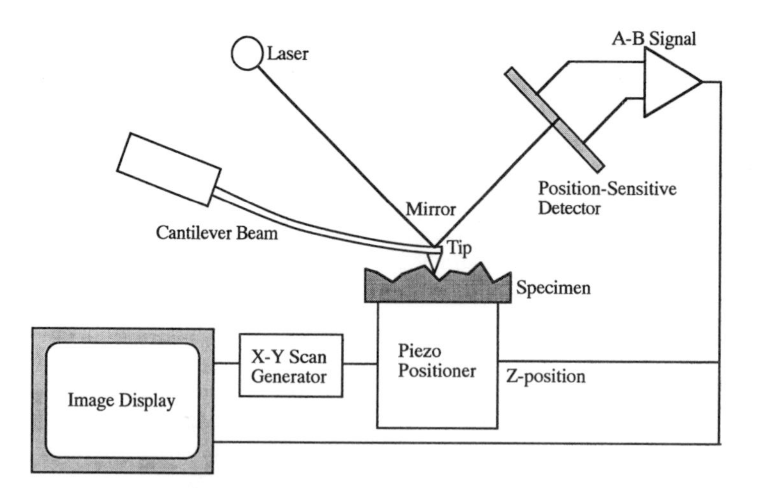

Figure 19. Schematic diagram of an AFM.

Although the AFM has in principle a lateral resolution of a few angstroms, as well as vertical sensitivity in this range, it is not always possible to realize that performance. For one thing, the adjustment of scan speeds and amplifier time constants for the visually best picture may eliminate some of the fine-scale roughness and thus bias subsequent analysis, or conversely may introduce additional "noise" from electrical or mechanical sources. Elimination of vibration, which would appear in the 2DFT power spectrum as a few discrete peaks or dips, requires special care for the AFM. The most common installation procedure is to suspend the entire instrument on "bungee" cords to damp out all but the lowest frequency vibrations.

Most of the attention to the MTF of the AFM and related instruments such as the STM has been concerned with the high-resolution limit (Denley 1990a; Denley 1990b; Grigg, Russell et al. 1992). This is generally set by the shape of the tip, which is not all that easy to characterize. Some authors (Reiss, Schneider et al. 1990; Reiss, Vancea et al. 1990; Pancorbo, Aguilar et al. 1991; Aguilar, Anguiano et al. 1992) have suggested that it may be possible to deconvolve the tip shape and improve the image sharpness, in exactly the same way that this is done for other imaging systems. If the MTF is not known directly, it can be estimated from the image itself (actually from the power spectrum) and subtracted. This presupposes that some portion of the power spectrum contains only tip-related information, so that the tip contribution can be determined. Logically, this should be the high-frequency portion, but as noted above many other factors including the instrument and above all the fractal nature of the surface may interfere. The apparent sharpening of the image would then be due to artificial tampering with the fractal roughness data.

One of the first attempts to measure the fractal dimension of surfaces with the STM was carried out by Mitchell and Bonnell (1990), who examined a deposited gold film, a fatigue fracture in copper, and a single-crystal silicon cleavage surface, at lateral resolutions down to 1.2 nm (limited by STM tip geometry). The problem of noise was dealt with by attempting to measure it. This was done by recording an image with no sample present, in other words just the electronic contribution to the image. This is somewhat analogous to recording a "blank" image from a video camera by covering the lens. It is not quite the same as the preferable method of recording an image with a uniformly illuminated grey card in front of the camera, which would be equivalent to measuring the microscope response to a perfectly flat sample (if one were available).

The authors found that the noise which they could measure was two orders of magnitude less than the magnitude of the signal from their rather rough surfaces, and so they felt justified in ignoring it in performing subsequent fractal analysis of the surfaces. For much flatter surfaces, or when considering all of the possible contributions to the instrument MTF, a more comprehensive method for determining the instrument contribution will be needed. The fractal dimension was determined from the power spectrum of profiles along the STM fast scan direction. They concluded that the surfaces were fractal, but no attempt was made to interpret the dimensions. Comparison of FT measurement of dimension with the Minkowski (variational) dimension was made using Mandelbrot–Weierstrass-generated profiles (not the actual measured data), and showed poor agreement, underestimating the value for profiles with a dimension greater than 1.5 and overestimating it for lower dimensions.

Aguilar, Anguiano et al. (1992) have commented on the use of STM to measure fractal surfaces. They also record a "blank" image to determine the instrument noise, and find that it is mathematically a Brownian fractal along scan lines. The effect of mixing noise into an image was discussed in Chapter 3, where it was shown that the frequency analysis method of determining a fractal dimension is much less sensitive to this than the slit-island dimensional analysis plot of log P vs. log A used by Aguilar and by Gomez-Rodriguez et al. (Gomez-Rodriguez, Baro et al. 1990; Gomez-Rodriguez, Asenjo et al. 1992; Gomez-Rodriguez, Baro et al. 1992).

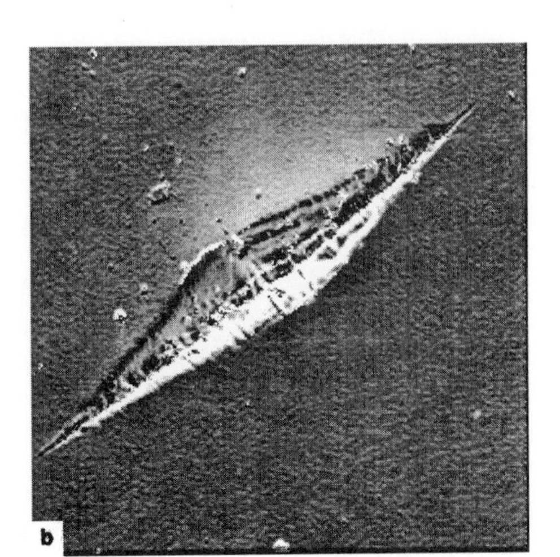

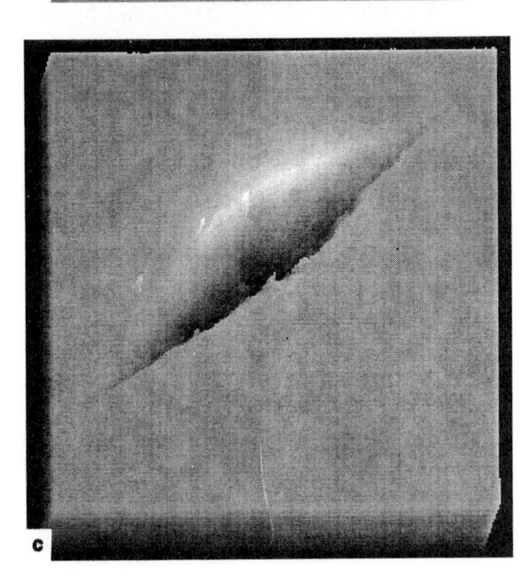

Figure 20. AFM image of a Knoop hardness indentation, shown as **(a)** range, **(b)** rendered, and **(c)** isometric presentations.

Most of the reports of fractal measurements from AFM and STM have used either Fourier analysis of line profiles along orthogonal directions in the array, or slit-island and dimensional analysis of the elevation array (citations above plus Chesters, Wen et al. 1989; Chesters, Wang et al. 1991; Files-Sesler, Randall et al. 1991). Miller and Reifenberger (1992) used the variation of standard deviation of elevation values with window size to determine a dimension. Anguiano, Vazquez et al. (1992) have compared several different fitting techniques for determining a fractal dimension from the FT power spectrum of STM images of surfaces, including different averaging techniques to combine data from multiple scan lines or directions and the effect of the number of points in each profile. They conclude that the FT results are slightly different from the numeric values calculated by methods such as dimensional analysis, but that meaningful and precise results can be obtained.

An evaluation of AFM performance was conducted using a sample of polished silicon (traditional roughness values indicate 0.2–0.3 nm magnitude, near the nominal vertical resolution limit for interferometry and STM) with a hardness indentation shown as a range image in Figure 20. One limitation of the STM can be seen by generating a rendered or an

isotropic display of the indentation, which shows that the left side of the indentation appears to be smoother than the right. This is an artifact of the scanning, as the tip response dynamics are different when following a surface down (where it may lag behind the actual surface and fail to record deviations) or up (where contact forces it to follow irregularities). In addition, the measured depth of the indentation is much less than the actual depth, because the tip cannot follow the deepest part of the indentation, and because the calibration of the AFM in the vertical direction is less precise than in the *x*, *y* directions.

A series of five AFM images was obtained (Eby 1993) with physical rotation of the sample through a range of 90 degrees, using optimal conditions, an extremely slow scan (1 Hz) and appropriate amplifier time constants, on the indentation and on the flat surface nearby. Each range image was analyzed using a 2DFT. The Fourier images were rotated to compensate for the specimen rotation, and averaged to enhance the sample-determined information while reducing instrument-dependent effects. The procedure was carried out using both a conventional AFM setup (referred to as a "tube" geometry) and with a tripod arrangement of independent piezo drivers for the *x*, *y*, and *z* directions. The former design is generally used for the very high spatial resolution images of atomic surfaces, and is considered by the manufacturers to produce sharper images.

For the indentation images, subtracting the mean image from each individual image (with appropriate rotation) leaves the noise image. Averaging all of these results gives an indication of the instument noise. As shown in Figure 21, the instrument noise left after removing the signal in this manner from the images of the indentations is radially symmetrical and consists of white noise with a magnitude nearly two orders of magnitude smaller than the signal. This indicates that because the surface has sufficient relief, the remaining noise is negligible. The noise magnitude from this measurement is less than 5 nm, which is a factor of 100 less than the total depth of the indentation.

When the same rotation procedure is used with the flat surface, the images (Figure 22) show a pattern of horizontal scan lines that dominate the measurements, with little if any indication of surface information. Summing the FT results isolates no significant information from the surface, leaving only the instrument noise. Except for images of occasional dust

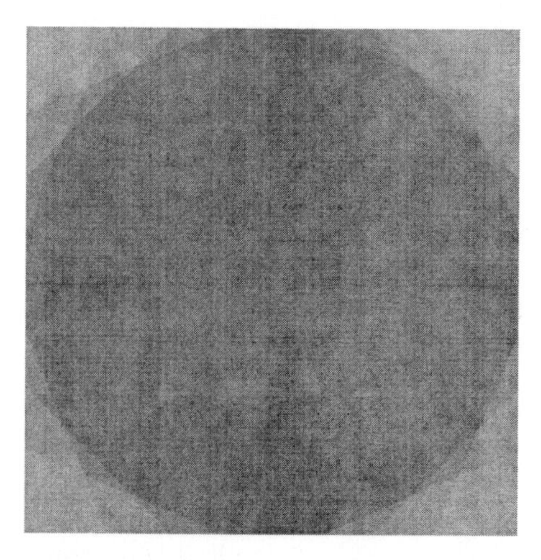

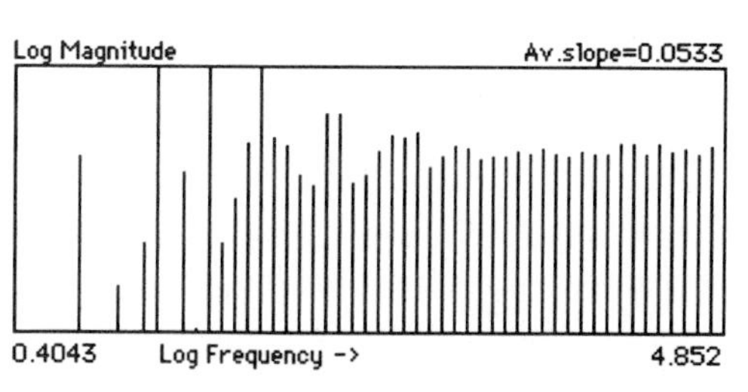

Figure 21. 2DFT image and log (Magnitude^2) vs. log (Frequency) plot for the instrument noise in the indentation images. The slope of zero corresponds to white noise, and a fractal dimension of 3.

particles, the image shows none of the small subnanometer steps or roughness in the surface that should follow crystallographic directions in the silicon. The instrument noise for the conventional "tube geometry" AFM design is shown in Figure 23, and has considerable anisotropy. Both the slope and intercept vary with direction relative to the scan, independent of the sample orientation, and arise from the complex scan mechanics of the AFM design. The magnitude of the intercept is slightly over 1 nm in the lowest noise directions. This reflects the accuracy with which the vertical position of the surface can be determined, but does not by itself indicate the ability to detect small variations in surface elevation, which depend (as discussed above) on the sharpness of the transition. In this case, it also depends strongly on direction.

On the other hand, with the use of independent x, y, z piezo drives the image noise is greater but also more isotropic. Figure 24 shows that the rotationally averaged noise has a fractal signature whose mean corresponds to Brownian noise ($D = 2.5$). While the pattern still shows some anisotropy, the variation of slope and intercept with direction is more gradual than for the conventional design. However, the magnitude is greater, with the intercept correspond-

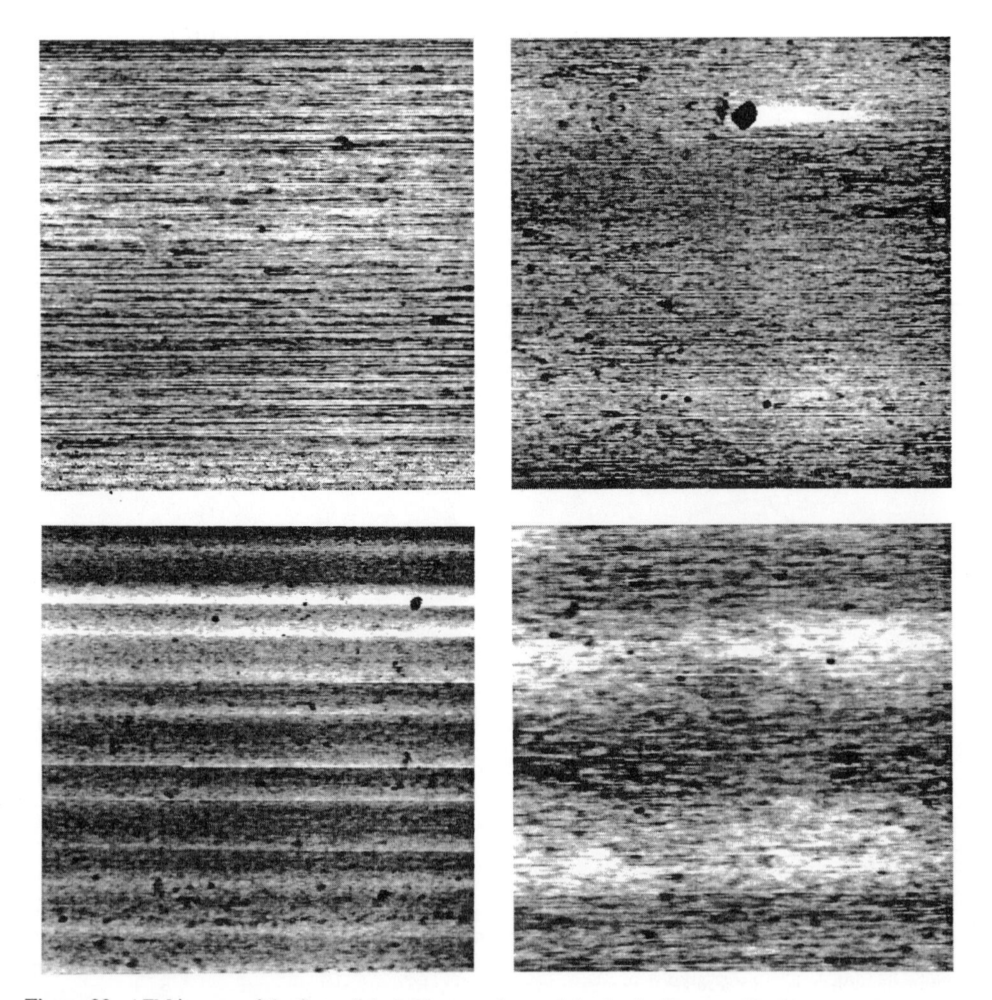

Figure 22. AFM images of the flat polished silicon surface as it is physically rotated in 22.5 degree steps. The apparent surface markings remain aligned with the raster direction of the instrument.

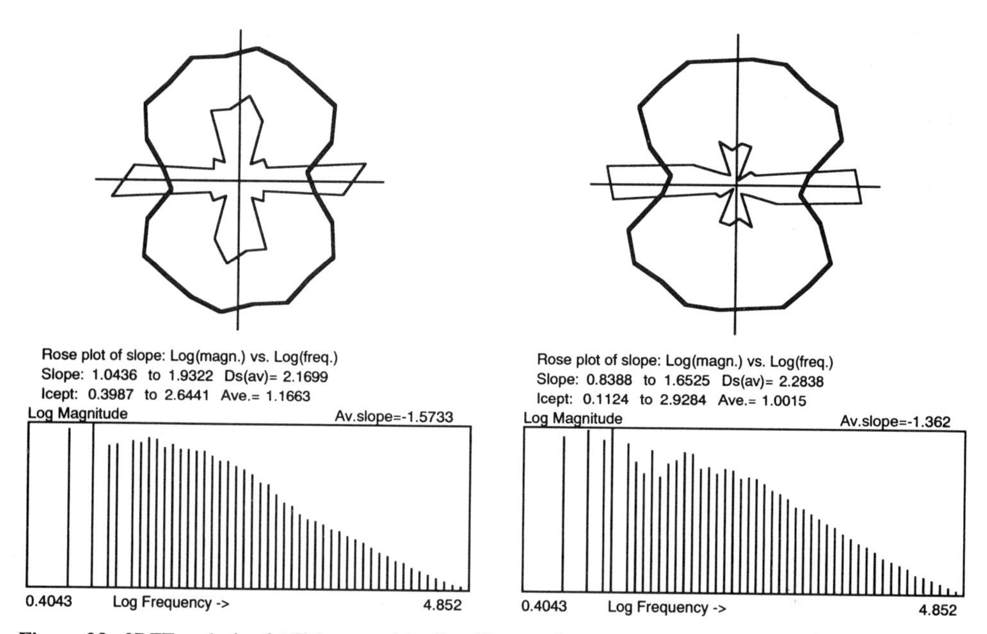

Figure 23. 2DFT analysis of AFM scans of the flat silicon surface rotated by 45 degrees. The anisotropy arises from the AFM scan mechanics and does not rotate with the sample.

ing to about 2.5 nm. This is larger than the 1 nm value shown above for Figure 23, and the difference probably indicates the relative vertical sensitivity of the two AFM designs. These results suggest that even with extreme care in selecting scan conditions, the measurement of very flat surfaces with the STM is strongly impacted by noise.

In conclusion, it seems that tools for surface characterization are available with sufficient resolution, both lateral and vertical. The interferometer is more convenient to use than the

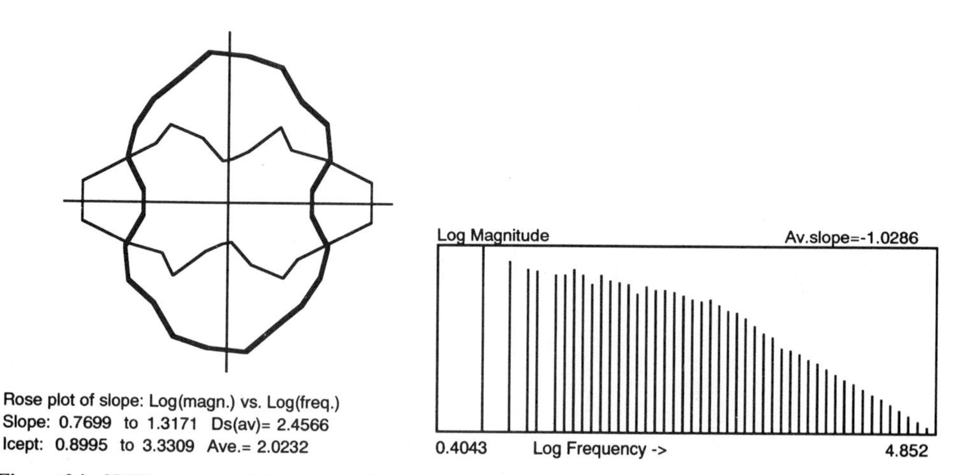

Figure 24. 2DFT analysis of the rotationally averaged flat-surface instrument response. A mean slope of –1 corresponds to Brownian noise and a fractal dimension of 2.5. The anisotropy in the surface is aligned with the raster orientation and does not rotate with the sample.

AFM, operates over a wide range of magnifications and accepts large samples. It also introduces less directional anisotropy due to the instrument characteristics. The AFM, on the other hand, has higher lateral resolution which may be required for the metrology of very fine features, now becoming commonplace in integrated circuit fabrication and nanotechnology. A complete understanding and measurement of the modulation transfer function is not available for either instrument as yet. This will be needed before the apparent changes in surface fractal dimension of very flat surfaces at very small dimensions can be interpreted.

References

Adams, H. M. and J. C. Russ (1992). "Chaos in the Classroom: Exposing Gifted Elementary Children to Chaos and Fractals." *J. of Science Education and Technology* **1**(3): 191–209.

Adler, P. M. (1989). Flow in Porous Media. *The Fractal Approach to Heterogeneous Chemistry*. New York, John Wiley & Sons. 341–360.

Aguilar, M., E. Anguiano, et al. (1992). "Digital filters to restore information from fast scanning tunnelling microscopy images." *J. Microscopy* **165**(2): 311–324.

Aguilar, M., E. Anguiano, et al. (1992). "Study of the fractal character of surfaces by scanning tunnelling microscopy: errors and limitations." *J. Microscopy* **167**(August): 197–213.

Aharony, A. (1985). "Fractals in Statistical Physics." *Ann. N. Y. Acad. Sci.* **452**: 220–225.

Aharony, A. (1986). "Fractals in Physics." *Bulletin of the European Physical Society* **17**(4, April): 41–43.

Aharony, A. (1991). "Fractal Growth." *Fractals and Disordered Systems*. Berlin, Springer Verlag. 151–173.

Aharony, A., Y. Gefen, et al. (1984). "Magnetic correlations on fractals." *J. Stat. Phys.* **36**: 795.

Ait-Keddache, A. and S. Rajala (1988). "Texture classification based on higher order fractals." *IEEE Trans* **CH2561**(Sept.).

Alexander, D. J. (1990). "Quantitative Analysis of Fracture Surfaces Using Fractals." *Quantitative Methods in Fractography*. Philadelphia, ASTM. 39–51.

Allain, C. and M. Cloitre (1986). "Optical diffraction on fractals." *Phys. Rev. B* **33**(5): 3566–3569.

Altus, E. (1991). "Fatigue, Fractals, and a Modified Miner's Rule." *J. Applied Mechanics* **58**(1): 37.

Anderson, T. L. (1989). "Application of Fractal Geometry to Damage Modeling in Advanced Materials." *Fractal Aspects of Materials—1989*. Pittsburgh, PA, Materials Research Society. 45–48.

Anguiano, E., F. Vazquez, et al. (1993). "Fractal characterization of surfaces: 1. Frequency analysis." *J. Microscopy* (*in press*).

Antolovich, S. D., A. M. Gokhale, et al. (1990). "Applications of Quantitative Fractography and Computer Tomography to Fracture Processes in Materials." *Quantitative Methods in Fractography*. Philadelphia, ASTM. 3–25.

April, G. V., M. Bouchard, et al. (1993). "Surface roughess characterization of dental fillings: A diffractive analysis." *Optical Engineering* **32**(2): 334–341.

Arbabi, S. and M. Sahmi (1990). "Fracture of Disordered Solid and Granular Media: Approach to a Fixed Point." *Scaling in Disordered Materials: Fractal Structure and Dynamics*. Pittsburgh, PA, Materials Research Society. 47–48.

Avnir, D., Ed. (1989). *The Fractal Approach to Heterogeneous Chemistry*. New York, John Wiley & Sons.

Avnir, D. and D. Farin (1984). "Molecular fractal surfaces." *Nature* **308**(15): 261–263.

Avnir, D. and D. Farin (1990). "Fractal Scaling Laws in Heterogeneous Chemistry: Part 1: Adsorptions, Chemisorptions and Interactions Between Adsorbates." *New Journal of Chemistry* **14**(3): 197–206.

Bak, P. and K. Chen (1989). "The physics of fractals." *Physica D, Nonlinear Phenomena* **38**: 5.

Bak, P. and K. Chen (1991). "Self-Organized Criticality." *Scientific American* (Jan.): 46–53.

Bak, P. and M. Creutz (1991). "Dynamics of Sand." *MRS Bulletin* (June): 17–21.

Baker, G. L. and J. P. Gollub (1990). *Chaotic Dynamics*. Cambridge, Cambridge University Press.

Baker, L., A. J. Giancola, et al. (1993). "Fracture and Spall Ejecta Mass Distribution: Lognormal and Multifractal Distributions." *J. App. Phys. (in press)* .

Balankin, A. S. and A. L. Bugrimov (1991). "Fractal dimensions of cracks formed during brittle fracture of model lattices and solids." *Soviet Technical Physics Letters* **17**(9): 630.

Bale, H. D. and P. W. Schmidt (1984). "Investigation of the fractal properties of porous materials by small-angle X-ray scattering." Symp. J, Boston, Materials Research Society.

Ball, R. C., M. J. Blunt, et al. (1990). "Diffusion-controlled growth." *Fractals in the Natural Sciences*. Princeton, NJ, Princeton University Press. 123–131.

Banerji, K. and E. E. Underwood (1984). "Fracture Profile Analysis of Heat Treated 4340 Steel." *Advances in Fracture Research*. Oxford, UK, Pergamon Press. 1371–1378.

Barabsi, A.-L. (1991). "Multifractality of self-affine fractals." *Phys. Rec. a[15] Statistical Physics* **44**(4): 2730.

Baran, G. R., C. Roques-Carmes, et al. (1992). "Fractal characteristics of fracture surfaces." *J. Amer. Ceram. Soc.* **75**(10): 2687–2691.

Barnsley, M. (1988). *Fractals Everywhere*. Boston, Academic Press.

Barnsley, M. F., V. Ervin, et al. (1986). "Solution of an inverse problem for fractals and other sets." *Proc. National Academy of Sciences* **83**: 1975–1977.

Barnsley, M. F. and L. P. Hurd (1993). *Fractal Image Compression*. Wellesley, MA, A. K. Peters.

Barnsley, M. F. and A. Sloan (1992). Method and apparatus for processing digital data. U. S. Patent. **5065447.**

Barth, H. G. and S.-T. Sun (1985). "Particle Size Analysis." *Analytical Chemistry* **57**: 151R.

Barton, C. C. (1989). "Fractal Characteristics of Fracture Networks and Fluid Movement in Rock." *Fractal Aspects of Materials–1989*. Pittsburgh, PA, Materials Research Society. 71–74.

Basingthwaighte, J. B. and J. H. G. M. van Beek (1988). "Lightning and the heart: Fractal behavior in cardiac function." *Proc IEEE* **76**: 693–699.

Beauchamp, E.K. and B.A. Purdy (1986). "Decrease of fracture toughness of chert by heat treatment." *J. Mater. Sci.* **21**: 1963.

Becker, K.-H. and M. Dörfler (1989). *Dynamical Systems and Fractals*. Cambridge, Cambridge University Press.

Beckmann, P. and A. Spizzichino (1963). *The Scattering of Electromagnetic Waves from Rough Surfaces*. New York, Pergamon.

Beddow, J. K., G. C. Philip, et al. (1977). "On relating some particle profile characteristics to the profile Fourier coefficients." *Powder Technology* **18**: 19–25.

Ben-Jacob, E., P. Garik, et al. (1990). "Adaptive Self-Organization in Nonequilibrium Growth." *Scaling in Disordered Materials: Fractal Structure and Dynamics*. Pittsburgh, PA, Materials Research Society. 11–14.

Berry, M. V. (1979). "Diffractals." *J. Phys. A* **12**(6): 781–797.

Berry, M. V. and J. Hannay (1978). "Topography of Random Surfaces." *Nature* **273**: 573.

Berry, M. V. and Z. V. Lewis (1980). "On the Weierstrass–Mandelbrot Fractal Function." *Proc. Royal Soc.* **A370**:459–484

Bhatt, V., P. Munshi, et al. (1991). "Application of fractal dimension for nondestructive testing." *Materials Evaluation* **49**(11): 1414.

Birgenau, R. J., R. A. Cowley, et al. (1987). "Static and dynamic properties of random magnets." *Fractal Aspects of Materials II*. Pittsburgh, Materials Research Society. 72–74.

Bishop, G. C. and S. E. Chellis (1989). "Fractal dimension: A descriptor of ice keel surface roughness." *Geophysical Research Letters* **16**(9): 1007.

Bittner, H. R. (1990). "Modelling of fractal vessel systems." 1st IFIP Conference on Fractals, Lisbon, Elsevier.

Bittner, H. R. (1991). "Limited Selfsimilarity." *Fractal Geometry and Computer Graphics*. Amsterdam, Elsevier.

Bittner, H. R. and M. Sernetz, Eds. (1990). "Selfsimilarity within limits: description with the log-logistic function." *Fractal 90*. Amsterdam, Elsevier.

Blumen, A. and G. H. Köhler (1990). "Reactions in and on fractal media." *Fractals in the Natural Sciences*. Princeton, NJ, Princeton University Press. 189–200.

Bouchard, E., G. Lapasset, et al. (1990). "Fractal dimension of fractured surfaces: A universal value?" *Europhysics Letters* **13**(1): 73.

Bouligand, G. (1929). “Sur la notion d’ordre de mesure d’un ensemble fermé.” *Bull. Sci. Math.* **2**: 185–192.

Broomhead, D. S. and R. Jones (1990). “Time-series analysis.” *Fractals in the Natural Sciences*. Princeton, NJ, Princeton University Press. 103–122.

Brown, C. A. and G. Savary (1988). “Fractal Aspects of Machined Surface Topography Determined by Stylus Profilometry.” *Fractal Aspects of Materials: Disordered Systems*. Pittsburgh, PA, Materials Research Society. 275–277.

Brown, H. R. (1990). “Relation between fracture and chain breaking in glassy polymers.” *Scaling in Disordered Materials: Fractal Structure and Dynamics*. Pittsburgh, PA, Materials Research Society. 37–38.

Brunauer, S., P. H. Emmett, et al. (1938). “Adsorption of gases in multimolecular layers.” *J. Am. Chem. Soc.* **60**: 309–319.

Bruno, B. C., G. J. Taylor, et al. (1992). “Lava Flows are Fractals.” *Geophysical Research Letters* **19**(3): 305.

Bueller, J. (1992). Institute of Archaeology, The Hebrew University of Jerusalem, Jerusalem, Israel, Private communication.

Bunde, A. and S. Havlin (1991). “Percolation.” *Fractals and Disordered Systems*. Berlin, Springer Verlag. 51–149.

Burrough, P. A. (1981). “Fractal dimensions of landscapes and other environmental data.” *Nature* **294**(19): 240-242.

Burrough, P. A. (1989). “Fractals and Geochemistry.” *The Fractal Approach to Heterogeneous Chemistry*. New York, John Wiley & Sons. 383–406.

Bush, A. W., R. D. Gibson, et al. (1978). “Strongly anisotropic rough surfaces.” *Journal of Lubrication Technology* **101**: 15–20.

Cahn, R.W. (1989). “Fractal dimension and fracture.” *Nature* **338**: 201.

Caldwell, C. B., S. J. Stapleton, et al. (1990). “Characterization of mammographic parenchymal pattern by fractal dimension.” *Phys. Med. Biol.* **35**: 235–247.

Campbell, D. K., Ed. (1990). *XAOC: Soviet American Perspectives on Nonlinear Science*. New York, American Institute of Physics.

Carr, J. R. and W. B. Benzer (1991). “On the practice of estimating fractal dimension.” *Mathematical Geology* **23**(7): 945.

Cates, M. E. and T. A. Witten (1986). “Family of Exponents for Laplace’s Equation near a Polymer.” *Phys. Rev. Letters* **56**(23): 2497–2500.

Chapman, R. (1992). “The Delineation of *p–n* functions by STM: Experimentation, Modeling and Computer Simulation.” Thesis, N. C. State Univ.

Charalampopoulos, T. T. and H. Chang (1991). “Agglomerate parameters and fractal dimension of soot using light scattering—effects on surface growth.” *Combustion and Flame* **87**(1): 89.

Chen, C.-C., J. S. Daponte, et al. (1989). “Fractal feature analysis and classification in medical imaging.” *IEEE Trans. Med. Imaging* **MI-8**: 133–142.

Chen, D., Z. Wang, et al. (1989). “The fractal analysis of fatigue fractures in dual-phase steels.” *Mater. Sci. Prog.* **3**(2): 115–120.

Chermant, J. L. and M. Coster (1978). “Fractal Object in Image Analysis.” Int’l Symp. Quant. Metall., Florence, Assoc. Italiana di Mettalurgica.

Chermant, J. L. and M. Coster (1983). “Recent developments in quantitative fractography.” *International Metals Review*. London, 28.

Chermant, L., J. L. Chermant, et al. (1987). “Fractal methods in profilometric analysis: Application to rupture of cold-worked brass.” *Acta Stereol.* **6**(III): 845–850.

Chesters, S., H.-C. Wang, et al. (1991). “A Fractal-based Method for Describing Surface Texture.” *Solid State Technology* **34**: 73–77.

Chesters, S., N. Y. Wen, et al. (1989). “Fractal-based characterization of Surface Texture.” *Appl. Surf. Sci* **40**: 185–192.

Chuang, K., D. J. Valentino, et al. (1991). “Measurement of fractal dimension using 3-D technique.” *Medical Imaging* **3**: 341.

Church, E. L. (1980). “Interpretation of high-resolution X-ray scattering measurements.” *Proc. SPIE* **257**: 254.

Church, E. L. (1982). "Direct comparison of mechanical and optical measurements of the finish of precision-machined surfaces." *Proc. SPIE* **429**: 105–112.

Church, E. L. (1983). "The Precision Measurement and Characterization of Surface Finish." *Proc. SPIE* **429**: 86–95.

Church, E. L. (1986). "Comments on the Correlation Length." *Proc. SPIE* **680**: 102–111.

Church, E. L. (1988). "Fractal surface finish." *Applied Optics* **27**(8): 1518–1526.

Church, E. L. and H. C. Berry (1982). "Spectral Analysis of the Finish of Polished Optical Surfaces." *Wear* **83**: 189–201.

Church, E. L., M. R. Howells, et al. (1982). "Spectral Analysis of the Finish of Diamond-Turned Mirror Surfaces." *Proc. SPIE* **315**: 202–218.

Church, E. L., H. A. Jenkinson, et al. (1979). "Relationship between surface scattering and microtopographic features." *Optical Engineering* **18**: 125.

Church, E. L., T. V. Vorburger, et al. (1985). "Direct comparison of mechanical and optical measurements of the finish of precision machined optical surfaces." *Optical Engineering* **24**: 388.

Clarke, K. C. (1986). "Computation of the fractal dimension of topographic surfaces using the triangular prism surface area method." *Computers and Geosciences* **122**(5): 713–722.

Clarke, K. C. and D. M. Schweizer (1991). "Measuring the fractal dimension of natural surfaces using a robust fractal estimator." *Cartography and Geographic Information Systems* **18**(1): 37.

Cook, R. F. (1988). "Effective-Medium Theory for Fracture of Fractal Porous Media." *Extended Abstracts: Fractal Aspects of Materials: Disordered Systems*. Pittsburgh, Materials Research Society. 47–49.

Corcuff, P. (1983). "Image analysis applied to the knowledge of human skin properties." *Acta Stereologica* **2** (Suppl.1): 85–88.

Corcuff, P., F. Chatenay, et al. (1984). "A fully automated system to study skin surface patterns." *International Journal of Cosmetic Science* **6**: 167–176.

Corcuff, P., J. de Rigal, et al. (1983). "Skin relief and aging." *J. Soc. Cosmet. Chem.* **34**(July): 177–190.

Corderman, R. R. and K. Sieradzki (1986). Fractal aspects of the selective dissolution of binary metal alloys. *Fractal Aspects of Materials*. Pittsburgh, Materials Research Society. 55–56.

Coster, M. (1978). Fracture, objet fractal et morphologie mathematique. *Sonderbände der praktischen Metallographie*. Stuttgart, Dr. Riederer Verlag. 61–73.

Courtens, E. and R. Vacher (1990). "Experiments on the structure and vibrations of fractal solids." *Fractals in the Natural Sciences*. Princeton, NJ, Princeton University Press. 55–69.

Daccord, G. (1989). "Dissolutions, Evaporations, Etchings." *The Fractal Approach to Heterogeneous Chemistry*. New York, John Wiley & Sons. 183–198.

Danielsson, P. E. (1980). "Euclidean Distance Mapping." *Computer Graphics and Image Processing* **14**: 227–248.

Daoud, M. and J. E. Martin (1989). "Fractal Properties of Polymers." *The Fractal Approach to Heterogeneous Chemistry*. New York, John Wiley & Sons. 109–130.

Dauskart, R. H., F. Haubensak, et al. (1990). "On the interpretation of the fractal character of fracture profiles." *Acta Metall. Mater.* **38**: 143–159.

Davidson, D. L. (1989). "Fractal surface roughness as a gauge of fracture toughness: Aluminium-particulate SiC composites." *J. Mater. Sci.* **24**: 681–687.

Dellepiane, S., S. B. Serpico, et al. (1987). "Fractal-based image analysis in radiological applications." *Proc. SPIE* **845**: 396–403.

Denley, D. R. (1990a). "Scanning tunneling microscopy of rough surfaces." *J. Vac. Sci. Techn.* **A8**(1): 603–607.

Denley, D. R. (1990b). "Practical applications of scanning tunneling microscopy." *Ultramicroscopy* **33**: 83–92.

Deutscher, G., Y. Lereah, et al. (1987). "Phase separation in metal-insulator mixture thin films." *Fractal Aspects of Materials II*. Pittsburgh, Materials Research Society. 2–4.

Devaney, R. L. (1989). "Fractal Patterns arising in chaotic dynamical systems." *The Science of Fractal Images*. New York, Springer Verlag. 137–218.

Devaney, R. L. and L. Keen, Eds. (1989). "Chaos and Fractals." Proceedings of Symposia in Applied Mathematics. Providence, American Mathematical Society.

Devreux, F., J. P. Boilot, et al. (1990). "NMR Determination of the fractal dimension in silica aerogels." *Phys. Rev. Letters* **65**(5): 614.

Diab, H. and N. Abboud (1991). "The engineering applications of fractals." *Simulation* **57**(2): 81.

Draper, N. R. and H. Smith (1981). *Applied Regression Analysis*. New York, John Wiley & Sons.

Dubuc, B. (1989). "On Takagi fractal surfaces." *Canadian Mathematical Bulletin* **32**(3): 377.

Dubuc, B., J. F. Quiniou, et al. (1989). "Evaluating the fractal dimension of profiles." *Physical Review A* **39**(3): 1500-1512.

Dubuc, B., C. Roques-Carmes, et al. (1987). "The variation method: A technique to estimate the fractal dimension of surfaces." *Fractal Aspects of Materials: Disordered Systems*. Pittsburgh, PA, Materials Research Society. 83–85.

Dubuc, B., S. W. Zucker, et al. (1989). "Evaluating the fractal dimension of surfaces." *Proc. Royal Soc. (London)* **A425**: 113.

Duval, E., A. Boukenter, et al. (1990). "Structure of Aerogels and Raman Scattering from Fractal Materials." *Scaling in Disordered Materials: Fractal Structure and Dynamics*. Pittsburgh, PA, Materials Research Society. 227–235.

Eby, R. K. (1993). Topometrix Corp., Bedminster, NJ. Private communication.

Ehrlich, R. and B. Weinberg (1970). "An exact method for characterization of grain shape." *J. Sediment. Petrol.* **40**: 205–212.

Elber, R. (1989). "Fractal Analysis of Proteins." *The Fractal Approach to Heterogeneous Chemistry*. New York, John Wiley & Sons. 407–424.

Evesque, P. (1989). "Energy Migration." *The Fractal Approach to Heterogeneous Chemistry*. New York, John Wiley & Sons. 81–104.

Fahmy, Y., J. C. Russ, et al. (1991). "Application of fractal geometry measurements to the evaluation of fracture toughness of brittle intermetallics." *J. Mater. Res.* **6**(9): 1856–1861.

Falconer, K. (1990). *Fractal Geometry: Mathematical Foundations and Applications*. New York, John Wiley.

Falconer, K. J. (1992). "The dimension of self-affine fractals II." *Mathematical Proceedings of the Cambridge Philosophical Society* **111**: 1.

Farin, D. and D. Avnir (1989a). "The Fractal Nature of Molecule–Surface Interactions and Reactions." *The Fractal Approach to Heterogeneous Chemistry*. New York, John Wiley & Sons. 271–294.

Farin, D. and D. Avnir (1989b). "Crystallite size effects in chemisorption on dispersed metals." *Journal of Catalysis* **120**: 55–67.

Farin, D., S. Peleg, et al. (1985). "Applications and Limitations of Boundary-Line Fractal Analysis of Irregular Surfaces: Proteins, Aggregates, and Porous Materials." *Langmuir* **1**(4): 399–407.

Feder, J. (1988). *Fractals*. New York, Plenum Press.

Files-Sesler, L. A., J. N. Randall, et al. (1991). "Scanning Tunneling Microscopy: Critical Dimension and Surface Roughness Analysis." *J. Vac. Sci. Technol. B* **9**(2): 659–662.

Flook, A. (1978). "Use of dilation logic on the Quantimet to achieve fractal dimension characterization of texture and structured profiles." *Powder Techn.* **21**: 295–298.

Flook, A. G. (1982). "Fourier Analysis of Particle Shape." *Particle Size Analysis* (N. G. Stanley-Wood, Ed.). London, Wiley Heyden.

Foley, J. D. and A. Van Dam (1984). *Fundamentals of Interactive Computer Graphics*. Reading, MA, Addison Wesley.

Fortin, C., R. Kumaresan, et al. (1992). "Fractal dimension in the analysis of medical images." *IEEE Engineering in Medicine and Biology* (June): 65–71.

Freniere, E., E. L. O'Neill, et al. (1979). "Problem in the determination of correlation function II." *J. Opt. Soc. Am.* **69**: 634.

Friel, J. J. and C. S. Pande (1993). "A Direct Determination of Fractal Dimension of Fracture Surfaces Using Scanning Electron Microscopy and Stereoscopy." *J. Mater. Res.* **8**(1):

Fripiat, J. J. (1989). "Porosity and Adsorption Isotherms." *The Fractal Approach to Heterogeneous Chemistry*. New York, John Wiley & Sons. 331–340.

Frisch, A. A., D. A. Evans, et al. (1987). "Shape discrimination of sand samples using the fractal dimension. " *Mathematical Geology* **19**: 131.

Gagnepain, J. J. and C. Roques-Carmes (1986). "Fractal approach to two-dimensional and three-dimensional surface roughness." *Wear* **109**: 119–126.

Gobel, I. R. (1991). "Strain and temperature dependence of the fractal dimension of slip lines in copper and aluminum." *Zeitschrift für Metallkunde* **82**(11): 858.

Goldberger, A. L. (1990). "Fractal electrodynamics of the heartbeat." *Ann. N. Y. Acad. Sci.* **591**: 402–409.

Goldberger, A. L., D. R. Rigney, et al. (1990). "Chaos and Fractals in Human Physiology." *Scientific American* **262**(2): 42–49.

Goldberger, A. L. and B. J. West (1987). "Fractals in Physiology and Medicine." *Yale Journal of Biology and Medicine* **60**: 421–435.

Gomez-Rodriguez, J. M., A. Asenjo, et al. (1992). "Measuring the fractal dimension with STM: Application to vacuum-evaporated gold." *Ultramicroscopy* **42–44**: 1321–1328.

Gomez-Rodriguez, J. M., A. M. Baro, et al. (1990). "Fractal Characterization of Gold Deposits by STM." *Journal of Vacuum Science and Technology B* **B9**(2): 495–499.

Gomez-Rodriguez, J. M., A. M. Baro, et al. (1992). "Fractal Surfaces of Gold and Platinum Electrodeposits. Dimensionality determination by STM." *Journal of Physical Chemistry* **96**: 347–350.

Goodchild, M. F. (1980). "Fractals and the accuracy of geographical measures." *Mathematical Geology* **12**(2): 85–98.

Gouyet, J.-F., M. Rosso, et al. (1991). "Fractal Surfaces and Interfaces." *Fractals and Disordered Systems*. Berlin, Springer Verlag. 229–262.

Grigg, D. A., P. E. Russell, et al. (1992). "Probe characterizataion for scanning probe metrology." *Ultramicroscopy* **42–44**: 1616–1620.

Guinea, F. and E. Louis (1991). "Fractures, Fractals and Foreign Physics." *Physics Today* (April): 13.

Guinea, F., O. Pla, et al. (1986). "Fractal Aspects of Fracture Propagation." Fragmentation, Form and Flow in Fractured Media. *Annals of Israel Physical Society* **8**: 587–594.

Guinea, F., O. Pla, et al. (1987). "Crack patterns in brittle materials, anisotropy and macroscopic properties." *Fractal Aspects of Materials II*. Pittsburgh, Materials Research Society. 8–10.

Hamblin, M.G. and G.W. Stachowiak (1993). "Comparison of boundary fractal dimensions from projected and sectioned particle images: I Technique evaluation"and "II Dimension changes." *Journal of Computer Assisted Microscopy* **5**(4): *(in press)*.

Haralick, R. M., K. Shanmugam, et al. (1973). "Textural features for image classification." *IEEE Trans SMC-3*: 610–621.

Harvey, D. P. I. and M. I. Jolles (1990). "Relationships between fractographic features and material toughness." *Quantitative Methods in Fractography*. Philadelphia, ASTM. 26–38.

Havlin, S. (1989). "Molecular Diffusion and Reactions." *The Fractal Approach to Heterogeneous Chemistry*. New York, John Wiley & Sons. 251–270.

Herrmann, H. J. (1988). "Fractal's physical origin and properties." *Fractals in Physics*. New York, Plenum Press.

Herrmann, H. J. (1991). "Fractures." *Fractals and Disordered Systems*. Berlin, Springer Verlag. 175–228.

Herrmann, H. J. and S. Roux, Ed. (1990). *Statistical Models for the Fracture in Disordered Media*. Amsterdam, North Holland.

Herrmann, R. A. (1989). "Fractals and ultrasmooth microeffects." *Journal of Mathematical Physics* **30**(4): 805.

Hibbert, D. B. and J. R. Melrose (1990). "Electrodeposition in support: Concentration gradients, and ohmic model and the genesis of branching fractals." *Fractals in the Natural Sciences*. Princeton, NJ, Princeton University Press. 149–158.

Hogrefe, H. and C. Kunz (1987). "Soft X-ray scattering from rough surfaces: Experimental and theoretical analysis." *Appl. Opt.* **26**: 2851.

Hornbogen, E. (1989). "Fractals in microstructure of metals." *International Materials Reviews* **34**(6): 277.

Huang, D. (1989). "Fractal Effects and Percolation of Material Fracture." *Fractal Aspects of Materials—1989*. Pittsburgh, PA, Materials Research Society. 35–36.

Huang, Z. H., J. F. Tian, et al. (1990). "A Study of the Slit Island Analysis as a Method for Measuring Fractal Dimension of a Fractured Surface." *Scripta Metallurgica et Materialia* **24**(6): 967.

Hurd, A. J., D. A. Weitz, et al., Ed. (1987). *Fractal Aspects of Materials: Disordered Systems*. Pittsburgh, Materials Research Society.

Hurst, H. E., R. P. Black, et al. (1965). *Long Term Storage: An Experimental Study*. London, Constable.

Imre, A. (1992) "Problems of measuring the fractal dimension by the slit island method." *Scripta Metallurgica et Materiala* **27**: 1713–1716.

Imre, A., T. Pajkossy, L. Nyikos (1992). "Electrochemical determination of the fractal dimension of fractured surfaces." *Acta Metall. Mater.* **40**(8): 1819–1826.

Ishikawa, K. (1990). "Fractals in dimple patterns of ductile fracture." *Journal of Materials Science Letters* **9**(4): 400.

Isichenko, M. B. and J. Kalda (1991). "Statistical topograpy. I. Fractal dimension of coastlines and number-area rule for islands." *Journal of Nonlinear Science* **1**(3): 255.

Jacquet, G., W. J. Ohley, et al. (1990). "Measurement of Bone Structure by use of Fractal Dimension." *IEEE Conference on Engineering in Medicine and Biology* **12**(3):

Jakeman, E. (1982). "Fresnel scattering by a corrugated random surface with fractal slope." *J. Opt. Soc. Am.* **72**(8): 1034–1041.

Jaroniec, M., K. Lu, et al. (1991). "Correlation between the fractal dimension and the microporous structure of a solid." *Monatshefte für Chemie* **122**(8/9): 577.

Jullien, R. and R. Botet (1989). "Geometrical optics in fractals." *Physica D, Nonlinear Phenomena* **38**: 208.

Jürgens, H., H.-O. Peitgen, et al. (1990). "The Language of Fractals." *Scientific American* (August): 60-67.

Kagan, Y. Y. (1991). "Fractal dimension of brittle fracture." *Journal of Nonlinear Science* **1**(1): 1.

Kapitulnik, A. and G. Deutscher (1982). "Percolation Characteristics in Discontinuous Thin Films of Pb." *Phys. Rev. Letters* **49**(19): 1444–1448.

Kardar, M., G. Parisi, et al. (1986). "Dynamic Scaling of Growing Interfaces." *Phys. Rev. Letters* **56**: 889–892.

Kaye, B. H. (1978a). "Sequential mosaic amalgamation as a strategy for evaluating fractal dimensions of a fineparticle profile." Institute for Fineparticle Research, Laurentian Univ.

Kaye, B. H. (1978b). "Specification of the ruggedness and/or texture of a fine particle profile by its fractal dimension." *Powder Techn.* **21**: 1–16.

Kaye, B. H. (1983). "Fractal descriptors of powder metallurgy systems." *Metals Handbook*. ASM International.

Kaye, B. H. (1984). "Multifractal Description of a Rugged Fineparticle Profile." *Particle Characterization* **1**: 14–21.

Kaye, B. H. (1985a). "Harmonious Rocks, Infinite Coastlines and Fineparticle Science." *Delightful Instruments and Moments in Applied Science*. Personal communication.

Kaye, B. H. (1985b). "Fractal dimension and signature waveform characterization of fineparticle shape." Personal communication.

Kaye, B. H. (1986). "Fractal Geometry and the Characterization of Rock Fragments." Fracture, Fragmentation and Flow, Conference, Israel.

Kaye, B. H. (1987a). "Morphological clues due to the formation dynamics of diesel soot and other fumed fineparticles." Powder Technology Conference, Rosemont, IL, Cahners Exposition Group.

Kaye, B. H. (1987b). "Fineparticle Characterization Aspects of Predictions Affecting the Efficiency of Microbiological Mining Techniques." *Powder Technology* **50**: 177–191.

Kaye, B. H. (1988). "Describing the structure of fineparticle populations using the Korcak fractal dimension." Size Characterization, Conference, Guildford, England,

Kaye, B. H. (1989a). *A Random Walk Through Fractal Dimensions*. Weinheim, VCH Verlagsgesellschaft.

Kaye, B. H. (1989b). "Image Analysis Techniques for Characterizing Fractal Structures." *The Fractal Approach to Heterogeneous Chemistry*. New York, John Wiley & Sons. 55–66.

Kaye, B. H., A. E. Beswick, et al. (1986). "Fractal description of the structure of concrete and other composite materials." Powder Technology Conference, Rosemont, IL, Cahners Exposition Group.

Kaye, B. H. and G. G. Clark (1986). "Fractal Description of Extra Terrestrial Fineparticles." *Particle Characterization* **2**: 143–148.

Kaye, B. H. and G. G. Clark (1987). "The fractal structure of filters and randomwalk modeling of fineparticle penetration of respirators." Powder Technology Conference, Rosemont, IL, Cahners Exposition Group.

Kaye, B. H., G. G. Clark, et al. (1986). "Image Analysis Procedures for Characterizing the Fractal Dimension of Fineparticles." ParTec 1986, Nürnberg.

Kaye, B. H., J. E. Leblanc, et al. (1983). "A study of the physical significance of three dimensional signature waveforms." Fineparticle Characterization Conference, Hawaii.

Khadivi, M. R. (1990). "Iterated Function System in Generating Fractal Fractures Models." *Scaling in Disordered Materials: Fractal Structure and Dynamics*. Pittsburgh, PA, Materials Research Society. 49–51.

Kirk, T. B. and G. W. Stachowiak (1990). "Development of fractal morphological descriptors for a computer image analysis system." International Tribology Conference, Brisbane.

Kirk, T. B. and G. W. Stachowiak (1991a). "Fractal computer image analysis applied to wear particles from arthritic and asymptomatic human synovial fluid: A preliminary report." *Journal of Orthopaedic Rheumatology* **4**: 13–30.

Kirk, T. B. and G. W. Stachowiak (1991b). "The applications of fractals to the morphological description of particles generated in tribological systems." *Transactions of the Institutions of Engineers, Au* **16**(4): 279.

Kirk, T. B. and G. W. Stachowiak (1991c). "Fractal Characterization of Wear Particles from Synovial Joints." *Journal of Computer Assisted Microscopy* **3**: 157–170.

Kirk, T. B., G. W. Stachowiak, et al. (1991). "Fractal parameters and computer image analysis applied to wear particles isolated by ferrography." *Wear* **145**: 347–365.

Kjems, J. K. (1991). "Fractals and Experiments." *Fractals and Disordered Systems*. Berlin, Springer Verlag. 263–296.

Klasa, S. L. (1991). "Fourier Fractals and Shapes." *Congressus Numerantium* **83**(3).

Klinkenberg, B. and K. C. Clarke (1990). "Exploring the Fractal Mountains." *Automated Pattern Recognition in Geophysical Exploration*.

Koch, H. v. (1904). "Sur une courbe continue sans tangente, obtenue par une construction geometrique elementaire." *Arkiv fur Matematik, Astronomie och Fysik* **1**: 681.

Kopelman, R. (1989). "Diffusion-controlled Reaction Kinetics." *The Fractal Approach to Heterogeneous Chemistry*. New York, John Wiley & Sons. 295–310.

Kubik, K. (1986). "Fractal behavior in ore deposits." *Fractal Aspects of Materials*. Pittsburgh, Materials Research Society. 57–59.

Kuklinksi, W. S., K. Chandra, et al. (1989). "Application of fractal texture analysis to segmentation of dental radiographs." *Proc. SPIE* **1092**: 111–117.

Kumar, S. and G. S. Bodvarsson (1990). "Fractal Study and Simulation of Fracture Roughness." *Geophysical Research Letters* **17**(6): 701–704.

LaBrecque, M. (1992). "To Model the Otherwise Unmodelable." *Mosaic* **23**(2): 13–23.

Laibowitz, R. B. (1987). "Electrical Conductivity in 2-D Percolating Thin Films." *Fractal Aspects of Materials: Disordered Systems*. Pittsburgh, PA, Materials Research Society. 17.

Lam, C.-H. and L. M. Sander (1992). "Fractals in surface growth with power-law noise." *J. Phys. A* **25**(3): 135.

Lam, N. S.-N. (1990). "Description and measurement of Landsat TM images using fractals." *Photogrammetric Engineering and Remote Sensing* **56**(2): 187.

Lanzerotti, M. Y. D., J. J. Pinto, et al. (1989). "Characterization of the Mechanical Failure Surfaces of Energetic Materials by Power Spectral Techniques." *Fractal Aspects of Materials–1989*. Pittsburgh, PA, Materials Research Society. 225–227.

Lanzerotti, M. Y. D., J. J. Pinto, et al. (1990). "Power Spectral Characterization of Fracture Surfaces of Energetic Materials." *Scaling in Disordered Materials: Fractal Structure and Dynamics*. Pittsburgh, PA, Materials Research Society. 133–136.

Larking, L. I. and P. J. Burt (1983). "Multi-Resolution Texture Energy Measures." Vision Patt. Recog, Washington, DC, IEEE Comput. Soc.

Le Méhauté, A. (1989). Fractal Electrodes. *The Fractal Approach to Heterogeneous Chemistry*. New York, John Wiley & Sons. 311–328.

Le Méhauté, A. (1990). *Les Geometries Fractales: L'espace-temps brisé*. Paris, Editions Hermes.

Le Méhauté, A. (1991). *Fractal Geometries: Theory and Applications*. Boca Raton, FL, CRC Press.

Lee, J. H. and J. C. Russ (1989). "Metrology of Microelectronic Devices by Stereo SEM." *Journal of Computer Assisted Microscopy* **1**: 79–90.

Lee, Y.-H., J. R. Carr, et al. (1990). "The fractal dimension as a measure of the roughness of rock discontinuity profiles." *International Journal of Rock Mechanics and Mining Sciences* **27**(6): 453.

Li, J. C. M. (1988). "A Theoretical Limit of Fracture Toughness." *Scripta Metallurgica* **22**: 837–838.

Lin, M. Y., H. M. Lindsay, et al. (1990). "Universality of fractal aggregates as probed by light scattering." *Fractals in the Natural Sciences*. Princeton, NJ, Princeton University Press. 71–88.

Ling, F. F. (1989). "The possible role of fractal geometry in tribology." *Tribology Transactions* **32**(4): 497–505.

Ling, F. F. (1990). "Fractals, engineering surfaces and tribology." *Wear* **136**(1): 141.

Linnett, L. M., S. J. Clarke, et al. (1991). "Remote sensing of the sea-bed using fractal techniques." *Electronics & Communication Engineering Journal* **3**(5): 195.

Liu, J., W.-H. Shih, et al. (1990). "Nonlinear Viscoelasticity and Restructuring in Colloidal Silica Gels." *Scaling in Disordered Materials: Fractal Structure and Dynamics*. Pittsburgh, PA, Materials Research Society. 43–46.

Long, Q. Y., L. Suqin, et al. (1991). "Studies on the fractal dimension of a fracture surface formed by slow stable crack propagation." *Journal of Physics d: Applied Physics* **24**(4): 602.

Louis, E., F. Guinea, et al. (1985). "The Fractal Nature of Fracture." *Fractals in Physics*. Amsterdam, North Holland. 177–180.

Louis, E., F. Guinea, et al. (1986). "The Fractal Nature of Fracture." *Fractal Aspects of Materials*. Pittsburgh, Materials Research Society. 43–45.

Lovejoy, S. (1982). "Area-perimeter relation for rain and cloud areas." *Science* **216**: 185–187.

Ludlow, D. K. and T. P. Hoberg (1990). "Technique for determination of surface fractal dimension using a dynamic flow adsorption instrument." *Analytical Instrumentation* **19**(2): 113.

Lundahl, T., W. J. Ohley, et al. (1986). "Fractional Brownian Motion: A Maximum Likelihood Estimator and Its Application to Image Textures." *IEEE Trans. Medical Imaging* **MI-5**(3): 152–161.

Lundahl, T., W. J. Ohley, et al. (1985). "Analysis and interpolation of angiographic images by use of fractals." *Comput. in Cardiol.* **24**: 355–358.

Lung, C. W. (1985). "Fractals and the Nature of Cracked Metals." *Fractals in Physics*. Amsterdam, North Holland. 189–192.

Lung, C. W. and Z. Q. Mu (1988). "Fractal dimension measured with perimeter-area relation and toughness of materials." *Phys. Rev. B* **38**: 11781–11784.

Luo, R. and H. H. Loh (1991). "Knowledge-Based Natural Scene Description." *IEEE IECON* **91**: 1555–1560.

Luo, R., H. Potlapali, et al. (1992). "Natural Scene Segmentations Using Fractal Based Autocorrelation." *IEEE IECON* **92**: 700–705.

Lynch, J. A., D. J. Hawkes, et al. (1991). "Analysis of texture in macroradiographs of ostroarthritic knees using the fractal signature." *Phys. Med. Biol.* **36**(6): 709–722.

Mackin, T. J., J. J. Mecholsky, et al. (1990). "Scaling Concepts Applied to Brittle Fracture." *Scaling in Disordered Materials: Fractal Structure and Dynamics*. Pittsburgh, PA, Materials Research Society. 33–36.

Mackin, T. J., D. E. Passoja, et al. (1987). "Fractal Analysis of Bond-Breaking Processes in Brittle Materials." *Extended Abstracts: Fractal Aspects of Materials: Disordered Systems*. Pittsburgh, Materials Research Society.

Maddox, J. (1986). "Gentle warning on fractal fashions." *Nature* **322**: 303.

Majumdar, A. and B. Bhushan (1990). "Role of Fractal Geometry in Roughness Characterization and Contact Mechanics of Surfaces." *Journal of Tribology* **112**(2): 205.

Majumdar, A. and B. Bhushan (1991). "Fractal Model of Elastic–Plastic Contact Between Rough Surfaces." *ASME Journal of Tribology* **113**(1): 1.

Majumdar, A. and C. L. Tien (1990). "Fractal characterization and simulation of rough surfaces." *Wear* **136**: 313–327.

Majumdar, A. and C. L. Tien (1991). "Fractal network model for contact conductance." *Journal of Heat Transfer* **113**(3): 516.

Malinverno, A. (1990). "A Simple Method to Estimate the Fractal Dimension of a Self-Affine Series." *Geophysical Research Letters* **17**(11): 1953.

Malozemoff, A. P. (1987). "Fractal structures in spin glasses." *Fractal Aspects of Materials II*. Pittsburgh, Materials Research Society. 86–88.

Mandelbrot, B. B. (1975). "Stochastic models for the earth's relief, the shape and the fractal dimension of coastlines, and the number-area rule for islands." National Academy of Science, USA.

Mandelbrot, B. B. (1982). *The Fractal Geometry of Nature*. New York, Freeman.

Mandelbrot, B. B. (1984). "Fractals in physics: Squig cluster, diffusions, fractal measures and the unicity of fractal dimensionality." *J. Stat. Phys.* **36**: 895.

Mandelbrot, B. B. (1985a). "Self-Affine Fractal Sets." *Fractals in Physics*. Amsterdam, North Holland. 3–28.

Mandelbrot, B. B. (1985b). "Self-Affine Fractals and Fractal Dimension." *Physica Scripta* **32**: 257–260.

Mandelbrot, B. B. (1986). "Self-affine fractals and fractal dimension." *Fractal Aspects of Materials*. Pittsburgh, Materials Research Society. 61–63.

Mandelbrot, B. B. (1989). "The Principles of Multifractal Measures." *The Fractal Approach to Heterogeneous Chemistry*. New York, John Wiley & Sons. 45–52.

Mandelbrot, B. B. (1990). "Fractals—A Geometry of Nature." *New Scientist* **127**(1734): 38.

Mandelbrot, B. B. and C. J. G. Evertsz (1991). "Exactly Self-similar Left-sided Multifractals." *Fractals and Disordered Systems*. Berlin, Springer Verlag. 323–344.

Mandelbrot, B. B. and J. A. Given (1984). "Physical properties of a new fractal model of percolation clusters." *Phys. Rev. Lett.* **52**: 1853.

Mandelbrot, B. B., D. E. Passoja, et al. (1984). "Fractal character of fracture surfaces of metals." *Nature* **308**: 721–722.

Mandelbrot, B. B. and J. W. Van Ness (1968). "Fractional Brownian motions, fractional noises, and applications." *SIAM Review* **10**: 422–437.

Mark, D. M. and P. B. Aronson (1984). "Scale-dependent fractal dimensions of topographic surfaces: An empirical investigation with applications in geomorphology and computer mapping." *Mathematical Geology* **16**(7): 671–683.

Markel, V. A., L. S. Murativ, et al. (1990). "Optical properties of fractals: Theory and numerical simulation." *Soviet Physics: JETP* **71**(3): 455.

Matshushita, M. (1985). "Fractal Viewpoint of Fracture and Accretion." *Phys. Soc. Japan* **54**: 857–860.

Matsushita, M. (1989). "Experimental Observations of Aggregations." *The Fractal Approach to Heterogeneous Chemistry*. New York, John Wiley & Sons. 161–180.

McMahon, T. A. and J. T. Bonner (1983). *On Size and Life*. New York NY, W. H. Freeman.

Meakin, P. (1983). "Diffusion-controlled cluster formation in two, three and four dimensions." *Phys. Rev.* **A27**: 604.

Meakin, P. (1984a). "Effects of cluster trajectories on cluster–cluster aggregation." *Phys. Rev* **A29**: 997.

Meakin, P. (1984b). "Diffusion-limited aggregation on two-dimensional percolation clusters." *Phys. Rev.* **B29**: 4327.

Meakin, P. (1986). "Fractal scaling in thin film condensation and material surfaces." *CRC Critical Reviews in Solid State and Material Sciences* **13**(2): 143–189.

Meakin, P. (1989a). "The Growth of Self-Affine Fractal Surfaces." *Fractal Aspects of Materials–1989*. Pittsburgh, PA, Materials Research Society. 107–110.

Meakin, P. (1989b). "Simulations of Aggregation Processes." *The Fractal Approach to Heterogeneous Chemistry*. New York, John Wiley & Sons. 131–160.

Meakin, P., P. Ramanlal, et al. (1987). "Ballistic deposition on surfaces." *Fractal Aspects of Materials II*. Pittsburgh, Materials Research Society. 55–58.

Meakin, P. and S. Tolman (1990). "Diffusion-limited aggregation." *Fractals in the Natural Sciences*. Princeton, NJ, Princeton University Press. 133–146.

Mecholsky, J. J. (1992). "Application of Fractal Geometry to Fracture in Brittle Materials." Scanning '92, Atlantic City, NJ, FAMS Inc.

Mecholsky, J. J. and S. W. Freiman (1991). "Relationship between Fractal Geometry and Fractography." *J. Am. Ceramic Soc.* **84**(12): 3136–3138.

Mecholsky, J. J. and T. J. Mackin (1988). "Fractal Analysis of Fracture in Ocala Chert." *J. Mater. Sci.* **7**: 1145–1147.

Mecholsky, J. J., T. J. Mackin, et al. (1987). "Crack propagation in brittle materials as a fractal process." *Fractal Aspects of Materials II*. Pittsburgh, Materials Research Society. 5–7.

Mecholsky, J. J., T. J. Mackin, et al. (1988). "Self-Similar Crack Propagation in Brittle Materials." *Fractography of Glasses and Ceramics*. American Ceramic Society. 127–134.

Mecholsky, J. J. and D. E. Passoja (1986). "Fractals and brittle fracture." *Fractal Aspects of Materials*. Pittsburgh, Materials Research Society. 117–119.

Mecholsky, J. J., D. E. Passoja, et al. (1989). "Quantitative analysis of brittle fracture surfaces using fractal geometry." *J. Am. Ceram. Soc.* **72**: 60–65.

Messier, R. and J. E. Yehoda (1985). "Geometry of thin-film morphology." *J. Appl. Phys.* **58**(10): 3739–3746.

Messier, R. and J. E. Yehoda (1986). "Morphology of ballistically aggregated surface deposits." *Fractal Aspects of Materials*. Pittsburgh, Materials Research Society. 123–125.

Miller, S. and R. Reifenberger (1992). "Improved method for fractal analysis using scanning probe microscopy." *Journal of Vacuum Science and Technology (in press)*.

Minkowski, H. (1901). "Über die Begriffe Länge, Oberfläche und Volumen." *Jahresbericht der Deutschen Mathematikervereinigung* **9**: 115–121.

Mitchell, M. W. and D. A. Bonnell (1990). "Quantitative topographic analysis of fractal surface by scanning tunneling microscopy." *J. Mater. Res.* **5**(10): 2244–2254.

Montagna, M., O. Pilla, et al. (1990). "Numerical study of Raman scattering from fractals." *Phys. Rev. Letters* **65**(9): 1136.

Mu, Z. Q. C.W. Lung, Y. Kang, and Q.Y. Long (1994). "A perimeter-maximum diameter method for measuring fractal dimension of fractured surface." *J. Computer Assisted Microscopy (in press).*

Mu, Z. Q. and C. W. Lung (1988). "Studies on the fractal dimension and fracture toughness of steel." *J. Phys. D* **21**: 848–851.

Nelson, J. A., R. J. Crookes, et al. (1990). "On obtaining the fractal dimension of a 3D cluster from its projection on a plane—application to smoke agglomerates." *J. Phys. d: Applied Physics* **23**(4): 465.

Noques, J., J. L. Costa, et al. (1992). "Fractal dimension of thin film surfaces of gold sputter deposited on mica: A scanning tunneling microscopic study." *Physica A* **182**(4): 532.

Normant, F. and C. Tricot (1991). "Method for evaluating the fractal dimension of curves using convex hulls." *Physical Review a[15] Statistical Physics* **43**(12): 6518.

O'Neill, E. L. and A. Walther (1977). "Problem in the determination of correlation functions." *J. Opt. Soc. Am.* **67**: 1125.

Obert, M., P. Pfeifer, et al. (1990). "Microbial Growth Patterns Described by Fractal Geometry." *Journal of Bacteriology* **1990**(March): 1180-1185.

Oliver, D. (1992). *Fractal Vision*. Carmel, IN, SAMS Publishing.

Orbach, R. (1986). "Dynamics of Fractal Networks." *Science* **231**(21 February): 814–819.

Osborne, A. R. and A. Provenzale (1989). "Finite Correlation Dimension for Stochastic Systems with Power-Law Spectra." *Physica D* **35**: 357–381.

Panagiotopoulos, P. D. (1992). "Fractals and fractal approximation in structural mechanics." *Meccanica* **27**(1): 25.

Pancorbo, M., M. Aguilar, et al. (1991). "New filtering techniques to restore scanning tunneling microscopy images." *Surface Science* **251/252**: 418–423.

Pancorbo, M., E. Anguiano, et al. (1993). "Fractal characterization of surfaces II: Effect of noise." *Journal of Microscopy (in press)*.

Pande, C. S., L. E. Richards, et al. (1987a). "Fractal characterization of fractured surfaces." *Acta Metall.* **35**: 1633–1637.

Pande, C. S., L. E. Richards, et al. (1987b). "Fractal characteristics of fractured surfaces." *Journal of Materials Science Letters* **6**: 295–297.

Passoja, D. E. (1988). "Fundamental Relationships between Energy and Geometry in Fracture." *Fractography of Glasses and Ceramics*. American Ceramic Society. 101–126.

Passoja, D. E. and J. A. Psioda, Ed. (1981). "Fourier Transform Techniques—Fracture and Fatigue." *Fractography and Materials Science*. Philadelphia, American Society for Testing and Materials.

Peitgen, H.-O. and D. Saupe, Ed. (1988). *The Science of Fractal Images*. New York, Springer Verlag.

Peitgen, H. O., H. Jürgens, et al. (1992). *Fractals for the Classroom Part One: Introduction to Fractals and Chaos*. New York, Springer Verlag.

Pekala, R. W., L. W. Hrubesh, et al. (1990). "A Comparison of Mechanical Properties and Scaling Law Relationships for Silica Aerogels and their Organic Counterparts." *Scaling in Disordered Materials: Fractal Structure and Dynamics*. Pittsburgh, PA, Materials Research Society. 39–42.

Peleg, S., J. Naor, et al. (1984). "Multiple Resolution Texture Analysis and Classification." *IEEE Trans. Patt. Anal. Mach. Intell.* **PAMI-6**(4): 518–523.

Pentland, A. P. (1984). "Fractal-Based Description of Natural Scenes." *IEEE Trans. Patt. Anal. Mach. Intell.* **PAMI-6**(6): 661–674.

Pfeifer, P. (1984). "Fractal Dimension as Working Tool for Surface-Roughness Problems." *Applications of Surface Science* **18**: 146–164.

Pfeifer, P. and D. Avnir (1983a). "Chemistry in noninteger dimension between two and three. Part 1: Fractal theory of heterogeneous surfaces." *J. Chem. Phys.* **79**: 3558.

Pfeifer, P. and D. Avnir (1983b). "Chemistry in Non-integer Dimensions between 2 and 3: Part 2: Fractal surface of adsorbents." *J. Chem. Phys.* **79**: 3566.

Pfeifer, P. and D. Avnir (1983c). "Fractal Dimension in Chemistry: An Intensive Characteristic of Surface Irregularity." *Nouveau Journal de Chemie* **7**(2): 71.

Pfeifer, P., D. Avnir, et al. (1983). "New Developments in the Application of Fractal Theory to Surface Geometric Irregularity." *Symposium on Surface Science 1983*. Vienna, Technical University of Vienna.

Pfeifer, P., D. Avnir, et al. (1984a). "Dynamical aspects of fractal surfaces at the molecular range." Symp. J, Boston, Materials Research Society.

Pfeifer, P., D. Avnir, et al. (1984b). "Scaling behavior of surface irregularity in the molecular domain: From adsorption studies to fractal catalysts." *J. Stat. Phys.* **36**: 699.

Pfeifer, P. and M. W. Cole (1990). "Fractals in surface science: Scattering and thermodynamics of adsorbed films II." *New Journal of Chemistry* **14**(3): 221.

Pfeifer, P. and M. Obert (1989). "Fractals: Basic Concepts and Terminology." *The Fractal Approach to Heterogeneous Chemistry*. New York, John Wiley & Sons. 11–44.

Pfeifer, P., M. Obert, et al. (1990). "Fractal BET and FHH theories of adsorption: A comparative study." *Fractals in the Natural Sciences*. Princeton, NJ, Princeton University Press. 169–188.

Pickover, C. A. (1990). *Computers Pattern Chaos and Beauty*. New York, St. Martin's Press.

Piscitelle, L. and R. Segars (1992). "Effect of particle size distribution in determining a powder's fractal dimension by single gas BET: A mathematical model." *Journal of Colloid and Interface Science* **149**(1): 226.

Pouligny, B., G. Gabriel, et al. (1991). "Optical Wavelet Transform and Local Scaling Properties of Fractals." *J. Applied Crystallography* **24**(5): 526.

Preuss, S. (1990). "Some remarks on the numerical estimation of fractal dimension." *Bulletin of the Geological Society of America*.

Prusinkiewicz, P. and A. Lindenmayer (1990). *The Algorithmic Beauty of Plants*. New York, Springer Verlag.

Przerada, I. and A. Bochenek (1990). "Microfractographical aspects of fracture toughness in microalloyed steel." *Stereology in Materials Science*, Kraków, Poland, Polish Society for Stereology.

Rammal, R. (1984). "Random walk statistics on fractal structures." *J. Stat. Phys.* **36**: 547.

Rarity, J. G., R. N. Seabrook, et al. (1990). "Light scattering studies of aggregation." *Fractals in the Natural Sciences*. Princeton, NJ, Princeton University Press. 89–100.

Ray, K. K. and G. Mandal (1992). "Study of correlation between fractal dimension and impact energy in a high strength low alloy steel." *Acta Metallurgica et Materialia* **40**(3): 463.

Ray, K. K., G. Mandal, et al. (1990). "Quantitative fractographic investigation of subzero impact tested fracture surfaces of a microalloyed steel." *HSLA Steels '90*. Beijing, Chinese Soc. Metals. 152–154.

Recht, J. M. (1993). Fractal patterns in PAP smears. Private communication.

Rehr, J. J., Y. Gefen, et al. (1987). "Friction between self-affine surfaces." *Fractal Aspects of Materials II*. Pittsburgh, Materials Research Society. 53–55.

Reiss, G., F. Schneider, et al. (1990). "Scanning tunneling microscopy on rough surfaces: Deconvolution of constant current images." *Appl. Phys. Lett.* **57**(9): 867–869.

Reiss, G., J. Vancea, et al. (1990). "Scanning tunneling microscopy on rough surfaces: Tip-shape-limited resolution." *J. Appl. Phys.* **67**(3): 1156–1159.

Richards, L. E. and B. D. Dempsey (1988). "Fractal characteristization of fractured surfaces in Ti-4.5Al-5.0Mo-1.5Cr." *Scripta Metall.* **22**: 687–689.

Richardson, L. F. (1961). "The problem of contiguity: An appendix of statistics of deadly quarrels." *General Systems Yearbook* **6**: 139–187.

Rigaut, J. P. (1990). "Fractal models in biological image analysis and vision." *Acta Stereologica* **9**(1): 37–52.

Rigaut, J. P. (1991). "Fractals, semifractals and biometry." *Fractals: Nonintegral Dimensions and Applications*. New York, John Wiley. 151–187.

Robinson, G. M., D. M. Perry, et al. (1991). "Optical Interferometry of Surfaces." *Scientific American* **265**(1): 66–71.

Romeu, D., A. Gomez, et al. (1986). "Surface fractal dimension of small metallic particles." *Phys. Rev. Lett.* **57**: 2552–2555.

Roques-Carmes, C., D. Wehbi, et al. (1987). "On the use of the Weierstrass–Mandelbrot function in the modelization of rough surfaces." *Fractal Aspects of Materials II*. Pittsburgh, Materials Research Society. 112–114.

Roques-Carmes, C., D. Wehbi, et al. (1988). "Modelling Engineering Surfaces and Evaluating their Non-integer Dimension for Application in Material Science." *Surface Topography* **1**: 237–245.

Rosenfeld, A. R. (1987). "Fractal Mechanics." *Scripta Metall.* **21**: 1359–1361.

Rovner, I. (1993). Anthropology Dept., North Carolina State University, Raleigh, NC. Private communication.

Roy, A., G. Gravel, et al. (1987). "Measuring the dimension of surfaces: a review and appraisal of different methods." Auto-Carto 8 (International Symposium on Computer-Assisted Cartography), 68–77.

Russ, J. C. (1986). *Practical Stereology*. New York, Plenum Press.

Russ, J. C. (1989). "A Simplified Approach to Harmonic Shape Analysis." *Journal of Computer Assisted Microscopy* **1**: 377–396.

Russ, J. C. (1990a). *Computer Assisted Microscopy*. New York, Plenum Press.

Russ, J. C. (1990b). "Surface Characterization: Fractal Dimensions, Hurst Coefficients and Frequency Transforms." *Journal of Computer Assisted Microscopy* **2**(3): 161–183.

Russ, J. C. (1990c). "Processing images with a local Hurst operator to reveal textural differences." *Journal of Computer Assisted Microscopy* **2**(4): 249–257.

Russ, J. C. (1990d). "Computer-Assisted Image Analysis in Quantitative Fractography." *Journal of Materials* (October): 16–19.

Russ, J. C. (1991a). "Measurement of the Fractal Dimension of Surfaces." *Journal of Computer Assisted Microscopy* **3**(3): 127–144.

Russ, J. C. (1991b). "Multifractals and Mixed Fractals." *Journal of Computer Assisted Microscopy* **3**(4): 211–231.

Russ, J. C. (1992a). *Image Processing Handbook*. Boca Raton, CRC Press.

Russ, J. C. (1992b). "Characterizing and Modeling Fractal Surfaces." *Journal of Computer Assisted Microscopy* **4**(1): 73–126.

Russ, J. C. (1993a). "Light scattering from fractal surfaces." *Journal of Computer Assisted Microscopy* **5**(2): 171–190.

Russ, J. C. (1993b). "Effects of Noise and Anisotropy on STM Determination of Fractal Dimensions." *Journal of Microscopy* **172** (*in press*).

Russ, J. C. and J. C. Russ (1987). "Feature-specific measurement of surface roughness in SEM images." *Particle Characterization* **4**: 22–25.

Russ, J. C. and J. C. Russ (1989). "Uses of the Euclidean distance map for the measurement of features in images." *Journal of Computer Assisted Microscopy* **1**(4): 343–376.

Salli, A. (1991). "On the Minkowski dimension of strongly porous fractal sets in Rn." *Proc. London Mathematical Society* **62**(2): 353.

Sander, L. M. (1986a). "Fractal Growth Processes." *Nature* **322**(August): 789–793.

Sander, L. M. (1986b). "Theory of ballistic aggregation." *Fractal Aspects of Materials*. Pittsburgh, Materials Research Society. 121–122.

Sapoval, B. (1991). "Fractal electrodes, fractal membranes, and fractal catalysts." *Fractals and Disordered Systems*. Berlin, Springer Verlag. 207–227.

Sapoval, B., M. Rosso, et al. (1989). "Fractal Interfaces in Diffusion, Invasion and Corrosion." *The Fractal Approach to Heterogeneous Chemistry*. New York, John Wiley & Sons. 227–246.

Sasajima, K. and T. Tsukada (1992). "Measurement of fractal dimension from surface asperity profile." *International Journal of Machine Tools and Manufac.* **32**(1/2): 125.

Saupe, D. (1989). "Algorithms for Random Fractals." *The Science of Fractal Images*. New York, Springer Verlag. 71–136.

Sayles, R. S. and T. R. Thomas (1977). "The spatial representation of surface roughness by means of the structure function." *Wear* **42**: 263.

Sayles, R. S. and T. R. Thomas (1978). "Surface topography as a nonstationary random process." *Nature* **271**: 431–434.

Schaefer, D. W., B. C. Bunker, et al. (1990). "Fractals and Phase Separation." *Fractals in the Natural Sciences*. Princeton, NJ, Princeton University Press. 35–53.

Schepers, H. E., J. H. G. M. van Beek, et al. (1992). "Four methods to estimate the fractal dimension from self-affine signals." *IEEE Engineering in Medicine and Biology* (June): 57–71.

Schmidt, P. W. (1982). "Interpretation of small angle scattering curves proportional to a negative power of the scattering vector." *J. Appl. Cryst.* **15**: 567.

Schmidt, P. W. (1989). "Use of Scattering to Determine the Fractal Dimension." *The Fractal Approach to Heterogeneous Chemistry*. New York, John Wiley & Sons. 67–80.

Schmidt, P. W., D. Avnir, et al. (1989). "Small-Angle X-ray Scattering Studies of Bone Porosity." *Fractal Aspects of Materials–1989*. Pittsburgh, PA, Materials Research Society. 155–158.

Schroeder, M. (1991). *Fractals, Chaos, Power Laws*. New York, W. H. Freeman.

Schwarz, H. and H. E. Exner (1980). "The implementation of the concept of fractal dimension on a semi-automatic image analyser." *Powder Technology* **27**: 207–213.

Scott, P. J. (1989). "Nonlinear dynamic systems in surface metrology." *Surface Topography* **2**: 345–366.

Scott, P. J. (1990). Some Recent Developments in the Analysis and Interpretation of Surface Topography. Private communication.

Scott, P. J. (1991). Fractal analysis of periodic profiles. Private communication.

Sernetz, M., H. R. Bittner, et al. (1991). "Fractal characterization of the porosity of organic tissue by interferometry." *Characterization of Porous Solids*. Amsterdam, Elsevier. 141–150.

Sernetz, M., H. R. Bittner, et al. (1989). "Chromatography." *The Fractal Approach to Heterogeneous Chemistry*. New York, John Wiley & Sons. 361–380.

Sernetz, M., B. Gelléri, et al. (1985). "The organism as bioreactor. Interpretation of the reduction law of metabolism in terms of heterogeneous catalysis and fractal structure." *J. Theoretical Biology* **117**: 209–230.

Sernetz, M., H. Willems, et al. (1989). "Fractal Organization of Metabolism." *Energy Transformations in Cells and Organisms*. Stuttgart, Georg Thieme Verlag. 82–90.

Sernetz, M., H. Willems, et al. (1990). "Dispersive Analysis of Turnover Rates of a CST Reactor by Flow-Through Microfluorometry under Conditions of Growth." *Enzyme Engineering* **10**: 333–337.

Sevick, E. M. and R. C. Ball (1990). "Dilute heteroaggregation: A description of critical gelation using a cluster–cluster aggregation model." *Scaling in Disordered Materials: Fractal Structure and Dynamics*. Pittsburgh, PA, Materials Research Society. 3–6.

Sinha, S. K. and R. Ball (1988). "Scattering from Hybrid Fractals." *Fractal Aspects of Materials: Disordered Systems*. Pittsburgh, PA, Materials Research Society. 287–289.

Solla, S. A. (1985). "Collapse of Loaded Fractal Trees." *Fractals in Physics*. Amsterdam, North Holland. 185–188.

Sreenivasan, K. R. (1991). "Fractals and Multifractals in Fluid Turbulence." *Ann. Rev. Fluid Mechanics* **23**: 539–600.

Srinivasan, S., J. C. Russ, et al. (1990). "Fractal analysis of erosion surfaces." *J. Mater. Res.* **5**(11): 2616–2619.

Stanley, H. E. (1987). "Multifractals." *Time-Dependent Effects in Disordered Materials*. New York, Plenum Press. 145.

Stanley, H. E. (1991). "Fractals and Multifractals: The Interplay of Physics and Geometry." *Fractals and Disordered Systems*. Berlin, Springer Verlag. 1–50.

Stanley, H. E. and N. Ostrowsky, Ed. (1986). *On Growth and Form*. Boston, Martinus Nijhoff.

Stauffer, D. (1985). *Introduction to Percolation Theory*. London, Taylor and Francis.

Stauffer, D. and A. Aharony (1991). *Introduction to Percolation Theory (Second Edition)*. New York, Taylor and Francis.

Stupak, P. R. and J. A. Donovan (1990). "Fractal Characteristics of Worn Surfaces." *Scaling in Disordered Materials: Fractal Structure and Dynamics*. Pittsburgh, PA, Materials Research Society. 23–26.

Stupak, P. R., J. H. Kang, et al. (1991). "Computer-aided fractal analysis of rubber wear surfaces." *Materials Characterization* **27**(4): 231–240.

Su, H., Y. Zhang, et al. (1991). "Fractal analysis of microstructures and properties of ferrite-martensite steels." *Scripta Met.* **25**(3): 651–654.

Sun, X. and D. L. Jaggard (1989). "Electromagnetic and Optical Wave Scattering from Fractal Objects." *Fractal Aspects of Materials–1989*. Pittsburgh, PA, Materials Research Society. 21–24.

Takacs, P. Z. and E. L. Church (1986). "Comparison of profiler measurements using different magnification objectives." *Proc. SPIE* **680**: 132.

Takagi, T. (1903). "A simple example of the continuous function without derivative." *Proc. Phys. Math. Soc. Jap.*

Takayasu, H. (1985). "Pattern Formation of Dendritic Fractals in Fracture and Electric Breakdown." *Fractals in Physics*. Amsterdam, North Holland. 181–184.

Takayasu, H. (1990). *Fractals in the Physical Sciences*. Manchester, Manchester Univ. Press.

Tang, C., P. Bak, et al. (1990). "Self-Organized Critical Phenomena." *Scaling in Disordered Materials: Fractal Structure and Dynamics*. Pittsburgh, PA, Materials Research Society. 55–58.

Taylor, C. C. and S. J. Taylor (1991). "Estimating the dimension of a fractal." *Journal of the Royal Statistical Society* **53**(2): 353.

Termonia, Y. and P. Meakin (1986). "The formation of fractal cracks in a kinetic fracture model." *Nature* **320**: 6061.

Thomas, A. P. and T. R. Thomas (1986). "Engineering surfaces as fractals." *Fractal Aspects of Materials*. Pittsburgh, Materials Research Society. 75–77.

Thomas, T. R. and A. P. Thomas (1988). "Fractals and Engineering Surface Roughness." *Surface Topography* **1**: 1–10.

Thompson, K. A. (1987). "Surface characterization by use of automated stereo analysis and fractals." *Microbeam Analysis*. San Francisco Press. 115–116.

Thong, J. T. L. and B. C. Breton (1992). "In Situ Topography Measurement in the SEM." *Scanning* **14**: 65–72.

Tricot, C. (1989a). "Local convex hulls of a curve, and the value of its fractal dimension." *Real Analysis Exchange* **15**(2): 675.

Tricot, C. (1989b). "Fractal Measures of Porous Media." *Fractal Aspects of Materials—1989*. Pittsburgh, PA, Materials Research Society. 19–20.

Tricot, C. (1989c). "Porous Surfaces." *Constructive Approximation* **5**(1): 117.

Tricot, C., J. P. Champigny, et al. (1987). "Practical evaluation of the fractal dimension for 2-dimensional profiles." *Fractal Aspects of Materials II*. Pittsburgh, Materials Research Society. 115–117.

Tricot, C., J. F. Quiniou, et al. (1988). "Evaluation de la dimension fractale d'un graphe ("Evaluating the fractal dimension of a graph")." *Revue Phys. Appl.* **23**: 111–124.

Trottier, R. and A. Beswick (1987). "Fractal dimension of fracture surfaces and rock fragments." Powder Technology Conference, Rosemont, IL, Cahners Exposition Group.

Tsai, Y. L. and J. J. Mecholsky (1991). "Fractal Fracture in Single Crystal Silicon." *J. Mater. Res.* **6**(6): 1248–1263.

Tyner, J. (1993). Precision Engineering Center, North Carolina State University, Raleigh, NC. Private communication.

Underwood, E. E. and K. Banerji (1986). "Fractals in fractography." *Materials Science & Engineering* **80**: 1–14.

Uozumi, J., H. Kimura, et al. (1991). "Fraunhofer diffraction by Koch fractals: The dimensionality." *Journal of Modern Optics* **38**(7): 1335–1347.

Van Damme, H. (1989). "Flow and Interfacial Instabilities in Newtonian and Colloidal Fluids (or The Birth, Life and Death of a Fractal)." *The Fractal Approach to Heterogeneous Chemistry*. New York, John Wiley & Sons. 199–226.

Varnier, F., A. Llebaria, et al. (1991). "How is the fractal dimension of a thin film top surface connected with the roughness parameters and anisotropy of this surfaces?" *J. Vacuum Science and Technology* **9**(3 part 1): 563.

Vicsek, T. (1992). *Fractal Growth Phenomena (2nd Edition)*. Singapore, World Scientific.

Voss, R. F. (1984). "Multiparticle fractal aggregation." *J. Stat. Phys.* **36**: 861.

Voss, R. F. (1985a). "Random fractal forgeries." *Fundamental Algorithms for Computer Graphics*. Berlin, Springer-Verlag. 805–835.

Voss, R. F. (1985b). "Random Fractals: Characterization and measurement." *Scaling Phenomena in Disordered Systems*. New York, Plenum Press. 1–11.

Voss, R. F. (1988). "Fractals in Nature: From Characterization to Simulation." *The Science of Fractal Images*. New York, Springer-Verlag. 21–70.

Voss, R. F. (1989). "Random fractals: Self-affinity in noise, music, mountains and clouds." *Physica D, Nonlinear Phenomena* **38**: 362.

Voss, R. F., R. B. Laibowitz, et al. (1982). "Fractal (scaling) clusters in thin gold films near the percolation threshold." *Physical Review Letters* **49**(19): 1441–1444.

Wang, G. (1993). State Univ. of New York, Buffalo, NY. Personal communication.

Wang, X. W., L. K. Dong, et al. (1990). "The change of fractal dimensionality in the recovery and recrystallization process." *J. Phys. Condens. Matter.* **2**: 3879–3884.

Wang, Z. G., D. L. Chen, et al. (1988). "Relationship between Fractal Dimension and Fatigue Threshold Value in Dual-Phase Steels." *Scripta Metallurgica* **22**: 827–832.

Wehbi, D., C. Roques-Carmes, et al. (1987). "Interfacial Energy Transfers in Tribology." *Fractal Aspects of Materials: Disordered Systems*. Pittsburgh, PA, Materials Research Society. 185–187.

Wehbi, D., C. Roques-Carmes, et al. (1992). "The perturbation dimension for describing rough surfaces." *International Journal of Machine Tools and Manufacturing* **32**(1/2): 211.

Weismann, H. J. and L. Pietronero (1986). "Properties of Laplacian fractals for dielectric breakdown in 2 and 3 dimensions." *Fractal Aspects of Materials*. Pittsburgh, Materials Research Society. 7–9.

Werman, M. and S. Peleg (1984). "Multiresolution texture signatures using min-max operators." *IEEE Trans. Patt. Anal. Mach. Intell.* **PAMI-6**: 97–99.

West, B. J. and A. L. Goldberger (1987). "Physiology in Fractal Dimensions." *American Scientist* **75**(4): 354–365.

West, B. J. and M. Shlesinger (1990). "The Noise in Natural Phenomena." *American Scientist* **78**(Jan-Feb): 40–45.

West, P. E., S. Marchese-Rugona, et al. (1992). "Fractal Analysis with Scanning Probe Microscopy." Electron Microscopy Society of America, Boston, San Francisco Press.

Wickramasinghe, H. K. (1989). "Scanned-Probe Microscopes." *Scientific American* **261**(4): 98–105.

Wight, J. F. and J. W. Laughner (1989). "Weibull Plotting of a Scaling Flaw Size Distribution." *Fractal Aspects of Materials–1989*. Pittsburgh, PA, Materials Research Society. 41–43.

Wilcoxon, J. P., J. E. Martin, et al. (1986). "Light scattering from gold colloids and aggregates." *Fractal Aspects of Materials*. Pittsburgh, Materials Research Society. 33–35.

Williford, R. E. (1988). "Scaling similarities between fracture surfaces, energies and a structure parameter." *Scripta Metallurgica* **22**: 197–200.

Wlczek, P., H. R. Bittner, et al. (1989). "3-Dimensional Image Analysis and Synthesis of Natural Fractals." *Acta Stereologica* **8**(2): 315–324.

Wlczek, P., A. Odgaard, et al. (1991). "Fractal 3D Analysis of Blood Vessels and Bones." *Fractal Geometry and Computer Graphics*. Springer.

Wlczek, P. and M. Sernetz (1991). "3 Dimensional Image Analysis and Synthesis of Natural Fractals." 8th International Congress for Stereology, Irvine, CA,

Xiao, H., A. Chu, et al. (1992). "Biomedical image texture analysis based on higher order fractals." *Biomedical Image Processing and Three-Dimensional Microscopy.* San Jose, CA, SPIE.

Yehoda, J. E. and R. Messier (1985). "Are thin film physical structures fractals?" *Appl Surf. Sci* **22/23**: 590–595.

Yuhong, B., V. K. Berry, et al. (1989). "The Fractal Nature of Crazing Paths in Front of a Crack in S-ABS." *Fractal Aspects of Materials—1989.* Pittsburgh, PA, Materials Research Society. 283–286.

Zhou, X. Y., D. L. Chen, et al. (1989). "Fractal characteristics of pitting under cyclic loading." *Materials Letters* **7**: 473–476.

Appendix

About the Accompanying Disks and Programs

The "Fractals" program provides a comprehensive set of tools for the synthesis and analysis of fractal surface images. It is expected that the user will be able to select and interpret the various models and methods available. Only minimal description of the various methods is given here, but each of the procedures for generating and measuring fractals is described in detail in the body of the book. The program deals primarily with 2D arrays (surfaces) or 1D arrays (profiles) of elevation data. These may be read from disk, generated internally, processed and measured, or saved to disk. The tools for cluster and network models are less completely developed, and are included as supplemental to the main purpose of dealing with surfaces.

The Mac program opens a window in which three buttons appear (Clear, Draw, Done). Done is equivalent to Quit on the menu (or press command-Q, abbreviated as ⌘Q), closing the window and ending the program. Clear is equivalent to Clear on the menu (⌘W), erasing any current data (surface or profile) and erasing the window. Draw redraws the current display (updates the window); this may be useful after some methods which superimpose several drawings.

The Mac program was written in a shell originally created by Christian Russ, Analytical Vision, Inc., 213 Merwin Road, Raleigh, NC 27606. He also contributed important utilities and routines within the program, particularly the file and array handling. The PC (Windows) program is a faithful translation of the original Mac program to the Windows environment. Porting the Pascal program to Windows 3.1 was performed by M. S. Esterman (email: ccocsme@prism.gatech.edu), 430 Tenth Street NW, Suite S-109, Atlanta, GA 30309. The principal differences between the PC and Mac programs are:

1) The PC program is somewhat slower than the Mac, primarily owing to differences in the speed of drawing things on the screen, reading and writing files, and perhaps a less efficient compiler. The times shown below for most of the slower operations (e.g., creation and measurement of 2D "surface" arrays") are for a Mac running a 25 MHz 68040 chip. They are shown as *(minutes:seconds)* and should be considered as approximate indications only, as exact times will depend to some extent on the actual values in the array and on the settings used for the various algorithms and options (the times shown are generally for the default settings).

2) In the PC program, the Done button is replaced by Stop. This can be used to halt a lengthy procedure, just as the ⌘-period command is normally used in the Macintosh.

3) The storage format for data on the PC is tab-delimited ASCII text. This allows the data files to be accessed by other programs including spreadsheets and word processors (and vice versa), but it does make the files larger and their access slower than in the Mac version, which has a default format of packed reals and an alternative format of PICTs for 2D arrays.

4) The Mac program is self contained and can be run from the floppy or dragged to your hard disk and run there. The rules of grammar for the L-systems are contained in a resource of type CFGR. You may modify or add these resources with a utility such as Apple's ResEdit. The PC program keeps the grammar rules in a separate file, and allows you to edit them from within the program. Before running the PC program, install it on your hard disk as follows:

```
C> mkdir fractals {create a directory to hold the files}
C> copy a:\exe\fractals.exe c:\fractals {copy the object code to the disk}
C> copy a:\exe\rules.gmr c:\fractals {copy the grammar rules to the disk}
C> copy a:\exe\BWCC.DLL c:\windows\system {copy required library files}
C> win {run windows; you may then run the program or use Setup Applications to
      place its icon in your program manager}
```

Both the PC and Mac disks contain, in addition to the ready-to-run compiled program, the complete source code (in Pascal). This is stored as text and can be read with most compilers, editors and word processors. It is included to assist the interested reader in understanding the methods used. The code has been written more for clarity than for speed. Readers are encouraged to use and adapt portions of the code as needed for their own research. However, the author retains copyright and before using any of the programs or code for commercial purposes, you must contact him.

The programs have been run extensively, and in fact were used to create many of the figures in this book. They are believed to be "bug free" (but see the notes at the end of this appendix for some "peculiarities"). However, they are provided for tutorial purposes and with absolutely no warranties of any kind.

DrJohnRuss @AOL.com

The entries in the main menu are:

Clear	⌘W	Clears all current data arrays and erases the window.
Copy Graph	⌘G	Places the current graphic into the clipboard for pasting into other applications. Note - most graphics displays of log-log plots, clusters, etc., automatically place the graphic onto the clipboard and this command is not required except for surfaces and elevation profiles.
Generate Line		Generates an elevation profile - see submenu.
Generate Cluster		Generates a cluster fractal - see submenu.
Generate Network		Generates a network fractal - see submenu.
Generate Surface		Generates a surface fractal - see submenu.
Analyze Line		Measures an elevation profile - see submenu.
Analyze Surface		Measures a fractal surface - see submenu.
Save Line	⌘L	Saves the 1D array (elevation profile) to disk. Default format is a packed array of 1024 reals. Depress the Shift key to save as an ascii file (a simple column of real values). Depress the Option key to save the values as text in two columns, with x-coordinate and column headers (Cricket Graph or StatView format, or may be read by spreadsheet programs).
Save Surface	⌘S	Saves the 2D array (surface) to disk. Default format is a packed array of 256x256 reals. Depress the Shift key to save as a PICT2 (with optional autoscaling provided by a dialog box).
Page Setup		The usual printer control dialog
Print	⌘P	Prints the current screen - graph, image, etc.
Random Numbers		Allows resetting the random number generator with a fixed seed to repeat a sequence of values. The user can also select between a Gaussian random number generator with mean of zero, and one with a mean of ±0.5.
Quit	⌘Q	End the program (no warning is given for any unsaved data)

The Generate Line submenu presents the following options:

Midpoint Displacement	One of the oldest and most straightforward methods for generating a fractal profile is iterated midpoint displacement. Starting with the entire length, the midpoint is displaced up or down by a random amount. This procedure is repeated for each of the two segments to produce four, and this continues until individual pixels are reached. The magnitude of each displacement is reduced as the length of the segment is reduced, so that for a length scale $r=1/2^n$, the mean square variation is $r^{2\alpha}$, and the fractal dimension is $2-\alpha$. See the examples in Peitgen & Saupe (1988) p. 51 and p. 85
Fractal Brownian	Conventional Brownian motion produces a fractal dimension of 1.5 for a profile. Fractal Brownian motion allows for other dimensions, which requires some correlation (D>1.5) or anti-correlation (D<1.5) between values. The method employs successive random additions with a Gaussian random number generator. The fractal dimension D is $2-\alpha$. The routine is described in Peitgen & Saupe (1988) p.86.
Mandelbrot-Weierstrass	The Mandelbrot-Weierstrass function generates a line profile by adding together sine functions of geometrically increasing frequency (unlike Fourier in which frequencies increase arithmetically). The profile is the sum of terms of the form $(1/\beta^{((2-D)*n)}) \bullet \cos(2\pi \bullet \beta^n \bullet x)$ where n is the index, running from n=–15 (the lowest frequency needed for a profile length of 1024) upwards until the magnitude drops below 1.0e–5 or the frequency drops below a fraction of one pixel. β controls how quickly the frequencies increase - typical values reported in the literature are β from 1.5 to 2; smaller values take longer but superimpose more terms. The fractal dimension D is equal to $2-\alpha$. Each successive term is displaced by a random phase. See Feder, p. 27-30, or Majumdar & Bhushan.
Fourier	Generates a line profile via the Fast Fourier Transform (FFT). First creates the FFT-space transform using Gaussian random numbers for the magnitude with the specified slope for the log(mag) vs log(freq) data, and random phase, and then inverts it to produce the spatial domain result. The fractal dimension is $(2-\alpha)$.
Multifractal	Uses the midpoint displacement technique with two different α values, changing from α_1 to α_2 halfway thru (or continuously if the option key is depressed).

Read from disk ⌘R	Loads the profile from disk. If the file is of type 'FrcP' it is assumed to be in the packed format saved by the program. If the file is of type 'TEXT' it is treated as a series of simple ASCII representation of values in free numeric format which may be separated by tab, space or return. Strings of alpha information will be evaluated as zeroes, and files that are in unexpected formats may cause various kinds of errors. If the number of entries exceeds the 1024 size of the array, extra values are ignored. If the record is too short, interpolation is used to spread the data out to fill the array. Press the Shift key to add the loaded data to the current profile in memory (scaled by a user-entered factor).
Add sine function	Adds a sine function of the specified magnitude and wavelength (specified as a fraction of the total width) to the current profile.

The Generate Cluster submenu presents the following options:

2D Diff. Lim. Aggreg.	Generates a classic DLA on a square pixel grid in which touching can occur to any of eight neighbors, motion probability is uniform in all directions, and only the sticking probability (uniform in all directions) and number of particles are entered. Depressing the shift key produces a grey-scale coded image with the time-of-arrival for each particle. Depressing the option key changes the logic to four-neighbor instead of eight-neighbor connectivity.
2D DLA Weighted	Generates a weighted DLA on a square pixel grid with 4 neighbors, using the method described by Vicsek. The number of particles reaching a touching site is counted, and compared to a weight value. Only when it is exceeded is the point added to the cluster, in which case the new neighbor points have their counts cleared. Changing the weight value produces random (low M), dendritic (medium M) and needlelike (high M) clusters.
3D Diff. Lim. Aggreg.	Performs a 3D DLA simulation identical to the 2D case above, with the same probabilities for adjusting sticking probabilities and direction of motion. The program shows a projection of the 3D cluster on a 2D plane.
Line Deposit - 2D	This procedure models the DLA deposition of particles onto a line, with motion restricted to a vertical plane and the same parameters for sticking probability and isotropy of motion as the 2D cluster DLA algorithm above. Depressing the Option key changes the connectivity rules from 8- to 4-connected, and depressing the Shift key causes the motion of particles to be biased toward a downward direction (semi-ballistic) instead of uniformly random.
Surface Deposit-Section	Models the 3D deposition of particles on a surface as shown in a vertical plane through the 3D space in which motion takes place. The sticking probability and motion probability are isotropic, as in the examples above.
Surface Projection	This simulation of DLA onto a surface is similar to the one above except that it displays a projection of the deposited particles on a vertical plane instead of a section through it. It also creates a grey-scale range image for the resulting surface (which is saved directly to disk), hence the size is reduced to the 3D array.
Ballistic Line Deposition	This 2D deposition onto a line is similar to the DLA method above except that it uses ballistic (downward) motion and allows the surface to be inclined up to 45° from normal. Depressing the Option key changes the connectivity rules from 8- to 4-connected, and depressing the Shift key causes the motion of particles that have once scattered to be biased toward a downward direction (semi-ballistic) instead of uniformly random.

The Generate Network submenu presents the following options:

Chain	Simulates a path or molecule whose steps or links are all of the same length, and whose angular deviations are fixed in angle but uniformly random in azimuth. A projection of the path onto a plane is shown.
L-System	Generates an iterative Lindenmayer system using any of the rules saved as a resource in the program. Also known as a 'context-free grammar', these rules specify the probability that branching will occur in a particular order. The angle of branching and size of the structure can be specified, as can the number of iterations. Rules are contained in resources and can be entered using ResEdit and the template built into the program.
Fat Fractals	The same as the L-system except that the thickness of the lines is varied according to the distance from the root.

The Generate Surface submenu presents the following options:

Midpoint Displacement	Midpoint displacement for a surface is just like that for a profile except that the displacement of points on a finer grid occurs in two steps. First the center of each square is displaced up or down from the average of the four corners. This produces a finer grid consisting of points with spacing $1/\sqrt{2}$ of the original, which is oriented at 45° to the original. Displacing the centers of these squares produces a finer grid aligned with the original and with 1/2 the spacing. The fractal dimension is $3-\alpha$. The standard deviation can be varied to control the lacunarity, as above. See Peitgen & Saupe (1988) p. 100 *(time=0:15)*
Fractal Brownian	Fractal Brownian motion in 2D by interpolation. This routine is similar to the 1D fractal Brownian profile above. The dimension of the surface is $3-\alpha$. The method starts with a coarse array and repeatedly expands it by interpolation into a finer one, adding Gaussian noise. R is the rate at which the size is varied. See Peitgen & Saupe (1988) p.104 *(time=0:30)*
Multifractal	The midpoint displacement algorithm from above, except that the α value varies with iteration (continuously if the option key is depressed)
Extrusion	Generates two fractal line profiles by the midpoint displacement method, and then 'extrudes' them across the surface at selected angles. For each one, the α ($D=2-\alpha$) value and the angle can be entered. It is not entirely clear what the fractal dimension of the resulting surface created by their superposition is, especially when the two alphas are different. *(time=0:15)*
Mandelbrot-Weierstrass	Mandelbrot-Weierstrass functions for surface modelling are identical to those for profiles. The only difference is that each term in the series of increasing frequency sinusoids must also be randomized in orientation and phase. The resulting surface 'looks' fractal but is not a very anisotropic because so few terms have been used. In FFT space, the discrete terms appear by frequency and orientation. Smaller values of β increase the density of terms. *(time=1:40)*
Fourier Series	Generates a surface image by first constructing the FFT-space data with the proper slope of log(mag) vs. log(freq) and then inverting it (with random phase). This would produce isotropic fractal surfaces with dimension $D=3-\alpha$ except that the routine allows entering two α values corresponding to the max and min values (orthogonal to each other) for fractal dimension of an anisotropic surface. Depress the option key to generate an anisotropic surface with a single α value. *(time=1:30)*

Takagi Functions	Constructs a fractal surface by iteratively adding together tilings of pyramids. Unlike the midpoint displacement method, the 'phase' of the location of the pyramids on the image is shifted randomly at each step. Also, the pyramids do not necessarily have square bases. As the maximum height of each set of pyramids is decreased in each step (by a ratio=R), the base widths of the pyramid are decreased by $(1/r)^{1+\alpha}$ where α is a function of direction, varying elliptically between α_1 and α_2.*(time=0:20)*
Add Sine Function	Adds a sine function to the 2D elevation array, with specified magnitude, orientation, wavelength and phase. Wavelength and phase are specified as fractions of the image width.
Open from Disk ⌘O	Loads a surface array from disk. Files of type 'Frct' are loaded as a packed array of reals; depressing the Shift key adds the result to the current image (multiplied by a user-entered scale factor). If the file is of type PICT the pixel values are converted to reals (0…255). If the picture is larger than 256x256, the upper left corner of the image is used; if it is smaller, the remainder is cleared to zero.

The Analyze Line submenu presents the following options:

Richardson	⌘1	A Richardson plot shows the total perimeter length along a boundary as a function of the stride length used. The fractal dimension D is 1+abs(slope). However, this is correctly applied only to zerosets such as the intersection of a surface with a plane, and to self-similar data, not to self-affine plots such as elevation profiles. It is included here to allow comparison to the other methods.
Minkowski	⌘2	The Minkowski dimension for a boundary profile such as a zeroset formed by the intersection of a surface with a plane can be determined by using dilation and erosion to measure the mean width of the Minkowski sausage as a function of the dilation distance. However, for a self-affine plot such as an elevation profile, the 'structuring element' for the dilation must be a line segment parallel to the x- direction. This makes the Minkowski method identical to Dubuc's 'variational' or covering method for this case. The program plots the mean difference between max and min values as a function of the width of the region used to select max and min (equivalent to the radius of the disk used for constructing the 'comforter' of the surface, above) vs. the distance. The dimension D = 2–abs(slope)
Kolmogorov	⌘3	Counts the number of grid squares 'visited' by the line profile as a function of box size and plots the result on log-log axes to estimate the dimension as 1 + abs(slope). Like the Richardson technique, this is only applicable to self-similar, not self-affine profiles. It is included here for comparison to other methods.
Korcak	⌘4	Constructs a plot of log N(L>x) vs log x where N is the number of horizontal distances between zero-crossings of the line profile which exceed L. Zerosets or Poincare sections are formed at nine threshold levels from 10% to 90% of the full scale range. The fractal dimension of the profile is 1+abs(slope), and is correct for a self-affine vertical profile.
Hurst	⌘5	Hurst or rescaled range plots show the maximum difference between values along a profile (normalized by the standard deviation) as a function of the separation distance. The dimension D is related to the slope of the line K as D = 2–K
RMS vs. L	⌘7	Constructs a log-log plot of the rms deviation of points within a window of increasing width vs the width. The data in each window are detrended with a linear least squares fit line, and the window displaced by half its width across the entire data array to obtain an average rms value. The slope H = 2–D. This also demonstrates that the "rms" roughness value of a profile is an artefact of the measurement procedure.

Fourier ⌘F	Plots the power spectrum as log(magn) vs. log(freq.) of a profile from the FFT data. The fractal dimension D is related to the slope β as $D=(2-\beta)$
Histo & Stats	Displays a histogram of elevation values along the profile, plus the limits, mean value and standard deviation.

The Analyze Surface submenu presents the following options:

Minkowski Cover ⌘M	This routine measures the fractal dimension of a surface using the Minkowski cover or variational method. The volume of a 'blanket' or 'comforter' draped over the surface is computed as the difference between dilations and erosions of different sizes, equivalent to placing disks of different radii on the surface to locate min and max values. The plot shows the mean blanket thickness (volume divided by disk area) as a function of disk radius. This is analogous to the same method applied to a profile. The fractal dimension is 2+abs(slope). The interpretation of this value for anisotropic surfaces is not clear. *(time=7:00)*
Kolmogov Box ⌘9	The Kolmogorov dimension is a box-counting dimension. The number of grid boxes 'visited' by the surface is counted as a function of box size and plotted on log-log axes to estimate the Minkowski dimension as 2+abs(slope). This should be applied only to self-similar and not self affine data (single-valued elevation maps are self-affine). It is included here for comparison to other methods. *(time=0:10)*
Hurst (isotropic) ⌘8	The Hurst method plots the largest elevation difference between any two points on the surface within a distance L of each other as a function of the distance L, on log-log axes. This is a direct analog to the Hurst measurement of a line profile, applicable to self-affine but isotropic surfaces. The surface fractal dimension is 3–Slope. *(time=4:30)*
RMS vs. Area ⌘7	Constructs a log-log plot of the rms deviation of points within a square window of increasing width vs the area. The window is displaced by half its width across the entire data array to obtain an average rms value. The slope H = 2.5–D. *(time=0:30)*
Draw Contours	Draws from 1 to 9 contour lines on the image, uniformly spaced in elevation. Depressing the Shift key erases the original grey scale image to leave just the contour lines.
Slit Island ⌘I	Draws an elevation contour (at a default height midway between max and min values) as a zeroset of the surface and calculates a fractal dimension using a Kolmogorov box count method (which is correct for the boundary line). For an isometric surface this is the fastest measurement procedure, and the surface dimension is 2+abs(slope). *(time=0:05)*
Korcak ⌘K	Measures the area of islands produced by thresholding the grey scale image at 20%, 40%, 60% and 80% of the elevation range for both brighter and darker (higher and lower) regions, and plots the Log of the number of features with area greater than A vs. Log A. *(time=1:00)*

Dimensional Analysis ⌘D	Measures the area and perimeter of islands produced by thresholding the grey scale image at 20%, 40%, 60%, 80% of the elevation range for both brighter and darker (higher and lower) regions, and plots the Log (Perimeter) vs. Log (Area) for each island which does not touch the edges. Perimeter is approximated by counting the number of pixels that touch the surrounding background. The fractal dimension is equal to 1 + 2 • slope. *(time=1:30)*
Fourier Transform ⌘T	Calculates the 2D FFT of a surface image, finally leaving the Log(magnitude) image in the display plane and showing a plot of the frequency of occurrence of each phase shift value (which should be uniformly random). *(time=1:45)*
Fourier Analysis ⌘A	Measures the anisotropy of surfaces from the 2D FFT image (which must be calculated separately beforehand). The slope of the log(magnitude) vs. log(frequency) plot is determined in many directions and plotted as a rose. Numerically, the slope β is related to the surface dimension D as $D=(6-\beta)/2$. Depressing the Shift key suppresses the intercept plot. Also shows a plot of the directionally averaged magnitude vs. frequency.
Hurst Orientation ⌘U	Determines the largest difference between values on a surface as a function of their separation distance and direction. From these data, multiple Hurst plots in different directions can be used to determine the surface dimension and anistropy. *(time=58:00)*
Hurst Analysis ⌘Y	Uses the H O T transform data (calculated beforehand) to measure the slope of the Hurst plot (max difference vs. distance) as a function of direction and shows the result as a rose plot. Depressing the Shift key suppresses the intercept plot. Depressing the Option key suppresses drawing the ellipse axes.
Normalize Array ⌘N	Scales the elevation data in the surface array between user-entered limits, and optionally inverts the values.
Reorient Array	Allows selection from a submenu to reorient the surface elevation array in any of 5 modes: vertical and horizontal flip, 90 degree rotation to the left or fight, diagonal flip or transpose (which interchanges X and Y directions).
Histogram ⌘H	Displays the elevation histogram of the image.
Elevation Profile ⌘E	Extracts an elevation profile from the 2D array. Direction is controlled by the shift and option keys: nothing=|, shift=–, option=\, shift+option=/
Plot Projection	Plots the projected (max) value of the array in the vertical or (shift or option key) horizontal direction. The width of the projection (up to 100% meaning the full width of the array) can be entered.

NOTES on the use of shift/option/command keys for the Mac program

Startup	option key initializes random number generator with fixed constant instead of using internal clock for a unique value.
	shift key biases random number generator to have mean = ±0.5 instead of zero.
Draw button	shift or option key forces drawing profile instead of surface (if both are present); command key forces drawing of profile instead of graph (if both are present)
Save profile	By default, elevation profiles are saved as packed arrays of 1024xExtended (type 'FrcP'). Hold down the shift key to save the values in ASCII format, or the option key to add column headers and a second column with the x-coordinate.
Save surface	By default, surfaces are saved as packed arrays of 256x256xReal (type 'Frct'). Hold down the shift key to save as PICT2. A dialog will be presented showing the minimum and maximum elevation values and allowing the desired scaling to be entered (default is autoscaling from 0 to 255).
Open from disk (surface or line)	shift key adds loaded data to current array, multiplied by a user-entered scale factor.
Surface - Midpoint	option key adds Gaussian random values between iterations
Surface - Fourier	option key makes isotropic, bypassing anisotropic alphas
Surface - Extrusion	option key bypasses second direction, extruding profile along straight line
Multifractal (surface or line)	option key selects continuous alpha interpolation instead of discrete change.
Line: Fourier	shift key sets all phase values to zero; option key removes Gaussian random variation in magnitude.
Fractal Brownian motion, Mandelbrot, multifractal	shift key suppresses erasing window, shows each iteration
L-Systems	option key selects multiple display; shift key selects step-by-step display (shift has precedence over option) command key allows non-zero standard deviation for turn angle and step length
HOT Measure	option key suppresses axes on the graph. shift key suppresses intercept plot
FFT Measure	shift key suppresses intercept plot
Surface: Profile	shift key and option key in combination select direction
	nothing=|, shift=-, option=\, shift+option=/
Surface: Projection	shift or option key selects horizontal direction instead of vertical
2D DLA	shift key creates grey scale image instead of mass fractal plot.
	option key changes the connectivity rules to 4-neighbor instead of 8.
Line deposit (DLA and Ballistic)	shift key biases the motion direction downwards instead of being uniformly random. Option key changes the connectivity rules to 4-neighbor instead of 8.

Other general notes:

1. Many of the analysis routines perform linear least squares fits and report the slope. The ± value shown for the slope is 1 standard deviation. If a comparision of the reduced chi-squared for the linear fit results to that for a quadratic fit shows that a quadratic fit is statistically superior, the symbol § is shown after the standard deviation value as a warning. The test is based on the ratio of reduced chi-squared values (which will have an F distribution), and the critical value of F used corresponds to (p=0.25); this means that one time in four the warning will be given in error, and that the data called into question are in fact adequately described by a straight line relationship in accordance with fractal theory.
2. It is often useful to read the packed array of reals saved by the program with Spyglass Transform. The settings are "Float" (Spyglass' description of 4-byte reals), 256x256 array, and zero length header. But, when Transform reads a file it transposes X and Y and inverts the grey scale assigned to pixel values. You can check the "transpose X & Y" box and select the inverse grey scale to correct this. Transform will also read PICTs and display the numeric values, but these are then scaled to integers (0...255) and will have lost any original elevation meaning. Spyglass can save data as a PICT, but not in the packed array of reals format that Fractals uses.
3. Some program idiosyncracies are listed below:
   a. Selecting Clear from the menu is not quite the same as clicking on the button. The latter erases the screen; both clear the data.
   b. An effort is made to set flags to keep track of whether the internal data arrays contain "good" data or whether they have been corrupted by the use of the arrays to hold intermediate calculation data such as Fourier coefficients. There are still a few cases in which it is possible to ask for a calculation to be performed when the data are no longer valid because of the actions of a prior analysis operation. If in doubt, save the original data and reload it before performing a further analysis.
   c. By preference, if there is a 2D array loaded, it will be drawn instead of a 1D profile which may also be in memory. However, if you obtain the elevation profile from the 2D array then it is drawn instead. Likewise, a graph of results is drawn instead of the data once an analysis has been selected even if the original data are still intact and available. Sometimes you can depress the Option key (to draw a profile instead of the surface) or the Command key (to draw a profile instead of a graph) if both are present, but the settings of the various flags sometimes fool the program into thinking that the data present are invalid when in fact they are actually OK. The solution, as in the case above, is to reload the data when in doubt.
   d. The various graph drawing routines automatically place the graphic onto the clipboard. This is somewhat at variance with the usual Macintosh conventions, but it works and it speeds up the display of drawings. If you actually select Copy after this has been done, you may sometimes find that the clipboard appends the image to the graph and superimposes both drawings. Be sure to paste your graphics someplace such as the scrapbook or a word processor to make sure they are OK.
   e. The present version of Fractals includes diagnostic routines that increase the size of the program and sometimes slow it down (like displaying the progress of feature counting in the Korcak routines). Screen updates are particularly slow, since the image is redrawn each time from the data array (but you can press Command-Period to halt the drawing). Nothing in the program was written to be speed-optimized; instead I have opted for clarity in the code to minimize the occurrence of

substantive errors, and to keep track of things over the 2+ years during which this program has evolved. I find that it still takes much longer to acquire the data and think about the results than to perform the various analyses and simulations. Eventually, the code may be modified with speed in mind. For now, I would suggest you just a) be patient, or b) get a faster computer.

f. The compiled version of the program expects a Mac with at least a 68020 processor and a floating point math chip. This includes most of the current models. It works best with a grey scale or color display but can be used with a 1-bit (black and white) display, in which case it draws the grey scale images with the points above the midpoint elevation in white and those below it in black. The PC version of the program will run on a system with a 386 (or 486) and 387 FPU, and a VGA display.

Index

Acoustic microscopy, 143
Aerial photographs, 115, 121, 141, 142
AFM, *see* Atomic force microscopy
Agglomeration, 16, 17, 227–228, *see also* Ballistic agglomeration deposition; Diffusion-limited aggregation
  light scattering and, 138
  mixed fractals and, 198
  modeling of, 153, 154, 179, 181
Airy disks, 139
Allotropic relationships, 66
Alumina, 234
Aluminum, 244
Alzheimer's disease, 23
Anisotropy, *see also* Strong anisotropy; Weak anisotropy
  of deposited surfaces, 245
  Fourier transform analysis of, 5, 107–109
  Hurst orientation transform and, 51, 95, 96, 107–108
  Korcak islands and, 15
  light reflection/scattering and, 128, 129, 130–131, 135, 136, 138, 148
  Minkowski comforters and, 71, 73
  Minkowski sausages and, 41, 42
  mixed fractals and, 212, 217
  modeling of, 152, 164, 168–172, 178–179, 180, 183–190
  of machined surfaces, 239, 244
  rose plots of, 107–108
  surface boundary line relationship and, 64, 71, 73
  of very flat surfaces, 251, 256, 263
  of wrinkles, 250
Arthritis, 242
Atomic force microscopy (AFM), 73, 74, 141–142, 143, 144, 147, 183
  of machined surfaces, 238, 240
  of mixed fractals, 200, 214, 215, 216
  of very flat surfaces, 253, 258, 259–265

Ballistic agglomeration deposition, 16, 17, 246–247
  mixed fractals and, 198
  modeling of, 153, 158–159, 160, 161
Barnsley method, 25
Basic (language), 8
BET measurements, 25–26, 61–62, 246
Blood vessels, 19, 199, 228, 248
Boundary lines
  Fourier transform analysis of, 101–102
  fractal dimension of, *see* Fractal boundary line dimension
  fractal surface relationship with, *see* Surface boundary line relationship
  Korcak islands and, 15
  monster curves and, 3, 4
  random fractals and, 9
Box-counting methods, 174
  Euclidean distance transform compared with, 43, 45
  fractal boundary line dimension and, 40, 41, 43, 45, 48, 49
  light reflection/scattering and, 122
  mixed fractals and, 203
  surface boundary line relationship and, 70–71, 78
Brain cortex, 66
Brightness, 115–116, 117, 120, 121–122, 142, 242, 250
  pore structures and, 249
  rough surfaces and, 122–129, 139–141
  surface dimension related to, 129–138
  very flat surfaces and, 252, 254
Britain, west coast of, 27, 28, 29
Brittle fractures, 229–238, 252
Brownian motion, 52, 54, 167, 228, *see also* Fractal Brownian motion
  deposited surfaces and, 16
  Fourier transform analysis and, 99, 102, 112
  Hurst plots and, 85, 87, 88, 112
  machined surfaces and, 240, 241
  modeling of, 152
  very flat surfaces and, 258, 260
  zerosets and, 13, 14
Brownian noise, 244, 256, 263
Brown noise, 103
Butterfly, 108

C (language), 8
Calcium fluorite, 234
Cantor dust, 10, 12–13
  deposited surfaces and, 16
  Hurst plots and, 86
  mixed fractals and, 202
  surface boundary line relationship and, 64
  zerosets and, 13, 14, 64
Cantor sets, 192
Carbon steel, 231
Catalysts, 62–63, 115, 181
CCD detectors, 253
Central limit theorem, 228
Ceramics, 231, 232, 233, 234, 252
Chain codes, 69, 101–102
Chaos theory, v, 140
Charpy fractures, 230, 231
Chert tools, 244
Chi-squared values, 51
Chromatography, 249
Clouds, 30, 64, 149, 191
Clusters, 2, 51, 227–228
  deposited surfaces and, 16–17, 248
  light scattering and, 140, 141
  mixed fractals and, 191, 193, 195, 198, 201, 203, 224
  modeling of, 153–157, 158
  pore structures and, 18, 249
Coastlines, 29, 63, 101, 229, *see also* Shorelines
Collage Theorem, 24
Complexity theory, *see* Chaos theory
Computed tomography (CT), 108, 141, 230
Computer-aided design, 122
Computers, 3, 8–9, 17, 31–32
Concrete, 237
Confocal scanning light microscopy (CSLM), 143, 144, 252
Copper, 119, 241, 244, 246, 260
Correlation length, 145, 242
Corrosion surfaces, 2, 250
  brittle fractures and, 234
  Korcak islands and, 15
  mixed fractals and, 192–193
  modeling of, 153, 181
  pore structures and, 17
Corrugated surfaces, 183, 184, 242
Cratering, 243
Crushed rocks, 193
CSLM, *see* Confocal scanning light microscope
CT, *see* Computed tomography
Curves, 100
  Hilbert, 5
  monster, 1, 3–6, 41, 116
  well-behaved, 55

Dense objects, 2
Deposited surfaces, v, 18, 240, 241, 252
  described, 16–17
Deposited surfaces (*cont'd*)
  examples of, 244–248
  Korcak islands and, 15
  light reflection/scattering and, 117, 119, 144
  modeling of, 153, 157–159, 179
  very flat surfaces and, 251
Dielectric breakdown, 232–233
Diffractals, 241
Diffusion-limited aggregation (DLA) deposition, 16, 17, 245–246, 248
  mixed fractals and, 193, 194–195, 198, 201
  modeling of, 153, 154, 155, 156, 157, 158
Digitized images, 32–38
Dilations, 41–42, 44, 47, 48–49, 83
Dimensional analysis, 66–71, 78, 95, 107
  of brittle fractures, 229, 230, 231
  of deposited surfaces, 245
  of surface boundary line relationship, 66–71, 78
  of very flat surfaces, 257, 258, 260, 261
Directionality, 212–223
DLA, *see* Diffusion-limited aggregation
Dow-Jones industrial average, 83
Ductile fractures, 231–232, 233–234
Dusts, 10–13, 149, 238, *see also* Cantor dust
Dye adsorption, 224

Earth, size of, 30
Eden model growth patterns, 246
Electromagnetic radiation, 141
Electronic noise, 33
Electron microscopy, 121, 227
Electrons, 117–121
Elevation profiles, 48, 51–52, 73–75, 76, 251
  brittle fractures and, 229, 230, 231
  Euclidean distance transform and, 45
  Fourier transform analysis of, 99, 112, 229
  Hurst analysis of, 88–89
  Hurst orientation transform and, 95
  Hurst plots and, 84–85, 112
  light reflection/scattering and, 116, 141, 144, 145, 146, 147, 148
  machined surfaces and, 241
  Minkowski sausages and, 48
  mixed fractals and, 196–197, 200, 203
  modeling of, 152, 167, 183, 186
  simulated noise and, 79–81
  surface boundary line relationship and, 64
  very flat surfaces and, 252, 253
  wear surfaces and, 243
Epoxy, 229
Erosion surfaces, 2, 63, 238, 243
  Euclidean distance transform and, 42, 44
  Korcak islands and, 15–16
  Minkowski sausages and, 41–42
  surface boundary line relationship and, 71
Euclidean distance transform, 42–48, 56
Euclidean geometry, v, 2, 4, 5, 59, 61, 66
  dusts and, 10

Euclidean geometry (*cont'd*)
Fourier transform analysis and, 100, 103
fractal boundary line dimension and, 29, 30
limitations of, 3
lines defined by, 1, 3, 6
machined surfaces and, 238, 239
mixed fractals and, 191, 192, 198, 200, 225
modeling and, 153, 155–157, 163, 165–167
range images and, 144
Richardson plots and, 29, 30
Extraterrestrial particles, 193
Extrusions, 242
mixed fractals and, 202, 212–223
modeling and, 183–190

Fast Fourier transform (FFT) analysis, 77, 99, 104, 108, 114
Fat fractals, 247
Vicsek's, 197–200
Fatigue fractures, 229, 235, 260
Ferromagnetism, 234
FFT analysis, *see* Fast Fourier transform analysis
Figures, 100, 139, 242, 254
Fizeau optics, 256–258
Flint tools, 244
Florida shoreline, 63
Fortran (language), 8
Fourier expansion theorem, 167–168, 181
Fourier transform (FT) analysis, 83, 85, 86, 107–109, 110–112, 113, 250–251
of boundary lines, 101–102
of brittle fractures, 229, 234, 235, 237
described, 97–101
fast, 77, 99, 104, 108, 114
of fractal boundary line dimensions, 49, 51, 54
Hurst orientation transform compared with, 96–97
inverse, *see* Inverse Fourier transform
of light reflection/scattering, 122, 130–131, 132, 145
of machined surfaces, 241
of mixed fractals, 203, 208, 210, 211, 212, 215, 224
modeling and, 163, 173–174
noise and, 77, 78, 79, 99–100, 102–103, 108, 109
of range images, 89
surface boundary line relationship and, 76–77, 78, 79, 81
two-dimensional, *see* Two-dimensional Fourier transform analysis
of very flat surfaces, 254–256, 260, 261
of wear surfaces, 241
Fractal boundary line dimensions, 27–57
comparisons of, 55–57
computer measurement of, 31–32
digitized images of, 32–38
Euclidean distance transform and, 42–48, 56
Minkowski sausage measurement of, 41–48, 49, 51
mosaic amalgamation and, 38–41
Richardson plot measurement of, 27–32, 35–36, 37–38, 40, 41, 42, 47, 48, 51, 55

Fractal Brownian motion, 52
Hurst plots and, 87–88
light reflection/scattering and, 116
modeling of, 165–167, 173, 174, 177
Fractal dimensions, v, 4–5, 69, 70–71
of boundary lines, *see* Fractal boundary line dimensions
of brittle fractures, 230–231, 237, 238
defined, 4
of deposited surfaces, 17, 245
dusts and, 11
fractal Brownian motion and, 116
Korcak islands and, 15
light reflection/scattering and, 115, 118–120, 121, 137–138, 139, 146, 148
L-systems and, 25, 26
of machined surfaces, 238
of mixed fractals, 194, 195, 196–197, 201, 203, 204–205, 209, 212
modeling of, 154
of pore structures, 248
of random fractals, 9
two-dimensional Fourier transform analysis of, 107
of very flat surfaces, 261
of wear surfaces, 244
zerosets and, 14
Fractal Geometry of Nature, The (Mandelbrot), 1, 149
Fractal lines, 13–14
Fractal profile modeling, 149–190
described, 149–153
Mandelbrot–Weierstrass functions in, 152, 167–172, 173, 174, 175, 177
Takagi functions in, 152, 177–180
Fractal profiles, 59, 228
machined surfaces and, 239
mixed fractals and, 202–203, 206–208, 212–223
noise and, 76
random fractals and, 9
Fractal surface modeling, v, 149–190
described, 159–165
Mandelbrot–Weierstrass functions in, 152, 167–172, 173, 174, 175, 177
Takagi functions in, 152, 177–180
Fractures, v, 2, 15, 144, 227
brittle, 229–238, 252
dimensional analysis of, 68
ductile, 231–232, 233–234
fatigue, 229, 235, 260
mixed fractals and, 198, 201, 209
modeling of, 153, 181, 183
Fringes, 142, 143, 250
FT analysis, *see* Fourier transform analysis

Gas adsorption, 61–62, 74, 224
Gaussian noise, 116
Gaussian random number distribution, 112, 228
fractal Brownian profiles and, 116

Gaussian random number distribution (*cont'd*)
light reflection/scattering and, 122, 145, 146
modeling and, 150, 151–152, 162, 170, 173
random fractals and, 6, 9
wear surfaces and, 242
Geometric modeling, 122
Glass, 231, 232, 234
Gold, 247, 248, 260
Greenwood–Williamson contact model, 240
Grinding, 239
Gulf of Elath, 237

Hausdorf dimension, 4, 41, 107
fractal boundary line dimension and, 36, 40, 51, 55, 56
mixed fractals and, 203
Hawaiian islands, 28, 63
Heartbeat irregularities, 140
Henon maps, 13
Hilbert curves, 5
HOT, *see* Hurst orientation transform
Hubble space telescope, 139
Hurst analysis, 88–89, 103
of fractal boundary line dimensions, 49, 51, 54
of light reflection/scattering, 117, 121, 122
of mixed fractals, 203, 208, 210, 211, 212, 216
of range images, 89–95
Hurst orientation transform (HOT), 107–108
described, 95–97
fractal boundary line dimension measured with, 49, 51
mixed fractals and, 212, 216
modeling and, 164, 165, 169, 170, 179, 180, 181, 182, 185–190
Hurst plots, 96, 110–112
Brownian motion and, 85, 87, 88, 112
described, 83–87
fractal Brownian motion and, 87–88
light reflection/scattering and, 121
mixed fractals on, 210, 211, 212
modeling and, 173, 179, 181, 184
range images and, 89–95
surface boundary line relationship and, 77

Icebergs, 251
Intercepts, 112, 186
Interferometry, 73, 74, 76, 113, 114, 141, 143
of machined surfaces, 238, 240
of mixed fractals, 214, 215, 216
noise and, 75
range images and, 144, 147
of very flat surfaces, 251–258, 264–265
Intersections, 63, 202
Inverse Fourier transform (FT), 105
of fractal boundary line dimensions, 52
of light reflection/scattering, 122
modeling and, 152, 173–174, 175, 176, 189, 190

Islands, 64
brittle fractures and, 229, 231
dimensional analysis of, 66–69
Koch, *see* Koch islands
Korcak, 10, 15–16, 87, 89, 203
machined surfaces and, 240
mixed fractals and, 201
modeling of, 149
pore structures and, 249
Iso-elevation contours, 162
Isotropy, 89, 96
digitized images of, 37–38
Euclidean distance transform and, 42–43, 44
fractal boundary line dimension of, 37–38
Kolmogorov dimension and, 40
Korcak islands and, 15
light reflection/scattering and, 122, 135, 141
of machined surfaces, 239
Minkowski sausages and, 41–42
mixed fractals and, 195, 196, 200, 201, 214, 217
modeling of, 168–172, 178, 181, 182, 190
rose plots of, 107
simulated noise and, 77
surface boundary line relationship and, 62, 63, 64, 66, 68, 71, 79, 80
of very flat surfaces, 256, 262, 263

Jupiter, 30

Kaye, Brian (mixed fractals defined by), 193–195
Knoop hardness, 261
Koch islands, 3, 4, 6, 9, 59
fractal boundary line dimension and, 44, 56
light reflection/scattering and, 116
L-systems and, 19, 20
modeling of, 149, 150, 177, 178
wear surfaces and, 241
Kolmogorov dimension
fractal boundary line dimension and, 40–41, 51, 55, 56
Minkowski dimension and, 41, 42
mixed fractals and, 203
modeling and, 172, 174
surface boundary line relationship and, 70–71, 78–79
Korcak analysis, 15, 51, 55
brittle fractures and, 229
Fourier transform analysis compared with, 109
Hurst plots compared with, 85, 86, 87
of light reflection/scattering, 122, 128, 130, 132, 134, 135
machined surfaces and, 240
modeling and, 173, 174
range images and, 89
surface boundary line relationship and, 64, 65, 68, 69, 76, 77, 78, 79, 81
wear surfaces and, 242
zerosets and, 14

Korcak islands, 10, 15–16, 87, 89, 203
Korcak plots, 52, 64, 65, 69, 95, 132, 251
Kurtosis, 145

Lacunarity, 116, 151–152, 205
Lakes, 64, 66, 68, 229, 242
Lake Sudbury, 29
Lambertian laws, 115, 122
Landsat images, 251
Laplace equations, 232–233
Lapping, 239
Lasers, 139
Lattices, 233, 249
Lava flow, 251
Least-squares fit, 107
Light diffraction, 241–242
Light microscopy, 143, 227, 236
Light reflection, 115–148, *see also* Brightness
  fractal Brownian profiles and, 116
  range measurement methods in, 141–144
Light scattering, 115–148
  described, 138–139
  fractal Brownian profiles and, 116
  range measurement methods in, 141–144
  from rough surfaces, 139–141
  wear surfaces and, 241–242
Lindenmayer systems, *see* L-systems
Linear least-squares fit, 50
Lipshitz-Hölder exponents, 193, 224
Lorenz functions, 13
L-systems, 10
  described, 19–26
  mixed fractals and, 198, 199
  modeling and, 149
  pore structures and, 25–26, 249
Lubrication, 242, 243, 244

Machined surfaces, v, 2, 238–241, 244, 253
  light reflection/scattering and, 115, 144, 147
  mixed fractals and, 191, 200, 201, 212, 215
  modeling of, 153, 181, 183, 186, 190
  very flat surfaces and, 251
MacIntosh computers, 8–9
Mammograms, 249
Mandelbrot's conjectures, 63
  described, 201–203
  mixed fractals and, 191, 192, 195, 196, 201–203, 204, 212, 214, 215–217
Mandelbrot–Weierstrass functions, 260
  brittle fractures and, 235
  machined surfaces and, 240
  mixed fractals and, 203
  modeling and, 152, 167–172, 173, 174, 175, 177
Manitoulin Island, 71
Martensite, 235
Mass dimensions, 17, 203
Mass fractals, 2, 51, 55, 153, 154, 156, 245–246
Mecholsky relationships, 232
Menger sponges, 10, 11, 12, 16, 149, 198, 228
Metals, 30, *see also* specific types
  brittle fractures of, 231, 232, 233, 234, 235, 236, 237
  machining of, 244
  mixed fractals and, 192–193
Meteorites, 228
Mica, 258
Midpoint displacement, 59–60
  mixed fractals and, 196, 203, 204, 205, 206, 209–212
  modeling and, 150, 151, 161, 162, 173, 174, 176, 178, 179, 184
  random, 179
Minkowski analysis, 83
  Fourier transform analysis compared with, 109
  of fractal boundary line dimension, 41–42, 55, 56
  Hurst analysis compared with, 88, 89
  of light reflection/scattering, 117, 118, 122, 128, 130, 134–135, 139
  of machined surfaces, 241
  mixed fractals and, 203, 208, 217, 224
  modeling and, 173, 174, 178
  range images and, 89
  of surface boundary line relationship, 71–73, 76, 77, 81
  of very flat surfaces, 258, 260
Minkowski comforters
  Hurst orientation transform compared with, 95
  light reflection/scattering and, 121, 130, 133, 135
  mixed fractals and, 215
  range images and, 89
  surface boundary line relationship and, 71–73
Minkowski plots
  Hurst plots compared with, 83, 85, 86
  mixed fractals and, 210, 211, 212, 215, 218, 219, 220, 221, 223
  modeling and, 154, 155, 179, 181, 184, 185, 189
Minkowski sausages, 71, 83
  brittle fractures and, 235
  Euclidean distance transform and, 42–48
  fractal boundary line dimension measured with, 41–48, 49, 51
  light reflection/scattering and, 121
Mirau optics, 256–258
Mirrors, 138–139, 241, 242, 253
Mixed fractals, 30, 50, 107, 185, 191–225
  addition in, 204–206
  brittle fractures and, 235
  directionality in, 212–223
  Kaye's definition of, 193–195
  Mandelbrot's conjectures and, *see* under Mandlebrot's conjectures
  simulation of, 195–197
  splicing and, 208–212
  variation with scale in, 206–208
Modulation transfer function (MTF), 254, 256, 257–258, 259, 260

Monster curves, 1, 3–6, 41, 116
Mosaic amalgamation, 38–41
Motion pictures, 9, 122
MTF, *see* Modulation transfer function
Multifractals, 30, 191–193, 198, 205, 206, 224, *see also* Mixed fractals

Nanometry, 253, 254
Natural fractals, 4, 12
Natural terrain, 181
Network percolation, *see* Percolation theory
Networks, 2, 51, 238
Neurons, 19, 23, 249
Nile river flooding, 83, 87
NMR, *see* Nuclear magnetic resonance
Noise, 76, 77, 228
  brown, 103
  Brownian, 244, 256, 263
  dimensional analysis and, 68
  electronic, 33
  Fourier transform analysis and, 77, 78, 79, 99–100, 102–103, 108, 109
  Gaussian, 116
  light reflection/scattering and, 141, 142, 143
  machined surfaces and, 240
  mixed fractals and, 195, 202
  pink, 102–103
  simulated, 77–81
  surface boundary line relationship and, 75–81
  very flat surfaces and, 256, 260, 262, 263–264
  white, 100, 102–103
Norwegian coastline, 29, 63
Nuclear magnetic resonance (NMR), 246

Oil recovery, 249
Optical surfaces, 242
Osteoporosis, 140, 249

Pap smears, 249
Particle agglomeration, *see* Agglomeration
Pascal (language), 8
Patterned fractals, 150
Percolation theory, 233, 234
  brittle fractures and, 232
  deposited surfaces and, 248
  pore structures and, 17–18, 249
Perimeters
  brittle fractures and, 229
  dimensional analysis of, 66, 68, 69
  mixed fractals and, 203
  surface boundary line relationship and, 78
Phong shading, 122
Photometric stereo, 142
Photon scattering, 74, 115, 138
Pink noise, 102–103
Pixels, 48, 56, 74–75, 78, 178, 252
  brightness of, 115, 117, 121, 122, 134
  digitized images and, 32–36, 37
Pixels (*cont'd*)
  dimensional analysis and, 68–69
  Euclidean distance transform and, 42, 43
  Fourier transform analysis and, 101–102, 108
  Hurst orientation transform and, 95–96
  Hurst plots and, 91, 112
  light scattering and, 138
  L-systems and, 25
  Minkowski comforters and, 71–73
  Minkowski sausages and, 41–42, 47
  mosaic amalgamation and, 38, 39, 41
  range images and, 144
Poincaré sections, 13, 15, 64, 239
Polishing, 238, 239, 240, 241
Polygons, 74
Polymers, 229, 231, 249
Porcelain fractures, 231
Pore structures, 2, 16
  brightness and, 249
  described, 17–19
  dusts and, 11
  examples of, 248–250
  light scattering and, 138
  L-systems and, 25–26, 249
  mixed fractals and, 198
Porod scattering, 191
Power-law distributions, 66, 228
  brittle fractures and, 235
  deposited surfaces and, 246, 248
  machined surfaces and, 240
  mixed fractals on, 224–225
  pore structures and, 18, 249
  wear surfaces and, 243
Profilometry, 73, 141, 144, 146, 241, 253, 259
Pseudo-isometric displays, 162
Pyrex glass, 240
Pythagorean theorem, 32, 37

Radar, 73, 74, 144, 241
  light reflection/scattering and, 117–121
  synthetic aperture, 138, 144
  of very flat surfaces, 251
Radiography, 140
Radio waves, 31, 241
Raman scattering, 138
Random fractals, 4, 6–9, 59, 195
Random midpoint displacement, 179
Random number generators, *see also* Gaussian random number distribution
  dusts and, 10
  L-systems and, 20–21
  mixed fractals and, 204
  modeling and, 150, 179
  random fractals and, 9
Random walks, 102, 154–155, 157, 224, 228
Random Walk through Fractal Dimensions, A (Kaye), 193

Range images, 162, 253
Hurst analysis of, 89–95
light reflection/scattering and, 144–148
mixed fractals and, 214
Range measurement methods, 141–144
Regular fractals, *see* Structured fractals
Relative dispersion analysis, 54
Rendered surfaces, 162
Richardson plots, 107, 250, 251
brittle fractures and, 229–230, 231
computer construction of, 31–32
digitized images and, 35–36, 37–38
Fourier transform analysis compared with, 109
fractal boundary line dimension measured with, 27–32, 35–36, 37–38, 40, 41, 42, 47, 48, 51, 55
Korcak islands and, 15
light reflection/scattering and, 117
machined surfaces and, 241
Minkowski analysis compared with, 41, 42
mixed fractals and, 30, 193–194, 195, 196, 197, 198, 200, 203, 206–208
mixed fractals on, 204, 205, 212
modeling and, 152
surface boundary line relationship and, 60–61, 62, 63, 64, 70, 71, 73–74
wear surfaces and, 242
RMS measures, 54, 85, 86, 89, 112
of light reflection/scattering, 122, 128, 129, 130, 132, 135, 144–145
machined surfaces and, 240, 244
mixed fractals and, 203
modeling and, 173
range images and, 89
wear surfaces and, 242
Robotics, 91
Rocks, fracture of, 235
Root-mean-square measures, *see* RMS measures
Rose plots, 107–108
light reflection/scattering and, 128, 131, 136
mixed fractals and, 212, 216
modeling and, 164, 169, 181, 187
Roughness, 15, 54, 250, 251
brightness patterns from, 122–129, 139–141
brittle fractures and, 229–230, 231, 235
deposited surfaces and, 245
digitized images and, 33, 36
Fourier transform analysis of, 110–112, 113–114
in fractal boundary line dimensions, 29, 30–31, 32, 33, 36
Hurst analysis and, 91, 93, 94
Hurst plots and, 110–112
light reflection and, 115, 117–121, 134, 135, 137–138, 143, 144, 145, 146–147
light scattering from, 139–141
machined surfaces and, 238, 239, 240, 241, 244
mixed fractals and, 194, 195–196, 200, 201–202, 206, 212, 214
Roughness (*cont'd*)
modeling of, 153, 167, 181, 183, 186
Richardson plots and, 29, 30–31
rose plots and, 107
surface boundary line relationship and, 63, 64
very flat surfaces and, 252, 253, 258, 260
wear surfaces and, 242
Rubber, 244
Run-length encoding, 36–37

SAR, *see* Synthetic aperture radar
Satellite imagery, 115
Saturn rings, 13
Scanning electron microscopy (SEM), 142, 250
of brittle fractures, 231, 237
light reflection/scattering and, 117–120, 121
of mixed fractals, 197
surface boundary line relationship and, 61
of very flat surfaces, 252
Scanning tunneling microscopy (STM), 74, 114, 141, 142
of brittle fractures, 229–230
noise and, 75–76, 103
of very flat surfaces, 259, 260, 261–262, 264
Sedimentology, 250
Self-affinity, 2, 83, 85, 89, 229
box-counting methods and, 41
brittle fractures and, 229, 230, 231
of deposited surfaces, 246, 247
dimensional analysis of, 68, 69–70, 71
Euclidean distance transform and, 45
fractal boundary line dimension and, 30, 36, 41, 45, 48, 51
fractal Brownian motion and, 116
Korcak islands and, 15
light reflection/scattering and, 120, 122, 137
L-systems and, 25
of machined surfaces, 241
Minkowski comforter and, 71
Minkowski sausages and, 48
mixed fractals and, 196, 200, 203, 205
modeling of, 152, 159, 181, 184
mosaic amalgamation and, 41
noise and, 76
Richardson plots and, 30
surface boundary line relationship and, 62, 64, 68, 69–70, 71, 73, 76
of very flat surfaces, 251
Self-similarity, 1, 2, 83, 228–229
box-counting methods and, 41
brittle fractures and, 230
of deposited surfaces, 16
dimensional analysis of, 68, 69–70
dusts and, 10
Euclidean distance transform and, 45
fractal boundary line dimension and, 29, 30, 36, 41, 45, 51
Korcak islands and, 15

Self-similarity (*cont'd*)
  light reflection/scattering and, 120, 122, 137
  L-systems and, 23
  Minkowski comforters and, 71
  mixed fractals and, 191–193, 196, 198, 200
  modeling of, 152, 159, 181
  mosaic amalgamation and, 41
  pore structures and, 18–19
  of random fractals, 6, 9
  Richardson plots and, 29, 30
  surface boundary line relationship and, 62, 64, 68, 69–70, 71, 74
  of very flat surfaces, 251
SEM, *see* Scanning electron microscopy
Shaded grid displays, 162
Shorelines, 63, 70, 71, 254, *see also* Coastlines
Sierpinski fractals, 11
Sierpinski gasket, 228
Silica gels, 246
Silicon, 232, 234, 253, 256, 258, 260, 261
Similarity dimension, *see* Hausdorf dimension
Simulated noise, 77–81
Sinusoids, 174, 177
Skew, 145
Skin, 227, 250
Slit-island analysis
  of brittle fractures, 229, 230–231, 235
  of deposited surfaces, 245
  of light reflection/scattering, 117
  of mixed fractals, 195
  modeling and, 172, 174
  surface boundary line relationship and, 78
  of very flat surfaces, 258, 260, 261
Slopes
  light reflection/scattering and, 117, 121, 129, 130–131, 134, 141, 143, 145, 146, 148
  modeling of, 165
Solidification, 241
Sonar, 73, 74, 138, 143, 251
Soot, 153, 155, 193
Sponges, 238
  Menger, 10, 11, 12, 16, 149, 198, 228
Stainless steel, 240
Standard deviations, 51–52
Steel, 229, 230, 231, 232, 235, 240
Stereoscopy, 73, 74, 141, 142, 144
Sticking probabilities, 246
  deposited surfaces and, 16
  modeling of, 153–154, 155, 156, 157, 158, 159, 160
STM, *see* Scanning tunneling microscopy
Stride length, 68
Strong anisotropy
  Hurst orientation transform and, 95, 96
  modeling of, 181–183, 186
Structured fractals, 4, 6
  mixed fractals and, 194, 195, 197, 206, 224
  modeling of, 149, 150, 179
Structured light, 142, 144
Structured walks, 30, 36, 42, 196, 203
Surface albedo, 118, 121
Surface boundary line relationship, 59–81
  data formats used to determine, 73–75
  dimensional analysis of, 66–71, 78
  direct methods of measuring, 61–64
  noise and, 75–81
Surface dimensions
  brightness dimension related to, 129–138
  brittle fractures and, 234
  machined surfaces and, 240
  mixed fractals and, 215–217
Surface elevations
  brittle fractures and, 230, 234
  light reflection/scattering and, 115–116, 144
  modeling of, 162
Surface measurement techniques, 74
Surface parameters, 144–148
Synovial fluid, 242
Synthetic aperture radar (SAR), 138, 144

Takagi functions
  machined surfaces and, 239, 240
  modeling and, 152, 177–180
Takagi–Landisberg functions, 178
Textural fractals, 194, 195, 197, 206, 224
Texture
  light reflection/scattering and, 115, 117, 118, 121–122, 134, 138, 139, 143
  local measurement of, 121–122
  modeling of, 163, 165
Thin films, 2, 234, 240
Thresholding, 33, 64, 172
  digitized images and, 36
  dimensional analysis and, 70, 71
  Euclidean distance transform and, 43
  Hurst analysis and, 91
  mosaic amalgamation and, 39
Titanium alloys, 231
Tool wear, 136–138, 244
Topographic maps, 61
Topothesy, 138, 148, 152, 181, 186
  described, 112–114
  Fourier transform analysis of, 113–114
  machined surfaces and, 244
  mixed fractals and, 205
Transmission electron microscopy, 250
Tube geometry (in AFM), 262, 263
Turbulence, 193, 206
Two-dimensional Fourier transform (FT) analysis
  described, 104–107
  of light reflection/scattering, 128–129, 130, 133, 135, 136, 138
  of machined surfaces, 244
  of mixed fractals, 203, 212, 216
  modeling and, 164, 165, 166, 169, 170, 179, 180, 181, 182, 185–190
  of very flat surfaces, 256, 260, 262, 264

Unrolling, 101

Variational methods, 48, 73, 76, 81
Variogram methods, 54
Very flat surfaces
  atomic force microscopy of, 253, 258, 259–265
  interferometry of, 251–258, 264–265
Vicsek's fat fractal, 197–200
Video cameras, 32, 33
Visible light, 121
Void coalescence, 233–234

Weak anisotropy
  Hurst orientation transform and, 96
  modeling of, 181–183, 186
Wear surfaces, 2, 227, 238, 241–244
  Korcak islands and, 15
  mixed fractals and, 201, 212
  modeling of, 181
  on tools, 136–138, 244
Weathering, 63, 237
Weibull strength distribution, 235
Weighting, 154–155, 156
Well-behaved curves, 55
White noise, 100, 102–103
Wrinkling, skin, 250

X-ray scattering, 138, 140, 249

Zerosets, 15, 51
  brittle fractures and, 229
  described, 13–14
  dimensional analysis and, 66
  modeling and, 158, 172
  surface boundary line relationship and, 64–66, 71
  very flat surfaces and, 258
  wear surfaces and, 242
Zinc, 246